# Mössbauer Spectroscopy: Principles and Applications

"Every aspect of the world today - even politics and international relations - is affected by chemistry"

Linus Pauling, Nobel Prize winner for Chemistry, 1954, and
Nobel Peace Prize, 1962

"We chemists have not yet discovered how to make gold but, in contentment and satisfaction with our lot, we are the richest people on Earth."

Lord George Porter, OM, FRS
Nobel Prize winner for Chemistry, 1967

"God, you have giv'n us power to sound depths hitherto unknown: to probe earth's hidden mysteries, and make their might our own. So for your glory and our good may we your gifts employ, lest, maddened by the lust of power, we shall ourselves destroy.

G.W. Briggs (1875-1959)

*Hymns Ancient & Modern, New Standard*

# ALFRED MADDOCK

Alfred Gavin Maddock was born in Bedford Park, the first garden suburb of London which dated from the William Morris period. Educated at the Latymer Upper School, Hammersmith, he won a state scholarship to study chemistry at the Royal College of Science, part of Imperial College, London University. With the Professor Emeleus as supervisor he gained his PhD in 1942 for work on silicon hydrides. His work on World War II problems included protection against arsine, and the study of toxicity of volatile compounds of fluorine, from which he suffered acute fluorine poisoning. With Lord Rothschild he developed a mercuric chloride device which was used by parachutists into France.

In 1941 Maddock joined a Free French group, who had escaped from the Juliet-Curie Laboratory, Paris bringing results on a divergent nuclear chain reaction, and worked with them at the Cavendish Laboratory in Cambridge. In 1942 his group moved to North America and with a British-Canadian team constructed an uranium heavy water reactor at Ottawa. Back in England in 1945, he began the use of carbon dioxide as a coolant in a graphite uranium reactor, and at Harwell under Cockcroft he helped design the new Radiochemical Laboratory.

In 1946 Maddock returned to Cambridge University as Assistant Director of Research, where he helped to employ readily available reactive materials in a broad interdisciplinary spectrum leading to the geometry of solids in atoms. He visited Chile 1953 and organised nuclear science courses in the University of Concepcion in Chile, and was next sent by the International Atomic Agency to advise on nuclear science education and training to Greece, the Philippines, Portugal, Yugoslavia, Morocco, Poland, Rumania, Mexico etc.

In 1960 Cambridge University acknowledged his work with the award of Doctor of Science, and appointed him to a personal Readership. His Mossbauer studies there have included work with tellurium, gold, iridium, tungsten, and Technetium (used in medical practice). There was fruitful collaboration with EPFL, Lausanne and the National Institute for Nuclear Studies, Mexico.

Mossbauer discoveries by Bookhaven scientists in USA motivated Maddock towards this new spectroscopic technique. With a Canadian scientist he built a Mossbauer spectrometer and began to uncover Mossbauer elements of iron and tin.

The University of Louvain in Belgium made him an honorary DSc, and he was invited as Visiting Professor to the State University of New York at the Buffalo Campus. He also went to the Nuclear Research Centre, Puerto Rico, and at the Universite Pasteur Louis Pasteur in Strasbourg he studied isomeric transmission in cross-over systems spectroscopy in littoral deposits.

He was elected to the Brazilian Academy of Science, and in 1995 the Academy conferred the Grand Cross of the Order of Merit in Science. At St.Catharine's College, Cambridge from 1958 he was Director of Chemical Studies and Tutorial Fellow; and in 1981 became College President His published work numbers some 300 papers, reviews, and chapters in books, and he has edited three other books. The Royal Society of Chemistry awarded him the Becquerel medal in 1996.

In 1988 he spent nearly two years in hospital after fracturing a femur which resulted in the amputation of his left leg at the hip joint. He soon gained wheelchair mobility, and has been on lecture tours and conferences in Japan and Mexico and on frequent visits to France.

# Mössbauer Spectroscopy
# Principles and Applications

**Alfred Maddock**, BSc, DIC, PhD, ScD(Cantab), DSc(Louvain)
Department of Chemistry
University of Cambridge
*and*
Fellow of St. Catherine's College

**Horwood Publishing**
**Chichester**

First published in 1997 by
**HORWOOD PUBLISHING LIMITED**
International Publishers
Coll House, Westergate, Chichester, West Sussex, PO20 6QL
England

**British Library Cataloguing in Publication Data**
A catalogue record of this book is available from the British Library

ISBN 1-898563-16-0

Printed in Great Britain by Hartnolls, Bodmin, Cornwall

# Table of Contents

# 1

# Basis of Mössbauer Spectroscopy

## 1.1 RESONANCE ABSORPTION AND SCATTERING

The photons emitted when an electronically excited atom radiates might seem to have exactly the appropriate energy for excitation of the same kind of atom to the same excited state. Subsequent emission by the excited atom then takes place isotropically. Thus a process of resonant scattering of the radiation occurs and a beam of such photons passing through a gas composed of these atoms will be attenuated. Indeed such behaviour is well known for the radiation from a mercury vapour lamp. If a tube containing mercury vapour is exposed to this radiation, the beam is attenuated and the tube emits the same radiation in all directions, including that normal to the incident beam --- a process of resonant absorption and scattering.

Many years ago attempts were made to demonstrate similar behaviour with the more energetic photons from excited nuclei, that is to say to explore the fluorescent resonant scattering of gamma radiation. For a long time such experiments were unsuccessful [Ref.1.1].

Let us look at this matter of the absorption and emission of photons rather more closely. Firstly, one must note that there is not a unique photon energy involved. Although the energy of the ground state of the emitter, assuming it to be stable, is sharply determined, the excited state has a rather short half-life before emitting and the product of the uncertainty of the energy of the state and the uncertainty in the time, $\delta t$ is given by $\delta E..\delta t \geq \hbar$ where $2\pi\hbar$ = Planck's constant. The uncertainty in the time is the mean life-time of the excited state, $\tau$. [Ref.1.2]. Thus the photons emitted will have a distribution of energies as shown in Figure 1.1.

The form of the distribution is given by the expression
$W(E) = (\delta E)^2/4 /[(E-E_t)^2 + (\delta E)^2/4]$. $W(E)dE$ measures the probability of emission of a photon with energy between $E$ and $E + dE$. $E_t$ is the most probable energy available due to the de-excitation. The expression has been normalised to give $W(E_t) = 1$. Such a distribution gives a Lorentzian line shape.

For a half-life of the excited state of $10^{-8}$ s. the uncertainty in the energy is of the order of $5\times10^{-8}$ eV. The line width, or **FWHM** (<u>F</u>ull <u>W</u>idth at <u>H</u>alf <u>M</u>aximum) is the difference, $\delta E$ in Fig.1.1, between the energies for which $W(E) = 1/2$.
rest mass, and momentum must be conserved in the emission events. Thus the emitting species must acquire an equal amount of recoil momentum. Hence the emitted photon

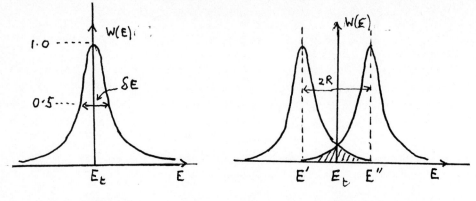

Fig.1.1                                    Fig.1.2

will have an energy less, by the the recoil energy of the emitter, R, than the energy available from the de-excitation process.    Similarly the photon energy needed for the excitation process must be greater, by the amount of the recoil energy, than the excitation energy.   Since the two recoil energies are very little different we will use the mean value, R.

So far it has been assumed that the emitter is at rest.   For a moving system, for example gaseous atoms, there will be a Döppler broading of the distribution of photon energies and further momentum conservation must be considered.

Let us represent the momentum of the emitter in the direction of the emitted photon by P and the photon momentum by p. Then the total recoil $R' = (P-p)^2/2M - P^2/2M$ where M is the mass of the emitter.

Hence $R' = p^2/2M - pP/M$. The second term in this expression is the Döppler contribution and the first the recoil without thermal motion, R. Suppose the mean kinetic energy of the emitter before emission is T then the second term can be written as $D \cos\varphi$, with $D = 2(TR)^{1/2}$ and $\varphi$ the angle between the directions of emission of the photon and movement of the emitter.    This angle ranges from 0 to $2\pi$.   So one finds that the emission and excitation distributions are displaced, as shown in Fig. 1.2.

With E' the emitted photon energy, E" the photon energy for excitation and $E_t$ the energy available in the de-excitation

$$E' = E_t - R + D \cos\varphi \quad \text{and} \quad E" = E_t + R + D \cos\varphi.$$

Resonance absorption and scattering only occur where the two distributions overlap. (Hatched area in Fig.1.2)  As can be seen in the figure the overlap can be anything from negligible to substantial depending on the value of R.

In the visible and near UV regions and for room temperature R is very small and D small, overlap is considerable and resonant scattering can take place.   The effect of the Döppler broadening is greatest in the tails of the distributions.

For the much more energetic photons involved in nuclear transitions R is very much greater, there is negligible overlap of the two distributions and resonant scattering would not be expected. Expressing the recoil, R, in eV, the photon energy, $E_\gamma$, in keV and the mass of the recoiling atom, M, in atomic mass units, $R = 5.36 \times 10^{-4} E_\gamma^2 / M$     or about 0.05 eV for a 100 keV photon from an atom of mass 100. The Döppler broadening term has a comparable magnitude at 300 K.

Comparing with atomic spectra, emission of the sodium D line by a gaseous atom at 300 K leads to a recoil of about $10^{-10}$ eV and the Döppler contribution to the line width amounts to about $3.9 \times 10^{-6}$ eV, while the uncertainty line width is $4.4 \times 10^{-8}$ eV. Hence resonant re-excitation by the emitted radiation can occur. For the emission of the 14.4 KeV photon from $^{57}$Fe the recoil amounts to $1.9 \times 10^{-3}$ eV, the Doppler contribution amounts to $1.6 \times 10^{-2}$ and the uncertainty line width is $4.7 \times 10^{-9}$ eV. Resonant re-excitation by this radiation seems most unlikely.

## 1.2 THE MöSSBAUER EFFECT

The above analysis applies to gaseous emitters and absorbers at low pressures. Mössbauer found that if the nuclei involved in the process were present in a solid material, in some fraction of the emission events R and D become effectively zero and the line width for the distribution of emitted photon energies is essentially that determined by the uncertainty relation, $\delta E = \hbar / \tau$. This is the FWHM or $\Gamma_t$ (See Fig.1.1). Thus for a life-time of the excited state of $10^{-8}$ s.the line width will be about $4.6 \times 10^{-8}$ eV, an extremely narrow line.

Since $\tau = t_{1/2}/\ln 2$, $\Gamma_t$ (eV) $= 4.56 \times 10^{-16} / t_{1/2}$ (s)    The recoil energy is generally too small to eject the emitting atom from its lattice site. Such events need 10 eV or more. Indeed the recoil energy may even be smaller than the phonon energy, the energy of the quanta of the vibrational modes of the atoms in the lattice. Now the recoil can only transfer its energy to the vibrational modes in an integral number of such quanta, thus there is some probability that the vibrational mode will not be excited and the emitted photon carries the whole of the energy of the transition, $E_t$. These events are called **zero phonon events**. Of course, momentum is still conserved but the recoil now involves a very large number of the atoms in the solid, an effectively infinite mass leading to an infinitesimal recoil. The emitted photons will then have a most probable energy $E_t$ and a line width determined by the uncertainty relation.

Consider the rather simplified model of an Einstein solid, which is characterised by a single vibrational frequency $\upsilon$, or angular frequency $\omega = 2\pi\upsilon$. Suppose that a fraction of emission events, f, are zero phonon events, all the rest involve excitation of a phonon. The average energy lost in recoil is R, therefore $(1-f) \hbar \omega = R$ and $f = 1 - R/\hbar \omega$.

This already tells us something about the conditions for a substantial recoil free, or zero phonon, fraction. Clearly R should be as small as possible and the Einstein frequency of the solid as high as possible. This implies a hard, high melting point solid. Polar crystals and refractories seem appropriate. A small value of R implies as small as possible a photon energy. f is usually called the **Mössbauer fraction.**

This kind of behaviour may be familiar in another context. A certain fraction of the X-ray photons scattered by a solid also preserve their original energy This fraction, usually called the Debye-Waller fraction, arises in exactly the same way as the Mössbauer fraction [Ref.1.3].

A somewhat more realistic approximation to f can be made using a quantum mechanical treatment. The probability of a photon emission event with quantum mechanical treatment. The probability of a photon emission event with the nucleus and the lattice going from an initial state $i$ to a final state $f$ is given by Fermi's Golden Rule. $W = k|<f\,|\textbf{H}|i\,>|^2$, where the bra and ket terms refer to the final, $<f$, and initial, $i >$, wave functions of the system and **H** is the interaction Hamiltonian operator. k is a constant.

Now the magnitudes of the nuclear, i.e. photon energy, and the vibrational energy are very different, as are the range of the nuclear and vibrational forces. Hence the nuclear and vibrational contributions to the above expression are separable and the nuclear term can be taken as constant for different changes in the vibrational state of the system. Thus we obtain $W = k'|<f\,|\textbf{H}|i\,>|^2$, where the wave functions relate only to the vibrational states and k' now includes the nuclear contribution.

In the zero phonon case $f = i$ so that $W = f = k'|<i\,|\textbf{H}|i\,>|^2$ .It can be shown that for translational and relativistic invariance **H** must have the form $\exp.ix(p/\hbar\,)$, with x the displacement of the nucleus from its mean position in the lattice along the direction of emission of the photon.

For an Einstein solid, with only one vibrational frequency one obtains
$f = \exp-(p/\hbar\,)^2<x^2>$, where $<x^2>$ is the mean square displacement of the emitting atom in the direction of photon emission and p is the recoil momentum. Now $p = E_\gamma\,/c$ therefore

the exponent is equivalent to $- E_\gamma^2 <x^2>/\hbar^2c^2$.

But $E_\gamma^2 = 4\pi^2 \hbar^2 v^2$ and $v^2 = c^2/\lambda^2$ so that $f = \exp -4\pi^2 <x^2>/\lambda^2$ or $K<x^2>$ with

$K = 2\pi /\lambda$. ( $v$ and $\lambda$ are the frequency and wavelength of the emitted photon)

With the rather more realistic Debye spectrum of vibrational frequencies, which has $N(v)$ proportional to $v^2$, $f = e^{-2\alpha}$ where

$$\alpha = -\frac{3R}{2k}\frac{1}{4}+\left(\frac{T}{\theta}\right)^2\int_0^{T/\theta}\frac{x\,dx}{e^x - 1}$$

$\theta$ is the Debye temperature of the solid and k is Boltzmann's constant. $\alpha$ corresponds to the Debye-Waller factor in X-ray crystallography.

If $T \ll \theta$, $\alpha = -\frac{3R}{2k\vartheta}\left[1+\frac{2\pi^2 T^2}{3\theta^2}\right]$ and as $T \to 0$ $f \to e^{-3R/2k\vartheta}$ and at

high temperature, $T \geq \theta/2$, $f \to e^{-6RT/k\theta^2}$

As with the cruder approximation to f the conditions favouring an optimum value are (i) a low value of R, (ii) a high value of θ, to which one now adds(iii) a low temperature.

As a result of this fraction of zero phonon events  resonant scattering of the energetic, gamma, radiation is possible with solid sources and scatterers. The solid sources will emit photons with a very narrow line width, determinedonly by the life-time of the excited state and the fraction of recoil-free events will increase as the temperature of the source is lowered.

But where is there any chemical information in this process?

## 1.3  INTERACTION OF NUCLEUS WITH ORBITAL ELECTRONS

The nuclear energy levels are modified by the electric and magnetic fields to which the nucleus is subjected. Since we are concerned with solids the nucleus will be present in an atom, which may be ionised by a few units. It will therefore experience the electric and magnetic fields arising from its orbital electrons. In addition there may be a smaller contribution to the electric field from more remote ions in the lattice. An externally applied magnetic field may also have an effect providedit is large enough. In principle the same is true of electric fields but at present large enough external electric fields cannot be produced.

These effects, called **hyperfine  interactions**, arise from the interaction of the magnetic dipole of the nucleus with the magnetic field and the electric quadrupole moment with the electric field. Such interactions had been observed in optical spectroscopy many years ago [Ref.1.4].

Nuclei with nuclear spin quantum numbers greater than zero possess a magnetic dipole moment. For these nuclei only certain orientations of the spin in relation to the field will occur, in the same way as the space quantisation of angular momentum with electrons. The ground state of nearly all nuclei with even mass number prove to have zero spin and these have no magnetic dipole moment.

The symmetry of nuclei precludes a nuclear dipole moment so that no similar electric interaction occurs. But if the spin quantum number is equal to, or greater than, one the nucleus will possess an electric quadrupole moment.  In a non-spherically symmetric electric field only certain orientations of this moment in relation to the field will be possible.

A magnetic field removes all the spin degeneracy of the nuclear levels; if the nucleus has an half integral spin an asymmetric electric field leaves each level doubly degenerate, a case of Kramers doublets [Ref.1.5]. In principle both higher magnetic and electric moments could be concerned, but the energies involved in these hyperfine interactions are extremely small and such higher order effects are hardly detectable.

Each of the above interactions will lead to a splitting of the nuclear energy levels into two or more levels.  It is only because of the very narrow line width in the zero phonon emission that we can explore these energy levels.  The data obtained tell us something about the electronic environment of the atom containing the photon emitting nucleus.

Another source of chemical information arises because the radius of the excited state of the nucleus usually differs from that of the ground state. The spherically symmetrical part of the electric field will then interact differently with the two levels, displacing both from their energies for the hypothetical bare nucleus situation. This displacement will depend on the electron density at the nucleus. Now the electron density at the nucleus will depend on the state of chemical combination of the atom containing the emitting nucleus, so that the same nucleus in two different compounds will emit slightly different energies of photons.

By deliberate use of the Döppler effect, moving the source of gamma radiation at a velocity v, one can change the energy of the radiation emitted in the direction of movement by an amount $\Delta E = (v/c)\ E_{\gamma}$. For the 14400 eV Mössbauer line from the excited state of $^{57}Fe$ a movement of 1 cm.s$^{-1}$ in the direction of propagation of the photons will lead to an increase in the photon energy of $4.82 \times 10^{-7}$ eV The same movement in the opposite direction will lead to a reduction in energy by the same amount. In this way a solid source can provide photons with a very narrow spread of energies over a small range of energies near to $E_t$.

Thus if a collimated beam of photons from a solid source, giving a single line emission, is directed at an absorber of the same composition, the number of photons per unit time recorded by a detector, situated on the far side of the absorber, in the line of the beam, can be measured as function of the velocity of movement of the source. If the velocity of the source is plotted as abscissa and the fraction of the photons removed from the beam as ordinate the plot will appear as an inversion of Fig. 1.1, with zero velocity at $E_t$.

The line width for the overall emission and absorption processes will be the sum of the values for the individual steps. The value found experimentally cannot be less than $2\delta E$ calculated from the mean life-time of the excited state. If the life-time, $t_{1/2}$ is expressed in seconds the overall line width is given by

$$2\Gamma_t = 2\delta E = 9.125 \times 10^{-16} / \ t_{1/2} \ eV.$$

## 1.4 INTERACTION WITH THE ELECTRIC FIELD

### 1.4.1 The Isomer or Chemical Shift

The nucleus is not a point charge, it has a finite radius and in most cases the excited and ground states have different radii.. Because of the electrostatic intreraction of the nucleus with the electric field due to the orbital electrons, the nuclear energy levels are slightly different in atoms of different compounds, as shown in Fig.1.3. Fig.1.3 a shows levels for a bare nucleus; 1.3 b levels in compound (i) and 1.3 c in compound (ii). The energy of interaction of the nucleus, of charge Ze, with the surrounding charges is

given by     $E_{elec} = \int_0^R \rho(r)V(r)d\tau$, where $\rho(r)$ is the nuclear sharge density at a point with coordinates $x_1, x_2, x_3$.

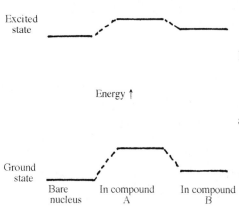

$r = (x_1^2 + x_2^2 + x_3^2)^{1/2}$ ,. V(r) is the potential at this point due to all exterior charges and $d\tau$ is the volume element $dx_1 dx_2 dx_3$. One can expand expand V(r) about the origin in a Taylors series and obtain:

**Fig.1.3**

$$E_{elec.} = V_0 \int \rho_n (r) d\tau + \sum_{i=1}^{3} (\partial V / \partial x_i)_0 \int \rho_n(r) x_i d\tau +$$

$$1/2 \sum_{i=1}^{3} (\partial^2 V / \partial x_i \partial x_j)_0 \int \rho_n(r) x_i x_j d\tau \qquad \text{Eq. 1.1}$$

Since $V_0 \int_0^R \rho(r) dr = Z e V_0$ this term simply contributes to the potential energy of the system. The second term represents the dipolar interaction, but since the symmetry of the nucleus precludes a nuclear electric dipole moment this term must be zero. The third term will require further consideration. In principle additional even terms of the expansion should be taken into account but they give rise to energy terms that are generally too small to detect.

### 1.4.2 The third term in the expansion

The $\partial^2 V / \partial x_i \partial x_j$ or $V_{ij}$ for compactness, form a 3 X 3 tensor and represent the electric field gradient at the nucleus (See 1.4.3).. With a suitable choice of the coordinate system all $V_{ij}$, $i \neq j$, can be made zero and one need only consider the $V_{ii}$. This coordinate system defines the principal axes of the electric field gradient.

Thus $E_{elec} = 1/2 \sum_{i=1}^{3} V_{ii} \int \rho_n (r) x_i d\tau$ whence adding and subtracting

$1/6 \sum_{i=1}^{3} V_{ii} \int \rho_n(r) r^2 d\tau$ one finds $E_{elec} = 1/6 \sum_{i-1}^{3} V_{ii} \int \rho_n(r) d\tau +$

$$1/2 \sum_{i=1}^{3} V_{ii} \int \rho_n(r) [x^2 - r^2 / 3] d\tau \qquad \text{Eq.1.2}$$

The first term gives the monopolar interaction and the second the quadrupolar interaction of the nucleus with its orbital electrons.

Now the Laplace relation requires $\nabla^2 V = -4\pi \rho_e$, where $\rho_e$ is the charge density, that $(\nabla^2 V)_0 = (\sum_{i=1}^{3} V_{ii})_0 = 4\pi e |\Psi(0)|^2$ where $|\Psi(0)|^2$ is the electronic charge density,

supposed constant over the nuclear volume. Hence the monopolar interaction energy is given by:

$$E_{elec.} = 2/3\ \pi e |\Psi(0)|^2 \int_0^R \rho_n(r) r^2 d\tau\ .$$

$$\text{Now} \int_0^R \rho_n(r) r^2 d\tau = 4\pi \int_0^R \rho_n(r) r^4 dr = Ze<R^2>$$

Thus $E_{elec.} = 2/3\pi Z e^2 |\Psi(0)|^2 <R^2>$, where $<R^2>$ is the expectation value of $R^2$.

The nuclear energy level will be displaced by this amount. Since different levels have different values of $<R^2>$ the levels will be displaced by different amounts. (See Fig.1.3.). The separation between the levels, determining the energy available for photon emission, will depend on the electron density at the nucleus, $|\Psi(0)|^2$, and the difference in the squares of the nuclear radii for the two levels.

A simpler approach can be made as follows. Let us suppose the electronic wave function in the nucleus is constant from $r = 0$, the centre of the nucleus, to R, the nuclear radius. This implies the electron density is constant over the nucleus and equals $|\Psi(0)|^2$ (Fig.1.4) The charge on the proton is denoted by e   Then the interaction energy of the nuclear charge with the orbital electron density in the nucleus is given by:

$$E_g = -\int_0^{R_g} V(r) e\ |\Psi(0)|^2\ 4\pi r^2 dr \qquad \text{for the nuclear ground state, of radius } R_g;$$

where V(r) is the potential due to the nuclear charge distribution.

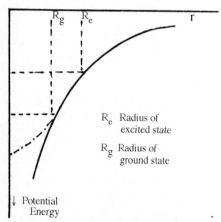

$R_e$   Radius of
      excited state

$R_g$   Radius of
      ground state

↓ Potential
  Energy

**Fig.1.4**

Unfortunately V(r) is not really known and we have to make, probably inaccurate, assumptions. A nucleus of radius $R_g$ with its charge Ze concentrated at its centre can centre can be used as a reference system.
. In this case the interaction energy for the ground state is,

$$E_g = -\int_0^{R_g} V(r) e |\Psi(0)|^2\ 4\pi r^2 dr$$

where $R_g$ is the nuclear radius for the ground state.
Hence $E_g = -2\pi Z e^2 |\Psi(0)|^2 R_g$. If we sup-
pose the electrostatic potential is constant
over $r = 0$ to $r = R_g$, at the value set by the intercept as shown in Fig.1.4
(horizontal lines) then $V(r) = Ze/R_g$ and $E_g = (4/3)\pi Z e^2 |\Psi(0)|^2 R_g$.

A better assumption is that the nucleus is a uniformly charged sphere, whence

$$V(r) = \frac{Ze}{R_g} \left[ \frac{3}{2} - \frac{r^2}{2R_g^2} \right]$$

Thus

$$E_g = -2Ze^2 \pi | \Psi(0)|^2 \int_0^{R_g} \left[ \frac{3}{R_g} - \frac{r^2}{R_g^3} \right] r^2 \, dr$$

$$= -2Ze^2 \pi | \Psi(0)|^2 \left[ \frac{r^3}{R_g} - \frac{r^5}{5R_g^3} \right]_0^{R_g}$$

$$= -(8/5)Ze^2 \pi | \Psi(0)$$

Hence the change in $E_g$, measured relative to the point nucleus, is $\Delta E_g = (2/5) \, Ze^2 \pi |\Psi(0)|^2 R_g$. Similarly for the excited state $\Delta E_e = (2/5) \, Ze^2 \pi |\Psi(0)|^2 R_e$. Thus the photon energy for emission, or absorption, will change from the value for the bare nucleus by the difference in these amounts. (Fig. 1.3). Now $|\Psi(0)|^2$ depends on the compound containing the atom with the Mössbauer nucleus. For a source made of a compound s, and an absorber of compound a, the peak of the absorption curve will move from zero velocity of source relative to absorber, found if s and a are the same compound, to a value given by : $\delta(v) = (2/5)\pi Ze^2 (R_e - R_g)(|\Psi(0)|^2_s - |\Psi(0)|^2_a)$

(See Fig. 1.6). This can also be written:

$$\delta(v) = (4/5)\pi e^2 R^2 (\Delta R/R)[|\Psi(0)|^2_s - |\Psi(0)|^2_a]$$ Where $R = (R_e + R_g)/2$ and

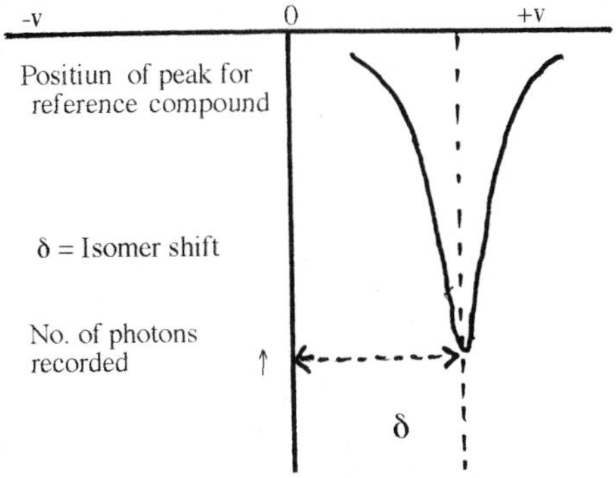

$\Delta R = R_e - R_g$ Hence $\delta(v) = K(|\Psi(0)|^2_s - |\Psi(0)|^2_a)$. K depends only on the nuclear characteristics and is a constant for a given Mössbauer transition. Taking the approximation $R = 1.2 \, A^{1/3}$, where A is the mass number of the nucleus, $K = k \, Ze^2 A^{2/3} \Delta R/R$, with k a purely numerical term. It can be seen that the magnitude of the this shift, called the **isomer shift** or sometimes, more inform-

Position of peak for reference compound

$\delta$ = Isomer shift

No. of photons recorded

$\delta$

**Fig. 1.5**

atively, the **chemical shift**, is determined by the difference in the electron densities at the nuclei of the atoms containing the Mössbauer nuclei in the source and absorber. (See Fig. 1.5) The sign of the shift depends on the sign of $\Delta R$, which may be positive or negative, the latter is the case for $^{57}Fe$ The excited nucleus is not necessarily larger than the ground state. For a positive $\Delta R$ and measurements with a given source, the values of $\delta$ increase with $|\Psi(0)|^2$ for different absorbers.

The value of $|\Psi(0)|^2$ is determined primarily by the s electrons, but electrons in other orbitals may influence $|\Psi(0)|^2$ by screening outer s electrons.

Since electron velocities near the nucleus are very large, comparable with the velocity of light, a more realistic calculation of $\delta$ demands the use of relativistic wave functions. This implies a contribution to the the electron density at the nucleus from $p_{1/2}$ electrons.[Ref.1.6]. It has been shown that this requirement can be taken into account by multiplying the supposed constant nuclear term, K, by a factor proportional to $Z^2$ for the Mössbauer element.[Ref.1.7].

A priori calculations of $\delta$ or of $\Delta R$ from observed values of $\delta$ are difficult and not very accurate, fortunately they are not needed for many of the applications.

### 1.4.3 Some electrostatic considerations

First we must digress on the electrostatic characteristics of an asymmetric electric field. A charge q located at polar coordinates $(r,\theta.\varphi)$ produces a potential, V, at the origin equal to q/r. Now $\mathbf{E} = \nabla V$.   $r = (x^2 + y^2 + z^2)^{1/2}$ where (x,y,z) are the Cartesian coordinates of q. (Fig.1.6) The component $E_x$ of $\mathbf{E} = -\partial V/\partial x$ therefore; $E_x = qxr^{-3}$. Similarly $E_y = qyr^{-3}$ and $E_z = qzr^{-3}$..Now the **electric field gradient,** or **EFG,** $= \nabla E = -\nabla^2 V$ which is represented by the tensor shown below..

$$\begin{vmatrix} V_{xx} & V_{xy} & V_{xz} \\ V_{yx} & V_{yy} & V_{yz} \\ V_{zx} & V_{zy} & V_{zz} \end{vmatrix}$$

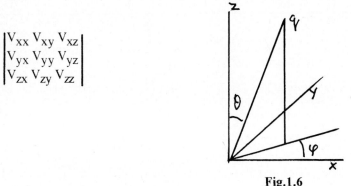

**Fig.1.6**

These terms can be obtained by differentiating the $E_x$, $E_y$, $E_z$ components with respect to x, y, and z. Thus one obtains:

$$V_{xx} = q(3x^2 - r^2)r^{-5} ,: V_{xy} = V_{yx} = 3qxyr^{-5}$$
$$V_{yy} = q(3y^2 - r^2)r^{-5} ; \; V_{xz} = V_{zx} = 3qxzr^{-5}$$
$$V_{zz} = q(3z^2 - r^2)r^{-5} ; \; V_{yz} = V_{zy} = 3qyzr^{-5}$$

Transforming to entirely polar coordinates, noting x = $r\sin\theta\cos\varphi$; y = $r\sin\theta\sin\varphi$; z = $r\cos\theta$; one obtains

$$V_{xx} = q(3\sin^2\theta\cos^2\varphi - 1)r^{-3} \qquad V_{yy} = q(3\sin^2\theta\sin^2\varphi - 1)r^{-3}$$
$$\text{and} \qquad V_{zz} = q(3\cos^2\theta - 1)r^{-3}$$

In SI units these expressions must be divded by $4\pi\varepsilon_0$ where $\varepsilon_0$ is the permittivity of the vacuum.

With appropriate axes all the off-diagonal terms, of the terms, of the type $\partial^2 V/ \partial x \partial y$, become zero and the EFG becomes:-

$$-\begin{vmatrix} V_{xx} & 0 & 0 \\ 0 & V_{yy} & 0 \\ 0 & 0 & V_{zz} \end{vmatrix}$$

In the above simple case of a single charge, q, diagonalisation of the tensor is easy if the point charge lies on the z axis.

In general there will be a distribution of charge, such that the charge density, $\rho$, is a function of $(r, \theta, \varphi)$. Then

$$V_{zz} = (1/4\pi\varepsilon_0) \int \rho(r, \vartheta, \varphi)(3 \cos^2 \vartheta - 1) d\tau$$ where $d\tau$ is the infinitesimal volume element $r^2 \sin\theta \, dr \, d\theta \, d\varphi$.

It is interesting to examine the case of a spherically symmetrical charge density function, $\rho = f(r)$. Then

$$V_{zz} = (1/4\pi\varepsilon_0) \int_0^\infty f(r) r^2 \, dr \int_0^{2\pi} d\varphi \int_0^\pi (3 \cos^2 \theta - 1) \sin \theta \, d\theta$$. The last integral is

zero, so that $V_{zz}$ is zero and there is no electric field gradient.

Now the electric field gradient can only arise from the non-spherically symmetric p,d, or f electrons and for these, for a single atom, the electron density at the nucleus is zero, so that $\nabla^2 V = 4\pi\rho = 0$, since $\rho = 0$. Hence $V_{zz} + V_{yy} + V_{xx} = 0$. Thus the electric field gradient can be characterised by two quantities; we identify the axes so that $|V_{zz}| > |V_{xx}| > |V_{yy}|$ and we define an **asymmetry parameter,** $\eta$, such that $\eta = (V_{xx} - V_{yy})/V_{zz}$. Thus the asymmetry parameter can range from 0 to 1 and the EFG can be characterised by the two parameters $\eta$ and $eq/4\pi\varepsilon_0$ or $V_{zz}$

The EFG tensor now becomes:

$$-V_{zz} \begin{vmatrix} -(1-\eta)/2 & 0 & 0 \\ 0 & -(1+\eta)/2 & 0 \\ 0 & 0 & 1 \end{vmatrix}$$

A positive $V_{zz}$ is found for a field such that points with equal magnitude of field form a surface defining a prolate spheroid with its axis along the z axis; a cross section of this solid including the z axis is elliptic. A cross section normal to the z axis is circular if $\eta$ is zero and elliptic if $\eta > 0$.

### 1.4.4  The Quadrupole Splitting

If the nucleus has spin, $I > 1$ it will possess a **quadrupole moment** arising from the non-uniform distribution of charge. This moment can, like the EFG, be specified by a 3 X 3 tensor. It can be diagonalised by a suitable choice of axes to give a traceless tensor.

As shown in Eq.1.2 the quadrupolar interaction energy of the nucleus with the electric field gradient is given by:

$E_q = 1/2 \sum_{i=1}^{3} V_{ii} \int \rho(r)(x_i - r^2/3)d\tau$ using coordinates $x_1$, $x_2$, $x_3$ for notational convenience. (The factor $1/4\pi\varepsilon_0$ which only adjusts units, will be omitted in this section.)

$$\text{Let } Q_{ii} = \int (3x_i - r^2)\rho_n(r)d\tau \quad \text{then } E_Q = 1/6 \sum_{i=1}^{3} (V_{ii}Q_{ii}))$$

or in terms of $\eta$ and eq or $V_{ZZ}$

$$E_q = eq/6 [(\eta - 1)/2] Q_{11} - [(\eta + 1)/2][Q_{22} + Q_{33}] \quad \text{(See 1.4.3)}$$

The quantum mechanical Hamiltonian operator for the quadrupole interaction can now be obtained by replacing the $Q_{ii}$ by the operators $\mathbf{Q}_{ii}$. We must now relate these quadrupole operators to the nuclear properties.

The nuclear angular momentum, like that of the orbital electrons, is characterised by a nuclear spin quantum number, I, and a quantum number, m, specifying the orientation of the spin in relation to the EFG axes. There are analogous nuclear spin operators to those used in the case of the orbital electrons.

Thus:- $\mathbf{I}^2 |I,m.> = I(I+1)|I,m.>$ and $\mathbf{I}_z|I,m.> = m|I,m.>$.

It is also useful to define the nuclear raising and lowering operators $\mathbf{I}_\pm = \mathbf{I}_x \pm i.\mathbf{I}_y$, such that; $<I,m.|\mathbf{I}_\pm|I,m.> = [(I(I+1) - m(m \pm 1)]^{1/2} |I,m \pm 1>$      Eq.1.3

(in each case taking all upper or all lower signs.) Now the $\mathbf{Q}_{ii}$ can give linear combinations $\mathbf{Q}_{\pm2}$, $\mathbf{Q}_{\pm1}$ and $\mathbf{Q}_0$ that behave as spherical harmonics on rotation of the coordinates. It is also possible to construct nuclear spin operators with the same transformation characteristics under rotation. Under these circumstances the Wigner - Eckart theorem [Ref.1.7] shows that: $<I,m.|\mathbf{Q}_{ij}|I,m.> = K.<I,m.|\mathbf{I}_i^2|I,m.>$. with $i = \pm 2$. $\pm 1$, where K is a function only of I. This enables one to express the $\mathbf{Q}_{ii}$ in terms of K, I and $\mathbf{I}_i$. Thus:     $\mathbf{Q}_{ii} = K[3\mathbf{I}_i^2 - I(I+1)]$ with $i = x,y$ or $z$.

We now define eQ, the quadrupole moment of the nucleus, as the average value of $Q_{ZZ}$ for the nuclear state $|I,I>$ where $m = I$.

$$\text{Thus } eQ = <I,I.|\mathbf{Q}_{ZZ}|I,I.> = K[3\mathbf{I}_z^2 - I(I+1)].$$

$$eQ = KI(2I-1) \quad \text{and } K = eQ/I(2I-1).$$

The Hamiltonian operator for the quadrupole interaction can now be obtained in a convenient form substituting for the $\mathbf{Q}_{ii}$ in

$$\mathbf{E}_q = eq/6 [1/2(\eta - 1)\mathbf{Q}_{xx} - 1/2(\eta + 1)\mathbf{Q}_{yy} + \mathbf{Q}_{zz}]$$

$$E_q = \frac{e^2 qQ}{6I(2I-1)} [1/2(\eta-1)\{3\ I_i^2 - I(I+1)\} - 1/2(\eta+1)\{3\ I_y^2 - I(I+1)\}} +$$

$$3I_z^2 - I(I+1)]$$

$$= \frac{e^2 qQ}{4I(2I-1)} [2I_z^2 + \eta(I_x^2 - I_y^2) - (I_x^2 - I_y^2)$$

$$= \frac{e^2 qQ}{4I(2I-1)} [3I_z^2 - I^2 + \eta(I_x^2 - I_y^2)].. \quad \text{But } I_+^2 + I_-^2 = 2(I_x^2 - I_y^2)$$

$$E_q = \frac{e^2 qQ}{4I(2I-1)} [3I_z^2 - I(I+1) + 1/2\eta (I_+^2 - I_-^2)]. \quad \text{Eq.1.4.}$$

It is not possible to derive a compact expression for the energy of the **quadrupole splitting** applicable to all values of I and $\eta$  But there are two solutions with certain restrictions that are very valuable.

　　a/ If $\eta = 0$, $V_{xx} = V_{yy}$, so that the electric field has rotational symmetry about the z axis, one obtains:

$$E_q|I,m.> = \frac{e^2 QV_{zz}}{4\ I(2I-1)} \quad [3I_z^2 |I,m.> - I^2|I,m.>], \quad V_{zz} = eq/4\pi\varepsilon_0$$

$$\text{whence } E_q|I,m.> = \frac{eQV_{zz}}{4I(2I-1)} [3m^2 - I(I+1)]|I,m.>$$

Therefore for the |3/2,3/2.> state   $E_q = + eQV_{zz}/4$ and
　　　　　for the |3/2,1/2.> state   $E_q = - eQV_{zz}/4$.

Thus in the case of an excited state with I = 3/2 and a ground state with I = 1/2, since the latter is not split, having no quadrupole moment, the energy level pattern will be as shown in Fig. 1.7(a). The separation of the lines is $eQV_{zz}/2$.

It must be emphasized the quadrupole splitting of the excited state is extremely small compared to its separation from the ground state. If the figure was drawn all to the same scale and the split levels placed 1 cm. apart the upper part of the figure would have to be located about $10^6$ km. above the ground state line!

Attention should be drawn to the following points:
　　(i) Since $E_q$ is a function of $m^2$, positive and negative m give the same value, therefore the split levels are still doubly degenerate.

　　(ii) When the ground state also has I ≥ 1 one must take into account that the value of Q will generally be different for the excited and ground states.

　　(iii) The baricentre of the split lines corresponds to the position at which the line would appear in the absence of any quadrupole interaction and determines the value of $\delta$ the chemical shift.

Fig. 1.7(b) shows a quadrupole split spectrum for a 3/2 ⇔ 1/2 Mössbauer nucleus

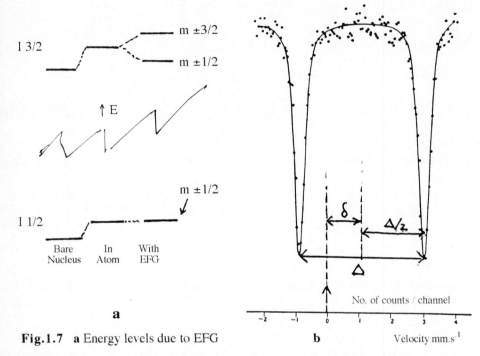

**Fig.1.7** **a** Energy levels due to EFG                    **b**                    Velocity mm.s$^{-1}$

**Fig.1.7** **b** shows a quadrupole split spectrum for a $3/2 \Leftrightarrow 1/2$ Mössbauer nucleus

Fortunately a very simple result can be obtained for this commonly occurring situation. Applying the full Hamiltonian operator from equation 1.4. to $|3/2,3/2\rangle$ it has already been shown that first two terms yield $3\beta|3/2,3/2\rangle$. Where $\beta = eQV_{zz}/12$

The third needs the evaluation of $1/2\eta(I_+^2 + I_-^2)|3/2,3/2\rangle$. Now $I_+|3/2,3/2\rangle = 0$ since

$m \leq 3/2$. Hence using Eq.1.3.

$$1/2\eta\ I_-^2\ |3/2,3/2\rangle = \{1/2\sqrt{3}\}\eta I\ |3/2,1/2\rangle = \{1/2\sqrt{3}\}\eta\ |3/2,-1/2\rangle$$

This gives the matrix                    $3/2,3/2\rangle$    $3/2,-1/2\rangle$

               $\langle3/2,3/2.$         $3\beta$          $3\beta\eta$

               $\langle3/2,-1/2.$      $3\beta\eta$       $-3\beta$

Which gives eigenvalues for the energy: $(3\beta - E)(-3\beta - E) - 3\beta^2\eta^2 = 0$. Whence the quadrupole interaction $E_q = \pm 6\beta(1 + \eta^2/3)^{1/2}$. But this is only valid for a.

$3/2 \Leftrightarrow 1/2$ change.

    Some important conclusions can be drawn from this result:

    i/ Since $\eta \leq 1$ the greatest effect the asymmetry parameter can have on the quadrupole splitting only amounts to about 15.5%

    ii/ The spectrum only yields one quantity, so that alone it cannot determine both $\eta$ and eq.

iii/ The spectrum does not tell one whether the |3/2,1/2.> or the |3/2,3/2.> state lies at the higher energy. That is to say that the spectrum does not give the sign of eq.

Ways by which these omissions can be made good will be considered later.

When $I > 3/2$ and $\eta > 0$ the calculations are more complicated but the Hamiltonian operator given in equation 1.4 can still be used to calculate the energy levels. The difficulties arise because, due to mixing of states, m is no longer a good quantum number. Some examples of higher I values will be considered later.

In the commonly occurring case of a $3/2 \Leftrightarrow 1/2$ transition the separation of the two lines in the quadrupole split spectrum, called the **quadrupole splitting**, will be denoted by $\Delta$,   $\Delta = eQV_{zz} (1+\eta^2/3)^{1/2}$.

## 1.5  INTERACTION OF NUCLEUS WITH A MAGNETIC FIELD

In the simpler situations this interaction is more straightforward to derive than the electric quadrupolar interaction. The magnetic field removes all the degeneracy of the nuclear levels, so that a nuclear spin value of 1 gives rise to 2I+1 energy levels. This splitting is sometimes called the nuclear Zeeman effect. Transitions between the levels produced in this way are concerned in nuclear magnetic resonance experiments.

The Hamiltonian for the interaction of the magnetic flux, B, with the nuclear magnetic moment, $\mu$, is $H_m = -\mu \bullet \mathbf{B}$.

The Hamiltonian operator for the magnetic interaction is:

$$\mathbf{H_m} = g_n\mu_n \mathbf{B} \bullet \mathbf{I}.$$

where $g_n$ is the nuclear Lande or gyromagnetic factor and $\mu_n = \mu/Ig_n$. Choosing the   z axis in the direction of the magnetic field   $\mathbf{H_m}|I.> = -g_n\mu_n\mathbf{B}$ m. Hence we obtain 2I+1 equally spaced levels arising from  m = I,(I-1), ---- (-I+1),-I.

In the important case of $^{57}$Fe both the 3/2 excited level and the 1/2 ground state are split; but the sign and magnitude of $\mu_n$ are different for the two states.

The pattern of energy levels that arises is shown in Fig.1.8 a.  The ground state has a positive moment   that is smaller than the absolute magnitude of the negative moment  for the excited state, but since the level separation depends on $\mu/I$ the separation of the ground state levels is greater than that of the excited state. For $^{119}$Sn the signs of the moments of the ground and excited states are the opposite of those for $^{57}$Fe, Fig.1.8 c.

As can be seen in Fig.1.8 two possible transitions are omitted; from -1/2 to +3/2 and from +1/2 to -3/2.   This is because selection rules for magnetic dipolar emission of a photon require that  $\Delta$m for the transition $= 0, \pm1$. Thus the nature of the nuclear photon emission is relevant to the spectrum. The less common instances of electric dipolar emission have the same selection rule, but for electric quadrupolar emission   m $= \pm2$ is also permitted.

Examination of Fig.1.8 a shows that if one numbers the lines in order of increasing energy, the separation of lines 2 and 4 or of lines 3 and 5 gives the magnetic splitting of the nuclear ground state. While the separation of lines 1 and 2, 2 and 3, 4 and

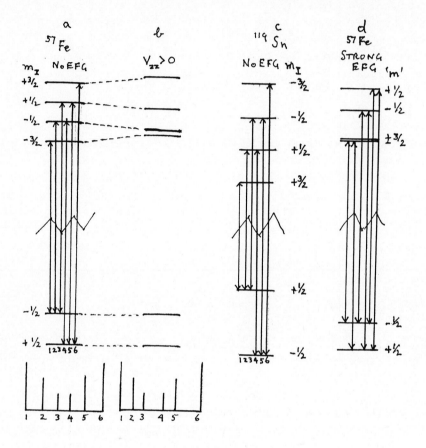

**Fig.1.8**

(a)Magnetic splitting of levels for $^{57}$Fe. (c) For $^{119}$Sn (b) As (a) but with moderate EFG. (d) As (a) with strong EFG.

5 or 5 and 6 give the magnetic splitting of the excited state. The magnetically split spectrum for a soft iron foil is shown in Fig.1.9.

In order that the magnetic dipolar splitting be well defined it is necessary that the field the nucleus experiences should remain the same for a period exceeding the inverse of the **Larmor precession frequency** of the excited nucleus. This time corresponds to the period of precession the nuclear magnetic moment around the direction of the magnetic field. An analogous condition requires that the EFG persists for a related time. However although the condition for magnetic splitting is often unfulfilled it is rare for quadrupole splitting to be absent for this cause.

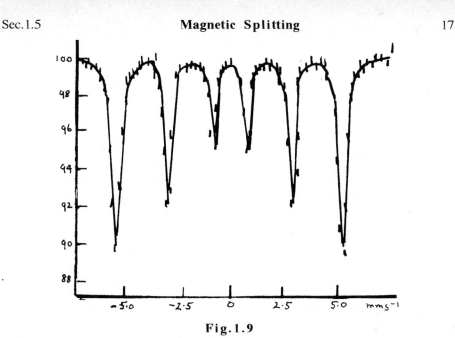

**Fig.1.9**

At first it might appear that all paramagnetic solids should yield magnetically split spectra. Certainly the magnetic field at the nucleus is sufficiently large, but the field is usually relaxing and changing its direction too quickly to fulfil the above mentioned condition and over the mean life of the excited state the field averages to zero.

This relaxation takes place by two mechanisms; **spin-spin** and **spin-lattice relaxation**. The first of these processes depends on the proximity of other like atoms and therefore on the concentration of paramagnetic Mössbauer atoms, for example iron atoms, in the solid. It will be slow in solids composed of a compound of high molecular weight possessing only a single iron atom. It can also be expected to be slow in quickly frozen dilute solutions of an iron salt. Free radicals present in the solid will generally enhance the rate of spin-spin relaxation.

The spin-lattice relaxation process takes place by the interaction of the paramagnetic moment with the lattice vibrations of the solid. If the Mössbauer atom has orbital angular momentum spin lattice relaxation is usually rather fast. It can be slowed down by cooling the absorber. At very low temperatures paramagnetic samples often begin to show magnetic splitting, but temperatures in the liquid helium region are usually necessary. For measurements down to liquid nitrogen temperature paramagnetic solids usually do not show magnetic splitting, although line broadening may take place.

The commoner cases of magnetically split spectra are with ferromagnetic or antiferromagnetic solids. In both kinds of solid the magnetic field at the nucleus remains the same for a sufficient period for splitting to occur.

An externally applied magnetic field will also lead to magnetic splitting but very strong fields are needed.

## 1.6  COMBINED MAGNETIC AND QUADRUPOLE INTERACTIONS

Very often a magnetic splitting is combined with an EFG. A combination of the Hamiltonian operators established in the last two sections can be used to obtain the energy levels that ensue, but unless the system complies with some rather severe restrictions the results are complicated and solutions can only be obtained by numerical methods using a computer.

There is, however, one set of conditions that appear in practice and where explicit results can be obtained. If (i) the EFG interaction is appreciably weaker than the magnetic interaction and (ii) $\eta = 0$; a perturbation treatment can be used to show that:

$$E_{MQ} = -g_n\mu_n \, mB + (-1)^{(|m| +1/2)} \frac{eQV_{zz}}{4}\frac{(3\cos^2\theta-1)}{2}.$$

where $\theta$ is the angle between the magnetic field and the z axis of the EFG. As a result the EFG displaces the the levels from their positions for magnetic splitting alone by

$$\pm \frac{eQV_{zz}}{4}\frac{(3\cos^2\theta -1)}{2}$$

The spectrum obtained (Fig.1.8.b) will not yield separate values of eq and $\eta$. It no longer has the symmetry of the simple magnetically split spectrum, except in the fortuitous circumstances when $\cos\theta = 1/\sqrt{3}$.

$eV_{zz}Q\dfrac{(3\cos^2\theta-1)}{2}$, or $\varepsilon$, can be obtained from the difference in the separations

of lines 5 and 6 and of lines 1 and 2 of the spectrum.
An example is shown in the spectrum of $\alpha$ $Fe_2O_3$, Fig.1.10.

Counts

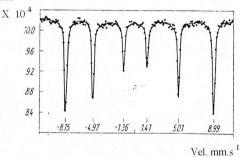

Note difference in separation of lines 1 and 2 and of 5 and 6.

Vel. mm.s$^{-1}$

**Fig.1.10**

In the general case the full Hamiltonian must be used to obtain the energy levels.

$$H = -g_n\mu_n \, \mathbf{I}\cdot\mathbf{B} + eQV_{zz} \, [3I_z^2 -I(I+1) +1/2\eta(I_x^2 - I_y^2 )]$$

If the magnetic interaction is small and can be regarded as a perturbation of the quadrupole interaction a splitting of levels such as is shown in Fig,1.8.d may ensue.

**References for Chapter 1**

1.1    See for example: Moon, P.B., (1951) *Proc.Royal Soc.* <u>64</u>, 76.

1.2    See Atkins, P.W., (1974) *" Quanta, a Handbook of Concepts"* Clarendon
Press, Oxford.

1.3    See for example: Glusker, J.P., Lewis, M. and Rossi, M., (1994)
*"Chemical Structure Analysis for Chemists and Biologists"* p.528,
Pub. VCH, N.Y.

1.4    Freeman, A.J. and Frankel, R.B., (1967) *"Hyperfine Interactions"*,
Academic Press, N.Y.

1.5    See for example: McWeeny, R., (1989) *"Methods of Molecular Quantum
Mechanics"* 2nd edition, Academic Press, London.

1.6    See for example: Moss, R., Chap.11 in *"Avanced Molecular Quantum
Mechanics"*, Chapman Hall, London.

1.7    Shirley, D.A., (1964) *Rev.Mod.Phys.*, <u>36</u>, 339.

1.8    See for example: Ziman, J.M., (1968*) "Elements of Advanced Quantum
Theory"* C.U.P., London.

**Acknowledgements**

Fig.1.10 reproduced with permission from Van der Woude, F., (1966)
*Phys.Stat.Solidi*, <u>**47**</u>, 417.

# 2

# Practical aspects of Mössbauer Spectroscopy

## 2.1   GENERAL NATURE OF TECHNIQUE

If the ground state produced in a Mössbauer transition occurs naturally, one can use absorption measurements to investigate the electronic environment of atoms of the element in an absorber.   A sample of the absorber compound is interposed between a source of the Mössbauer radiation and a detector.   The number of photons recorded by the detector is then measured as a function of the relative velocity of the source and absorber.   To ensure an essentially monochromatic emission from the source the radiating nuclei must be located in a solid at positions where they experience no electric field gradient or magnetic field.

A small number of systems can be studied in which the ground state is still radioactive, but with a long enough life-time that macroscopic amounts of compounds containing this state can be obtained.

Emission spectra can be examined by similar measurements using an absorber that has been shown to display a single absorption line. It will be shown later that the interpretation of the results of such measurements is sometimes rather difficult (See Chap.10).

## 2.2  SOURCES

The excited state in the Mössbauer transition generally has a half-life between $10^{-6}$ and $10^{-9}$ s.   Below $10^{-9}$ s the line width of the emission becomes so large that the tiny hyperfine interactions cannot be seen. Above $10^{-6}$ s the emission line is so narrow that extraordinary care has to be taken to avoid fortuitous line broadening due to vibration or other minute movements of the absorber in relation to the source. In principle the sensitivity of the measurement increases and very small changes in EFG at the Mössbauer atom in the absorber will change the spectrum. Such changes could arise from solid state defects in the absorber, and the preparation of absorbers for such measurement has to take this into account. These effects complicate the production of $^{181}$W sources for tantalum Mössbauer spectroscopy, the $2\Gamma_t$ being 0.0065 mm.s$^{-1}$. Nonetheless using specially designed spectrometers some useful results have been obtained using transitions with half-lives up to nearly $10^{-5}$ s.

### 2.2.1 **Nuclear requirements**

In Chapter 1 it was shown that a large Mössbauer fraction demands a low recoil and photon energy. A number of other nuclear factors have to be taken into account in developing a satisfactory Mössbauer source. Because of the short half-lives of the Mössbauer transitions it will be necessary that the Mössbauer excited state be fed continuously by the decay of some much longer lived parent species. The longer the half-life of this parent species the less frequent the need for its, usually very expensive, replacement.

To ensure the highest proportion of Mössbauer scattering events in the absorber it is desirable that the abundance of the ground state species in the natural element be high. For instance, in the case of $^{57}$Fe this isotope comprises only about 2% of natural iron. This means that a great gain in sensitivity ensues if one uses compounds for the absorber made from separated $^{57}$Fe. But separated stable isotopes are very expensive.

Another desirable feature is that a high proportion of parent decay events lead to Mössbauer photon emission. Decay of radioactive isotopes usually takes place by a number of channels, some missing the Mössbauer excited level entirely.

More important in most cases is the competition between photon emission and **internal conversion** in the decay of the Mössbauer excited state. When internal conversion occurs instead of photon emission, the atom ejects one of its orbital electrons with energy determined by the difference between the alternative photon energy and the binding energy of the electron in the orbital involved. This is always a rather probable event when the photon energy is small. The proportion is given by the total **internal conversion coefficient**, $\alpha$, which is the ratio of the number of conversion electrons emitted to the number of photons. A low value of the internal conversion coefficient is desirable.

The orbital electron vacancy created by internal conversion fills with the emission of a characteristic X-ray and this radiation can also be used to detect the resonant absorption events. The cross section presented by the ground state nucleus for resonant absorption of a Mössbauer photon of energy $E_\gamma$ is given by the Breit Wigner expression:

$$\sigma_0 = \frac{\lambda^2 (2I_{ex.} + 1)}{2\pi(2I_{gr.} + 1)} \quad \frac{1}{1 + \alpha} \qquad \text{Where } \lambda \text{ is the wavelength of the Mössbauer}$$

photon.

$$\sigma_0 = \frac{(hc)^2 (2I_{ex.} + 1)}{2\pi E_\gamma^2 (2I_{gr.} + 1)} \quad \frac{1}{1 + \alpha}$$

$$\sigma_0 = \frac{2.446 X 10^{-15} (2I_{ex.} + 1)}{E_\gamma^2 (keV)(2I_{gr} + 1)} \quad \frac{1}{1 + \alpha}.$$

Since this cross section should be as large as possible, it is important that $E_\gamma$ and $\alpha$ be as small as possible.

The fewer the photons of other energies emitted by the parent species the lower the background and the easier the measurement of the spectrum.

The radioactive decay process feeding the excited Mössbauer level should produce the minimum disturbance of the source lattice. Radioactive decay in a solid generally leads to recoil of the affected atom, as well as ionisation of the daughter and its surroundings.

The recoil energy will be dissipated in a very short time, well before the Mössbauer photon emission, but if the recoil has been sufficient to eject the nascent atom from its lattice site, it is likely to thermalise in an abnormal position where it will probably experience an electric field gradient. It is better therefore that the feed decay process is one that produces a very small recoil. Isomeric transition, orbital electron capture, and the emission of not too energetic beta particles fall into this category . The ionisation due to the decay events can lead to disturbing effects in insulators, but can be avoided in a metallic matrix (See Chap.10).

An alternative, but generally rather inconvenient, method of continuously generating the excited state is by **Coulomb excitation**. In principle it is applicable to all Mössbauer species. When a solid containing the ground state species is bombarded by heavy ions, not energetic enough to penetrate the Coulomb barrier presented by the ground state nuclei, for example 20 Mev $O^{8+}$ ions, the passage of the ion near the ground state nucleus leads to its excitation.

Clearly the bombardment has to continue throughout the spectroscopic measurement, usually a matter of some hours, and the measurements have to be made close to the accelerator producing the energetic ions. The bombarded solid suffers considerable radiolytic damage, the nascent excited atom recoils, and to obtain a satisfactory line width it is necessary that there is a high probability of the excited atom thermalising in a normal lattice site before the Mössbauer emission. The method is really only to be recommended when no reasonably long-lived parent is available.

A useful application of this technique uses the recoil to implant the Mössbauer atom into a solid which may not contain the same chemical species as the Mössbauer atom in its composition. Thus $^{57m}$Fe can be implanted in diamond. The emission spectrum obtained from such a source reflects the environment of the implanted atom.

It may eventually prove possible to use excitation by synchrotron radiation to provide sources for Mössbauer spectroscopy.

### 2.2.2 **The Source Matrix**

The parent species must be incorporated in a non-magnetic solid in sites possessing cubic symmetry. The matrix should have a high Debye temperature so as to ensure a large Mössbauer fraction.

The element involved in the Mössbauer transition is not usually a satisfactory matrix for the parent specie. This is because the presence of the ground state species in the source leads to resonant absorption within the source and thence to line broadening. For this reason the line width found with very strong sources, initially free from the ground state, increases with the age of the source.

During the life of the source the matrix will be subjected to a considerable dose of ionising radiation. It is necessary that this radiolytic action shall not lead to chemical decomposition of the matrix, nor shall large numbers of defects be formed.

The incorporation of the parent atom in a high melting point cubic metal is most likely to give a source providing an emission spectrum of a single line of width close to the value set by the half-life for the Mössbauer transition. For instance, incorporation of $^{57}Co$ in rhodium provides a very satisfactory source for iron Mössbauer spectroscopy, giving a line width very little more than the theoretical value. If this is not possible incorporation of the parent in a cubic refractory compound may be satisfactory.

Not surprisingly, in view of the number of these desirable features, most sources make some compromises.

Figs. 2.1.a to 2.1.j give the principle features of the decay schemes of a number of the more important Mössbauer species. Minor modes of decay are omitted since these only make a small contribution to the back ground radiation in the measurements.

Tables 2.1.a & 2.1.b give the relevant nuclear characteristics of several Mössbauer species suitable for chemical studies.

The theoretical line width $\Gamma_t$ in $mm.s^{-1}$ is given by the expression $136.8 / E_\gamma t_{1/2}$ with $E_\gamma$ in keV and $t_{1/2}$, the half-life of the excited state, in nanoseconds.

Some selected examples of sources will be described.

Iron is the most favourable element for Mössbauer spectroscopy, although the natural abundance of the ground state is low ($\approx 2\%$) and the internal conversion coefficient for the excited state is rather high ($\approx 8.2$). Satisfactory absorption measurements can be made with source and absorber at room temperature if the Debye temperature of the absorber is not too low.

The parent species is made by the deuteron irradiation of iron, $^{56}Fe(d,n.)^{57}Co$. 9.5 MeV deuterons are necessary and so, like all cyclotron products, the $^{57}Co$ is expensive compared to reactor irradiation products. The $^{57}Co$ is separated chemically from the iron target, electroplated onto a suitable cubic metal and diffused in by heating in vacuo. A thin aluminium foil filter can be used to attenuate the intense 6.3 keV X radiation from the source.

Tin is another element that can be studied at room temperature, although the low Debye temperatures of the organometallic compounds demand that these absorbers be

## Table 2.1 a

| A | B | C | D | E | F | G | H | I |
|---|---|---|---|---|---|---|---|---|
| $^{57}$Fe | 2.14 | $^{57}$Co ec | 270 d | 14.41 | 141.1 | 8.21 | -3/2 | -1/2 |
| $^{61}$Ni | 1.19 | $^{61}$Co β | 99 m | 67.4 | 7.6 | 0.12 | -5/2 | -3/2 |
| $^{67}$Zn | 4.11 | $^{67}$Ga ec | 78 h | 93.2 | 13201 | 0.89 | -3/2 | -5/2 |
| $^{99}$Ru | 12.72 | $^{99}$Rh ec | 16.1d | 90.0 | 29.7 | 0.47 | +3/2 | +5/2 |
| $^{119}$Sn | 8.58 | $^{119m}$Sn IT | 250 d | 23.87 | 26.5 | 5.12 | +3/2 | +1/2 |
| $^{121}$Sb | 57.25 | $^{121m}$Sn β | 76 y | 37.1 | 5.05 | 10 | +7/2 | +5/2 |
| $^{125}$Te | 6.99 | $^{125m}$Te IT | 58 d | 35.48 | 2.31 | 12.7 | +3/2 | +1/2 |
| $^{§17}$I | 100 | $^{127m}$Te β | 109 d | 57.6 | 2.68 | 3.3 | +7/2 | +5/2 |
| $^{129}$I | * | $^{129m}$Te β | 33 d | 27.7 | 24.24 | 5.3 | +5/2 | +7/2 |
| $^{§19}$Xe | 26.44 | $^{§19}$I β | 17. 10$^7$y | 39.58 | 1.46 | 11.8 | +3/2 | +1/2 |
| $^{149}$Sm | 13.8 | $^{149}$Eu ec | 106 d | 22.5 | 10.96 | 12 | -5/2 | -7/2 |
| $^{151}$Eu | 47.7 | $^{151}$Gd ec | 129 d | 21.6 | 13.0 | 29 | +7/2 | +5/2 |
| $^{157}$Gd | 15.7 | $^{157}$Eu β | 15.4 h | 64.0 | 664 | 0.8 | +5/2 | -3/2 |
| $^{161}$Dy | 18.9 | $^{161}$Tb β | 6.9 d | 25.65 | 41.8 | 3 | -5/2 | +5/2 |
| $^{181}$Ta | 100 | $^{181}$W ec | 140 d | 6.25 | 9810 | 46 | +9/2 | +7/2 |
| $^{182}$W | 26.4 | $^{182}$Ta β | 115 d | 100.1 | 1.98 | 3.85 | +2 | 0 |
| $^{189}$Os | 16.1 | $^{189}$Ir ec | 13.3 d | 69.6 | 2.37 | 8 | -5/2 | -3/2 |
| $^{193}$Ir | 62.7 | $^{193}$Os β | 31 h | 73.0 | 9.09 | 6.5 | +1/2 | +3/2 |
| $^{197}$Au | 100 | $^{197}$Pt β | 18 h | 77.3 | 2.71 | 4.3 | +1/2 | +3/2 |
| $^{237}$Np | 100 | $^{257}$Am α | 458 d | 59.54 | 63 | 1.06 | -5/2 | +5/2 |

A Isotope of element concerned, B Natural abundance, C Parent feeding Mössbauer level, ec Orbital electron capture, IT Isomeric transition, D Half life of parent, E Energy of Mössbauer photons in keV, F Mean lifetime of Mössbauer excited state in ns, G Internal conversion coefficient, H Nuclear spin of excited state, I Nuclear spin of ground state.
* Fission product of very long half life.

cooled to liquid nitrogen temperature. The natural abundance of the ground state is again not very high, ≈8.6%, and the internal conversion coefficient, ≈5.1, is rather high.

The $^{119m}$Sn is made by the neutron irradiation of tin, $^{118}$Sn(n,γ.)$^{119m}$Sn. The abundance of $^{118}$Sn is satisfactory, ≈24%, but its capture cross section for thermal neutrons to give $^{119m}$Sn is very small, so that the use of separated $^{118}$Sn is advantageous. Long irradiations are needed. Even if $^{118}$Sn is used the product still contains the ground state $^{119}$Sn formed by the competing $^{118}$Sn(n,γ)$^{119}$Sn reaction.The irradiated β-tin can be used after annealing as a source, but the Mössbauer fraction is low and the line width substantially greater than the theoretical value, owing to unresolved quadrupole splitting.

. A much better Mössbauer fraction and line width is possible by converting it to $CaSnO_3$ or $BaSnO_3$.

## Table 2.1 b

| A | J | K | L | M | N | P | Q | R |
|---|---|---|---|---|---|---|---|---|
| $^{57}$Fe | 256 | 0.194 | M1 | - | 0 | +0.213 | +0.091 | -0.155 |
| $^{61}$Ni | 72.1 | 0.77 | M1 | - | +0.16 | -0.2 | -0.75 | +0.48 |
| $^{67}$Zn | 4.96 | 0.0003 | E2 | | +0.16 | | +o.88 | |
| $^{99}$Ru | 13.9 | 0.149 | E2/M1 | + | +0.12 | +0.34 | -0.62 | -0.28 |
| $^{119}$Sn | 140 | 0.628 | M1 | + | 0 | -0.08 | -1.04 | +0.67 |
| $^{121}$Sb | 21.5 | 2.10 | M1 | - | -0.26 | -0.36 | +3.36 | +2.35 |
| $^{125}$Te | 28.4 | 5.01 | M1 | + | 0 | -0.2 | | |
| $^{127}$I | 22.8 | 2.56 | M1 | - | -0.79 | -0.71 | +2.81 | +2.92 |
| $^{129}$I | 38 | 0.59 | M1 | + | -0.55 | -0.68 | +2.62 | +2.84 |
| $^{129}$Xe | 24.4 | 6.83 | M1 | + | 0 | -0.4 | | |
| $^{149}$Sm | 28 | 1.60 | M1 | + | | | -0.66 | -0.62 |
| $^{151}$Eu | 23 | 1.41 | M1 | + | 1.16 | 1.50 | 3.46 | 2.59 |
| $^{157}$Gd | 50 | 0.0093 | E1 | | 1.67 | 2.97 | | |
| $^{161}$Dy | 93 | 0.37 | E1 | | 1.35 | 1.36 | -0.47 | +0.59 |
| $^{181}$Ta | 167 | 0.0064 | E1 | + | | | 2.36 | 5.22 |
| $^{182}$W | 25.2 | 1.99 | E2 | - | 0 | 0.51 | | |
| $^{189}$Os | 8.4 | 2.41 | E2/M1 | - | +0.8 | -0.6 | +0.66 | +0.98 |
| $^{193}$Ir | 3.1 | 0.59 | E2/M1 | + | +1.5 | 0 | +0.47 | +0.16 |
| $^{197}$Au | 3.8 | 1.88 | E2/M1 | + | 0.59 | 0 | +0.14 | +0.42 |
| $^{237}$Np | 33.5 | 0.11 | E1 | | | | | |

$J = \sigma_0$, $K = 2\Gamma_t$, L Type of emission, M Sign of $\Delta R/R$, Quadrupole moment of ground state, N, and of excited state P, in units of $10^{-24}$ cm$^2$, Magnetic moment of ground state, Q, and of excited state, R, in nuclear magnetons.

Such a source gives a line width not much greater than the theoretical value.

The source always emits the tin X-rays because of the strongly internally converted 65.66 keV transition preceding the Mössbauer emission. Interference from this X-radiation can be considerably reduced by interposing a palladium foil filter between the absorber and the detector. The palladium has its K absorption edge at 24.35 keV, between the undesirable X-rays at around 25 keV and the desired 23.88 keV Mössbauer photons. With such a filter radiation of more than 24.35 keV will be attenuated much more strongly than the less energetic Mössbauer radiation.

Most of the other elements that can be studied require the source and absorber to be cooled to liquid nitrogen temperature or below, to obtain a satisfactory Mössbauer fraction . Inorganic compounds of the next two elements antimony and tellurium, give acceptable spectra at liquid nitrogen temperature with inorganic compounds, but a lower temperature is necessary for their organic derivatives.

**Fig.2.1.a.**          **Fig.2.1.b.**          **Fig.2.1.c.**

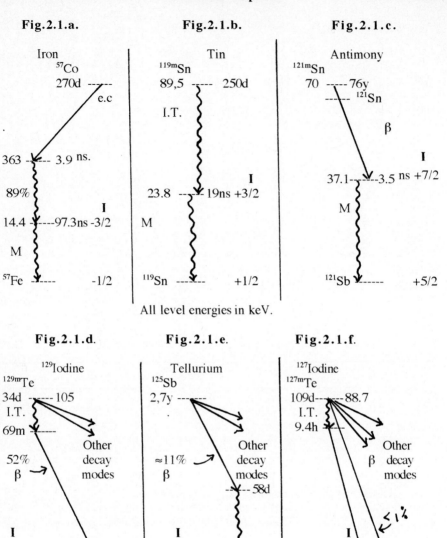

All level energies in keV.

**Fig.2.1.d.**          **Fig.2.1.e.**          **Fig.2.1.f.**

[127]I Excited state is fed mostly by β decay of 109d [127m]Te
[129]I Excited state is fed mostly by decay of 69m [129]Te
Other decay modes involve several photon emissions.
M Denotes Mössbauer emission.

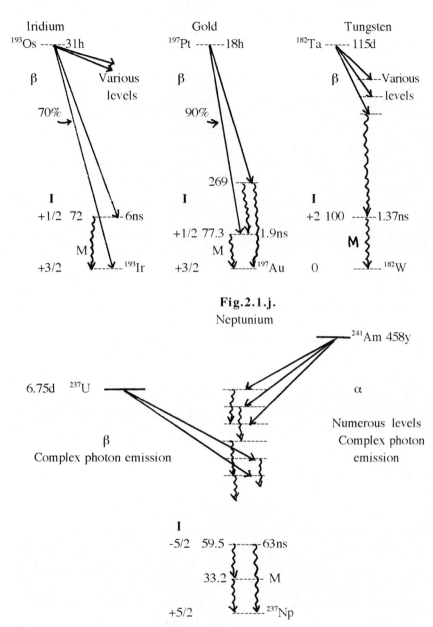

**Fig.2.1.g.**

Iridium

**Fig.2.1.h.**

Gold

**Fig.2.1.i**

Tungsten

**Fig.2.1.j.**
Neptunium

In the case of antimony the parent species, an isomeric state of a tin isotope, feeds the $^{121m}$Sb by β decay, but the maximum β particle energy is modest, ≈ 0.4 MeV and the recoil small. The $^{121m}$Sn is prepared by the neutron irradiation of tin metal, which contains ≈33% of $^{120}$Sn, but the capture cross section is very small and the half life of the tin isomer is very long: hence very long irradiations in a very high flux reactor are necessary. After irradiation the other tin activities are allowed to decay for about two years, the tin dissolved and separated from impurity activities, and converted to $BaSnO_3$. Such a source gives an acceptable Mössbauer fraction and line width. The source emits antimony K X-rays at about 26 keV but these are easily resolved from the 35.48 keV Mössbauer emission.

There are several ways of feeding the $^{125m}Te$ excited state. It is formed directly in the orbital electron capture decay of $^{125}I$. The latter isotope has the advantage of being readily commercially available because of its important medical applications. However the electronic disturbance associated with the decay process leads to line broadening unless the iodine is in an essentially metallic matrix. Diffusion of the iodine into copper produces a source with reasonable characteristics.

A better source is based on the use of the rather long-lived, 58 day, isomeric state of $^{125}Te$. This is formed by the decay of $^{125}Sb$, but it is inadvisable to try to use this directly because of the wide variety of its other photon emissions. The $^{125}Sb$ is formed by the β decay of $^{125}Sn$ formed by neutron capture by $^{124}Sn$. But the capture cross section is small, the abundance of $^{124}Sn$ low, and the half lives involved are rather long.

It is better to carry out a long irradiation of separated $^{124}Sn$, and after allowing time for the growth of $^{125m}Te$, to separate the isomer chemically. It can be deposited on platinum and diffused into the metal by heating in vacuo. A rather better source can be made by converting the $^{§15m}Te$ to $\beta TeO_3$ The line width and Mössbauer fraction of these sources are very satisfactory.

There is interference by X-rays but this can be avoided by recording the escape peak with a xenon filled proportional counter (See below). A considerable gain in Mössbauer absorption can be obtained by using absorbers made from separated $^{125}Te$.

As a final example of source selection the case $^{193m}Ir$ is considered. For this element both the source and the absorber must be cooled to the temperature of liquid helium to obtain a satisfactory absorption. Unfortunately the long lived $^{193}Pt$, which decays by orbital electron capture, does not feed the Mössbauer state of $^{193}Ir$ so that the source has to be based on the $31h.^{193}Os$, which undergoes β decay with a maximum β particle energy of 1.13 MeV. The parent species can be made by the neutron irradiation of osmium, $^{192}Os(n.\gamma)^{193}Os$. The abundance of $^{192}Os$, 41% is ample but the cross section is rather small and several other osmium activities are produced. Thus lower background counting rates and easier selection of the Mössbauer photons can be achieved by using separated $^{192}Os$. The $^{193}Os$ itself emits several energies of γ rays

However a single line emission cannot be obtained by simply irradiating osmium metal, because it is hexagonal and some unresolved quadrupole splitting of the 3/2 ground state of $^{193}Ir$ occurs. A satisfactory single line source has been made using a cubic alloy of separated $^{192}Os$ with niobium. Such a source can be re-irradiated repeatedly to regenerate the comparatively short-lived $^{193}Os$..

Measurements are still complicated by the complex γ rays of the osmium parent and, more seriously, by iridium X ray emission.


## 2.3    **DETECTORS**

The requisite properties are high efficiency and good energy resolution. As can be seen from the decay schemes shown in Fig,2.1 the Mössbauer photons are invariably accompanied by other γ rays and X-rays arising from internal conversion and photoabsorption events. The detector should be chosen to facilitate the selection of only the Mössbauer photons. Three kinds of detector are in common use.


### 2.3.1  **Scintillation detectors**

A sodium iodide crystal, activated by a small amount of thallium, observed by a photomultiplier tube, gives voltage pulses whose size distribution reflects the energies of the incident photons. For not too energetic photons, $E_\gamma < 150$ keV, the incident photons interact principally with the iodide ions suffering photo-absorption. The electrons ejected in these events, mostly from the 1s orbital of the iodine, if this is energetically possible, produce a burst of ultraviolet photons in the crystal which are converted by the photomultiplier tube into a voltage pulse. The height of these pulses depends on the energies of the ejected electrons. The filling of the orbital vacancy on the iodine produced in this way leads to the emission of an iodine X-ray. This may also suffer photo-absorption in the crystal. In some events internal conversion may occur, and this leads to another energetic electron and a vacancy in an orbital of higher principal quantum number. By a combination of events of this kind there occurs a transformation of the energy of the original photon into kinetic energy of electrons and thence to ultraviolet photons and a voltage pulse from the photomultiplier tube. The height of the resultant pulse is a measure of the original photon energy.

If the crystal is not very big the X-ray photon may escape from the crystal without becoming involved in another photo-absorption event and its energy will not contribute to the observed voltage pulse. Hence mono-energetic photons incident on the crystal will produce a voltage height spectrum with a photo-peak arising from complete absorption events; and an escape peak, due to events in which the X-ray photon escapes.

Both peaks show a range of pulse heights or widths, because of varying amplification by the photomultiplier tube. This spectrum is superposed on a continuous background of pulse heights due to Compton scattering and other events. The width of the photo-peak, which determines the energy resolution of the counter, usually amounts to about 8% of the pulse height at the photo-peak maximum intensity.

The efficiency of detection of such a counter is high for photons of $< 200$ keV. A scintillation counter is satisfactory when the photon energy is $> 40$ keV and there is unlikely to be interference from photons of not very different energy. It is not very satisfactory for detecting low energy photons in the presence of much higher energy photons, as is commonly needed in Mössbauer spectroscopy (e.g. for $^{57m}$Fe, see $^{57}$Co decay scheme, Fig. 2.1).

Some improvement can be obtained by using a thin crystal, so that the more energetic photons dissipate only part of their energy in the crystal. However a thin crystal will give a stronger escape peak. Indeed sometimes it is better to use the escape peak signal to count the Mössbauer photons. This technique may enable one to avoid interference from X-rays.

A novel type of detector, suitable for iron Mössbauer spectroscopy, has been developed in which an iron compound was incorporated in an organic scintillator. The internal conversion arising from the $^{57m}$Fe produced by resonant absorption in the absorber-cum-detector leads to electrons of short range that give a burst of ultraviolet photons detected by a photomultiplier tube. This process is almost ten times as likely as photon re-emission. Such a detector is relatively insensitive to γ and X-rays but very sensitive to the resonant absorption events. Similar counters can be made for other Mössbauer radiations.

### 2.3.2 **Proportional counters**

The proportional counter is usually a cylindrical ionisation chamber with a central wire anode. It is filled with an inert gas and a small percentage of methane. Over a certain range of operating voltages electrons produced in ionising events in the chamber cause further ionisation during their accelerating passage to the anode and give a voltage pulse proportional in height to the initiating ionisation. An internal amplification of $10^2$ to $10^4$ can be obtained. At too high a voltage a discharge occurs. Further amplification by an external amplifier is necessary before pulse-size selection and counting.

These counters are not really suitable if the photon energy much exceeds 40 keV. Their efficiency is lower than for the scintillation counter, but it can be increased for low photon energies by filling with the heavier inert gases and using a higher pressure of gas. The latter expedient demands higher operating voltages. The peak widths in the pulse height spectrum from the proportional counter are appreciably narrower than for the scintillation counter. The width arises because of statistical fluctuations in the multiplication events.

The interaction of the photons with the gas filling is essentially similar to that described for the scintillation counter. The filling is more transparent to the photons and for the normal size of counter the escape peak is more prominent. Indeed this peak is often suitable for counting the Mössbauer photons.

Filters with a convenient K-absorption edge can sometimes be used to attenuate the photo peaks from X-rays from the source and to facilitate measurement of the escape peak events.(See $^{119m}$Sn source above).

The proportional counter is especially well suited to the measurement of the rather low energy $^{57m}$Fe Mössbauer photons. The counter has to be fitted with a window permitting the photon radiation to reach the inside of the counter. For this purpose a thin beryllium foil is best, but care must be taken to ensure that the beryllium has a negligible iron content.

### 2.3.3 Semiconductor detectors

For the highest resolution Ge/Li or pure Ge and Si/Li are best. The older lithium drifted germanium detectors have the serious disadvantage of needing to be kept permanently at the temperature of liquid nitrogen. They become useless if allowed to warm to room temperature. Very pure,"intrinsic", germanium detectors give similar resolution but can be kept at room temperature when not in use. Both kinds must be operated at liquid nitrogen temperature.

The germanium behaves as a solid ionisation chamber. The voltage pulse it produces from ionising events must be amplified externally and the height of the final pulse is proportional to the number of ions produced in the initiating event. As with the proportional counter a thin window, usually of beryllium, is necessary to enable the less energetic photons to reach the germanium.

The interaction of the photons with the germanium is essentially similar to that described for the NaI/Tl crystal. Because of the lower atomic number of the germanium, and even more so of silicon, the efficiency of counting and the proportion of events contributing to the photo-peak is generally less than for the scintillation counter.

However the resolution and line width for the semiconductor counter are an order of magnitude better than with the scintillation counter so that separation of events due to Mössbauer photons is easier, even in the presence of X rays and more energetic photons. The importance of the latter can be minimized by an appropriate choice of thickness of the detector.

For the lowest photon energies, below about 15 keV and particularly below 5 keV, a small lithium drifted silicon detector is better. A small detector of this kind is comparatively insensitive to photons above about 70 keV.

The counting rate possible with these detectors is restricted to less than about 50 kHz., although the scintillation detector wll allow about double this rate.

A schematic representation of the detector ensemble is shown in Fig.2.2. The output from the amplifier is fed to a pulse analyser which sorts the pulses according to their height and records the number of detected events giving pulses of different heights, displaying the spectrum on an oscilloscope. Inspection of the display helps one to identify the Mössbauer photon peak. The amplifier output is also fed to a single channel pulse analyser to select only the pulses in this peak. This analyser rejects pulses above or below certain predetermined heights, but those within the chosen range produce output pulses of fixed height.

The correct settings for the single channel analyser are most easily chosen using a coincidence technique. Most multichannel analysers have a facility for gating the input pulses so that only those coinciding in time with some other event are accepted. The output from the single channel analyser is used to control the gate and the feed from the amplifier to the multichannel analyser is delayed slightly to allow for the operating time of the single channel analyser. In this way the pulse height analyser only displays pulses that fall within the acceptance channel of the single channel analyser. It is then a simple matter to change the opper and lower levels of

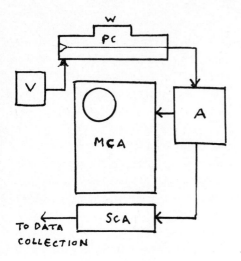

V Polarising voltage for detector, PC Proportional counter, A Linear amplifier MCA Multichannel - analyser, to monitor output polses. SCA Single channel analyser. to select Mössbauer photons. W Thin window on counter.

**Fig.2.2** Detector assembley

the latter analyser to bracket the range of pulse heights covered by the Mössbauer peak. Settings on either side of the peak at about 20% of the peak maximum intensity usually give good results.

## 2.4  DOPPLER MODULATION OF PHOTON ENERGY

The spectrum displays the extent of the absorption of the Mössbauer radiation as a function of the relative velocity of the source and absorber. For absorption spectra, a single source is used with various absorbers; for emission spectra various sources are explored with a single line absorber.

In the earliest work measurements were made, point by point, at various constant velocities. Such a procedure is now only used in a few special applications where very accurate velocities are necessary. Normally the source, or absorber, is driven repetitively through a cycle of motion, and the pulses from the single channel analyser recorded for each of a number of successive small time intervals in phase with the movement of the source or absorber.

Using the multichannel analyser in the multiscaler mode, or a small computer, which can serve the same purpose, the output from the single channel analyser is fed successively and repetitively from channel to channel in phase with the cycle of motion. In this way each multi-scaler channel accumulates data from the detector system while the source is moving at a fixed average velocity, the spread in velocities depending only on the number of channels used and the total velocity scan of the drive system. A small computer can serve the same function as the pulse analyser. [Ref.2.1].

### 2.4.1 **The Drive Waveform**

Three wave forms for the motion of the source are in common use, sinusoidal and two kinds of constant acceleration drives. These are shown in Fig.2.3 a, b & c

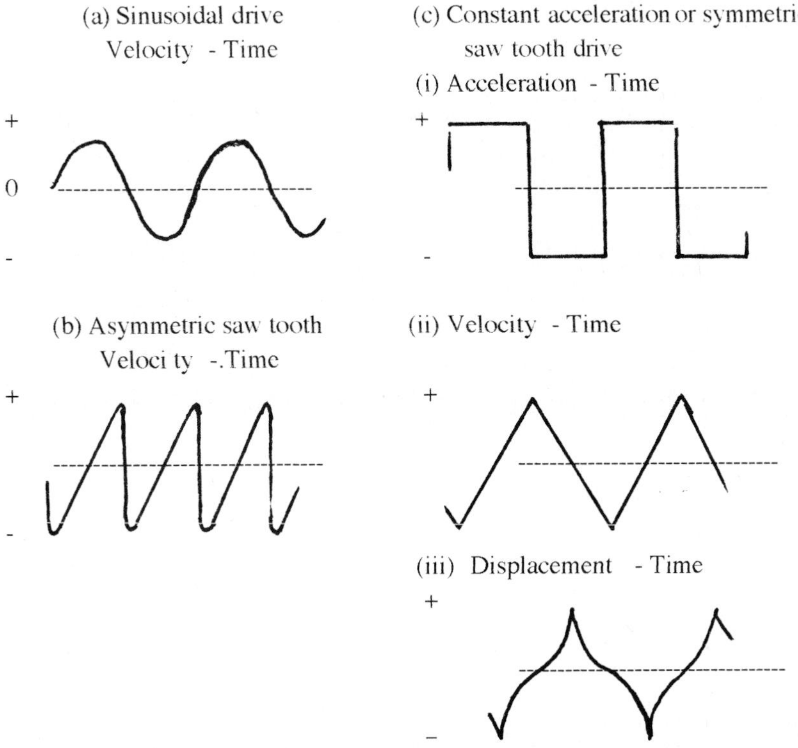

(a) Sinusoidal drive
Velocity - Time

(c) Constant acceleration or symmetric
saw tooth drive
(i) Acceleration - Time

(b) Asymmetric saw tooth
Veloci ty -.Time

(ii) Velocity - Time

(iii) Displacement - Time

**Fig.2.3a,b & c**

With the symmetric wave forms (a) or (c) the spectrum, which extends from -v to +v and back again, appears twice in the span of the multiscaler channels. However all the data is used and the computer program for analysis of the data can fold the two spectra to produce one set of results. The sinusoidal mode has the advantage of no discontinuities in the acceleration of the source which makes it easier for it to follow closely the idealised sinusoidal motion. The disadvantage is that the velocity coordinate is not linear, which makes it more difficult to make a visual assessment of the spectrum and estimate the line width and quadrupole splitting from the oscilloscopic presentation of the data.

The weakness of the symmetric constant acceleration mode is that for inertial reasons, and therefore the more so the heavier the source, it is not possible for the source to change instantly from positive to negative acceleration.

Hence the linearity of the velocity coordinate is impaired in the vicinity of the maxima and minima of the saw tooth drive. The asymmetric saw tooth drive gives only one spectrum in the cycle but it does not avoid the inertial problem.

### 2.4.2 **The Drive**

A schematic representation of the drive system is shown in Fig.2.4.

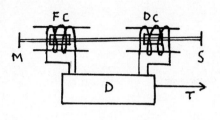

DC Drive coil, FC Sensor coil
supplies feedback, D Waveform
generator, S Source, T Synchronised pulses initiate
channel advance in data collection system,.

### Fig.2.4

The source is driven electro-mechanically by using a loud speaker coil or, better,
a specially designed vibrator coil. A similar, higher impedance, sensor solenoid,
rigidly attached to the drive rod, monitors the actual motion of the source and
provides feed back to the drive electronics to make the source follow the desired wave
form more closely.

The data collection system must be synchronised with the movement of the
source. This is usually done by the drive system generating sharp pulses at regular
intervals, in phase with the wave form. These are used to advance the input from
channel to channel in the analser or data collection system.  Multichannel analysers
usually provide for such switching. This technique is versatile and a wide range of
velocities is possible. The frequency of these pulses determines the dwell time for
each channel of the analyser.

In most cases the addition to the square wave drive of a pulse, at or near the
change of sign, and a parabolic component, improve the linearity of the acceleration
of the source.

To discover the best settings for the drive control one connects the monitor
signal and the drive signal to the inputs of a double beam oscilloscope, amplifies the
former signal so that its amplitude is the same as for the drive signal, inverts one
signal, and adds the two signals together. Many oscilloscopes provide facilities to
carry out these operations. The drive controls should then be adjusted so that the
trace of the addition signal is as flat as possible.

A schematic diagram of the complete spectrometer is shown in Fig. 2.5.
The vibrator system will have a resonant frequency,  often around 30 Hz. If driven at
a lower frequency the applied voltage opposes the restoring force due to the vibrator
springs, and the velocity of the source is in phase with the drive voltage. Above the
resonant frequency the source velocity and drive voltage are out of phase. Since the
maximum velocity and the frequency are inversely related for a given amplitude of
movement of the source, a low frequency demands a larger movement of the source
to provide a given maximum velocity. As shown below a large amplitude has some
undesirable implications. Below the resonant frequency a change in the mass of the
source assembly requires alterations to the controls of the drive electronics.

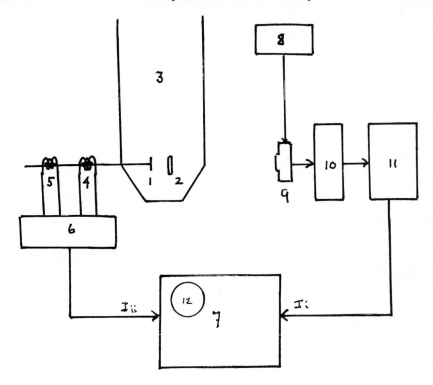

**Fig.2.5** Spectrometer ensemble. 1 The source. 2 The absorber. 3. Cryostat. 4. Drive coil

5. Sensor coil. 6. Drive system. 7. Data collector. 8.Power for detector. 9. Detector.

10. Amplifier. 11. Single channel analyser. 12. Oscilloscopic presentation of spectrum.

However, it is nearly always best to operate at a frequency of less than about 20 Hz. Operating at a drive frequency of 16 Hz the cycle time is 125 ms.; thus if the analyser has n channels (typically  n = 512 or 1024 ) the dwell time in each channel will be    (125 - ns)/n, where s is the time taken by the analyser to switch from channel to channel. The analyser should be chosen to have as small a value of s as possible. The dwell time is typically between 25 and 100 $\mu$s.

### 2.4.3  Geometric considerations

2.4.3.1 *Cosine  effect*   In the expression    $\delta E = vE_{\gamma}/c$  the v refers to the velocity of the source in the direction of emission of the photon. For photons not travelling parallel to a line joining the centre of the source to the centre of the detector the Doppler change becomes v $\cos\theta$ $E_{\gamma}/c$, where $\theta$ is the angle between the direction of emission and the above line. If the detector subtends too large an angle at the source this effect will  give  rise  to  appreciable  line  broadening. Assuming equal radii for a circular source and detector, r, the ratio r/l,  where l is the distance between them, should be as small as possible. A value < 0.2 is desirable.The cosine effect can also be reduced by the use of lead collimating stops in the photon beam. However, greater separation of source and detector or collimation by lead annuli reduce the counting rate for a given source so that some compromise is necessary.

2.4.3.2 *Inverse  square  effect*.  The value of l changes during the cycle of
motion of the source and this will lead to a change in counting rate independent of
any absorption. For a symmetric saw tooth drive a parabolic distortion of the base
line in the two spectra obtained will ensue. This can be allowed for in the computer
program used to process the results. The effect can also be avoided by movement of
the absorber rather than the source.

To avoid detecting too much secondary scattered radiation it is better to have the
absorber nearer to the source than the detector.

## 2.5  CALIBRATION

Absolute calibration, i.e. direct determination of the average velocity for each
channel of the analyser, is not often necessary. For most purposes one records the
six line spectrum given by a soft iron foil. The velocities at the peak positions for
these lines are accurately known in relation to the centre of gravity of the spectrum.
In this way one obtains six points on an average velocity versus channel number
plot.  The method has the advantage that the centre of gravity of the spectrum is
commonly used as the reference point for expressing chemical shifts for iron
compounds.  A foil enriched in $^{57}Fe$ is useful for this purpose and also enables one
to verify that the spectrometer is functioning correctly; the beginning of the six line
spectrum should be apparent on the pulse analyser display within a minute of
accumulating data.  But it should be noted that enriched foils lead to increased line
widths.

The lines in the soft iron spectrum lie at:
$$+ 5.328 \text{ mm.s}^{-1}, + 3.083 \text{ mm.s}^{-1}, \text{ and} + 0.839 \text{ mm.s}^{-1}.$$
and the other three lines at the corresponding negative velocities.

The velocity span needed to include all six lines, or even the middle four, is
rather large for some studies and an alternative method uses the two line quadrupole
split spectrum of sodium nitroprusside. This splitting is accurately known so that a
two point calibration can be made. The centre of this spectrum is also sometimes
used as a reference for chemical shifts, but this is not recommended, although the
compound is less variable in properties than soft iron.

The quadrupole splitting for sodium nitroprusside dihydrate is $1.705 \text{ mm.s}^{-1}$ and
the chemical shift relative to soft iron is $+ 0.258 \text{ mm.s}^{-1}$ ·

Both the above methods yield only the average velocity increment per channel
and fail to reveal small departures from linearity in the velocity-channel number
function. If an absolute calibration is needed and when higher velocities must be
covered an interferometric calibration can be used. The arrangement is shown in
Fig.2.6.

Light from a laser, L, is directed to a beam splitting prism, P.  One part travels
to the fixed mirror, $M_i$, the other to a mirror attached to the rear end of the source
rod, $M_{ii}$. The reflected beams interfere on combining beyond the beam splitter and

travel to the photodiode, D. As the path lengths differ, because of the movement of the source and the mirror $M_{ii}$, interference maxima will be experienced by the diode detector Fhe resultant pulses are recorded in the channels of the analyser, as when recording the spectra. Successive interference pulses correspond to movements of the

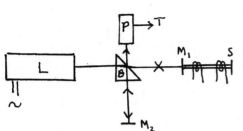

L Laser, S Source, B Beam splitter prism, $M_i$ Mirror rigidly attached to the source drive rod, $M_{ii}$ Another mirror.

P A photodiode, T Pulse lead to data collection system.

**Fig.2.6**. Interferometric calibration.

. Successive interference pulses correspond to movements of the mirror, and therefore of the source, by a quarter of the wavelength of the laser radiation. An oscillator can be used to provide an accurate time scale and the "spectrum" seen in the analyser gives the true average velocities for each channel.

## 2.6 THE ABSORBER

Mössbauer spectroscopy has the advantage that it is a non-destructive method of investigation. One might suppose that little attention need be paid to the preparation of the absorber, but for a variety of reasons this is not the case.

### 2.6.1 Thickness

The optimum thickness for the absorber is of some importance. It is convenient to express the absorber thickness in terms of the dimensionless quantity, $T_a = f_a t_a n_a \sigma_o$, where $f_a$ is the Mössbauer fraction for the absorber, $n_a$ is the number of atoms per unit volume in the absorber able to give rise to resonant absorption, $t_a$ is the thickness of the absorber and $\sigma_o$ the cross-section for Mössbauer absorption in the corresponding units of area. For the case of iron compound absorbers $T_a$ is given by $0.5907 f_a.m$ where m is the number of milligrams of natural iron per cm$^2$ in the absorber.

In the more general case for an absorber of molecules each containing one atom of the Mössbauer active element and with an abundance, r, for the Mössbauer isotope of this element one finds:

$T_a = f_a t_a N_o \sigma_o \rho_a r / M$. where $t_a$ is the actual thickness of the sample, $N_o$ is Avogadros number, $\rho_a$ is the density of the absorber, and M is the molecular weight of the compound.

For iron compound absorbers containing m milligrams of the compound per unit area $T_a = 32.99$ m $f_a/M$. In the general case if the absorber contains m mg/cm² of a compound of molecular weight M, which contains one atom of the Mössbauer element per molecule, $T_a = 6.02 \times 10^2 (m\ r\ f_a\ \sigma_0)/M$, where r is the isotopic abundance of the Mössbauer species.

Both the energy distribution of the incident photons and the cross-section, $\sigma_0$, are proportional to the function $1/[(E - E_\gamma)^2 + (\Gamma/2)^2]$, which gives the Lorentzian distribution. Thus in a thick absorber, in the first layers encountered by an incident Mössbauer photon beam, proportionately more photons will be absorbed from the beam in the region of energy close to $E_\gamma$ than in the wings of the distribution. Thus the observed absorption line becomes broader as the beam penetrates the absorber gradually losing its Lorentzian shape and tending to a Gaussian shape. If $T_a$ is not too large, < 4, the effective line width increases nearly linearly with thickness and departure from the Laurentzian shape is not very serious. In this region $\Gamma_{obs.} = (2 + 0.027T_a)\Gamma_t$, where $\Gamma_{obs.}$ is the experimental line width at half maximum absorption and $\Gamma_t$ is the theoretical line width, based on the mean life of the excited state.

Actual line widths tend to be a little larger, partly because of small differences in the environment of the atoms concerned in the absorber. Contributions from instrumental causes should be negligible.

Since larger line widths reduce the resolution of the measurement, this analysis would suggest that very thin absorbers might be best.

However, the extent of absorption increases with the thickness of the absorber while the number of photons recorded by the detector per unit time falls. There is therefore an optimum absorber thickness.

To establish the optimum thickness one must explore the absorption in the resonance region. The transmission through a resonant absorber, of thickness $t_a$, moving at a velocity vrelative to a source of thickness $t_s$, is given by:

$$p(v) = \exp.(-\mu_s\ \rho_s t_s)\ \{(1-f_s) \int_0^{t_s} N \exp.(-\mu_s\rho_s\ x)dx$$

$$+ \frac{f_s\Gamma}{2\pi} \int_{-\infty}^{+\infty} \frac{\exp.(-f_a n_a t_a \sigma_0 \Gamma^2)}{4(E - E_0)^2 + \Gamma^2}dE \text{ multiplied by}$$

$$\int_0^{t_s} \frac{N}{4[E - E_0(1 - v/c)]^2 + \Gamma^2} \exp \cdot \frac{-f_s n_s \sigma_0 \Gamma^2}{4[E-E_0(1-v/c)]^2} - \mu_s\ x\,dx$$

The subscripts a and s denote quantities relating to the absorber or source respectively. The $\mu$ are mass absorption coefficients for electronic absorption of the radiation at $E_0$, $\Gamma$ is the full line width at half maximum absorption, supposedly the same for both the emission and absorption processes. Other symbols are defined above.

For a uniformly populated source with N Mössbauer atoms per unit volume, $Nt_s$ is the total number of emitters per unit area of source. $t_s \geq x \geq 0$.

The term exp $-\mu_a \rho_a t_a$ represents the non-resonant electronic absorption of the incident radiation in the absorber, mainly due to photoelectric absorption, and is independent of v.

In the second term in the equation for p(v), the integral in E, allows for the change of resonant absorption with energy of the radiation and the integral in x for self absorption within the source. Since the resonant absorption greatly exceeds the electronic absorption the term $\mu_s$ can be neglected.

One can introduce a normalised transmission factor

$P(v) = p(v)/(Nt_s \exp.-\mu_a \rho_a t_a$

The first integral in x in the expression for P(v) becomes

$$(1-.f_s)\ \frac{(1-e^{-\mu_s \rho_s y_s})}{\mu_s \rho_s t_s}$$  This term represents the non-resonant transmission and it

goes to $(1-f_s)$. as $t_s \Rightarrow 0$.

$$f_s \Gamma/2\pi \int_{-\infty}^{+\infty} \exp.(-T_a \Gamma^2 / Q) dE \int_0^{t_s} (-N/Q') \exp.(T_s \Gamma^2 X/t_s Q'\ dX$$

Let $4(E-E_0)^2 + \Gamma^2 = Q$ and $4[E-E_0(1-v/c)]^2 + \Gamma^2 = Q'$. The second term in the expression for P(v) now becomes:

Now the last integral in the above expression is:

$[-t_s Q'/T_s \Gamma^2 \exp.-T_s \Gamma^2 x / t_s Q']_0 = (t_s Q'/T_s \Gamma^2)(1- \exp.-T_s \Gamma^2 /Q').$

Therefore the second term becomes:

$$(f_s / 2\pi T_s \int_{-\infty}^{+\infty} \exp.(-T_a \Gamma^2 / Q)(1 - \exp(-T_s \Gamma^2 / Q;')\ dE$$

Supposing the source is very thin and $T_s$ small, expanding the exponential in $T_s$ and taking only the first two terms one gets:

$$(f_s / 2\pi \Gamma \int_{-\infty}^{+\infty} \Gamma^2 / Q' . \exp.(-T_a \Gamma^2 / Q)\ dE$$

At resonance $Q = Q' = 4E^2 + \Gamma^2$ and one gets

$$P(v) = (1-f_s) + f_s / 2\pi \Gamma \int_{-\infty}^{+\infty} \Gamma^2 / Q\ \exp.(T_a \Gamma^2 / Q)\ dE.$$

The $(1-f_s)$ is a velocity independent term and will now be omitted.:

For the remaining term one obtains:

$$\frac{f_s}{2\pi\Gamma}\int_{-\infty}^{+\infty}\frac{\Gamma^2}{4E^2+\Gamma^2}\exp-\left[\frac{T_a\Gamma^2}{4E^2+\Gamma^2}\right]dE \quad \text{Put } E = \Gamma y/2$$

thus $dE = \Gamma dy/2$. The integral then becomes

$$f_s\ /\ 4\pi\int_{-\infty}^{+\infty}\frac{1}{1+y^2}\exp\ .-\frac{T_a}{1+y^2}\ dy$$

Now let $y = \tan\theta$, $dy = \mathrm{cosec}^2\theta\ d\theta$, $1 + y^2 = \mathrm{cosec}^2\theta$ and the

integral becomes $\dfrac{f_s}{4\pi}\displaystyle\int_{-\pi/2}^{+\pi/2}\exp(-T_a\cos^2\theta)d\theta$ .Finally, put

$\theta = 2\varphi$ and one gets $\dfrac{f_s}{2\pi}\exp(-T_a/2\displaystyle\int_{-\pi}^{+\pi}\exp(-T_a\cos\varphi)/2\ d\varphi$ .

This is a standard integral and gives:- $f_s\ \exp.(-T_a/2)\ J_0\ (iT_a/2)$, $J_0$ being the zero order Bessel function: $J_0(ix) = 1 + x^2/2^2 + x^4\ /(2.4)^2 + x^6\ /(2.4.6)^2$ +etc.

Thus the absorber does not show the usual exponential decrease in observed intensity with increase in thickness, because the Bessel function tends to flatten out more quickly.

The relative Mössbauer absorption is given by $1 - \exp.-(T_a/2)J_0(iT_a/2)$.

As the absorber gets thicker the total observed intensity decreases and $I = I_0\ \exp.-\mu_a\rho_a t_a$ Now the relative accuracy of the measurement, due to the statistical distribution of the decay events, is measured by $(I_0/I)^{1/2}$. In addition the observed line width depends on $(2+0.27T_a)$. Hence the most favourable thickness for the measurement is the value that maximises the function:

$$L(T_a)= 1- \exp.-(T_a/2)J_0(iT_a/2)\ /\ (I_0/I)^{1/2}.(2 + 0.27T_a).$$

Expressing $(I_0/I)$ in terms of $T_a$ one gets:

$(I_0/I) = \exp.\ T_a\mu_a M/\ f_a N_0\sigma_0 r$, where r is the abundance of the Mössbauer isotope. Suppose the absorber consists of a compound of the Mössbauer element, then $\mu_a$ becomes the mean mass absorption coefficient at the Mössbauer wavelength, $\displaystyle\sum_i m_i n_i \mu_i\ /M$ , where the compound has $n_i$ atoms per molecule of relative mass $m_i$ and mass absorption coefficient $\mu_i$. Taking $n_i = 1$ for the Mössbaue

element, then: $L(T_a) = 1 - \dfrac{\exp.-T_a\ /\ 2)\ J_0\ (iT_a\ /\ 2).\ \exp.-T_a\sum_i m_i n_i \mu_i\ /\ 2f_a\sigma_0 N_0 r}{(2+0,27T_a)}$

An explicit expression for the value of $T_a$ for which $L(T_a)$ is a maximum is difficult to derive, but a plot of $L(T_a)$ against $T_a$ for the case of $FeSO_4$ is shown in Fig.2.7, assuming various values of $f_a$. The optimum value of $T_a$ lies between 2 and 2.5.

Fortunately it is not very sensitive to the usually unknown value of $f_a$ and the region of the maximum is reasonably flat. A plot of this kind is a useful guide to the optimum absorber thickness but it does not allow for scattering by the

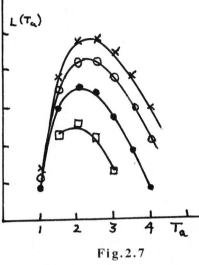

$L(T_a)$

**Fig.2.7**

absorber of the more energetic photons
emitted by the sourve and other factors.
The final choice of $T_a$ is best determined by experiment.

## 2.6.2  Texture

The intensities of the lines in a Mössbauer spectrum depend on the orientation
of the crystals in the absorber (See Chap.3).  It is convenient to use either single
crystal absorbers or randomly orientated small crystals in a powder. For most
purposes the latter choice is most convenient, but with many compounds
considerable care is necessary to ensure a random orientation of particles of the
absorber.  This is a particular problem with absorber materials such as mica that
crystallise in thin plates.  Sometimes it is necessary to grind the sample to a
powder, embed it in a plastic, preferably one composed of C, H and O, and to grind
the product to a fine powder to use in the absorber.  With a randomly orientated
sample a small rotation of absorber about an axis normal to the photon beam should
not affect the line intensities in the spectrum.(See also Section 3.4).

## 2.7  CRYOSTATS

In most cases the absorber, or the absorber and source, need to be cooled in
liquid nitrogen to obtain satisfactory spectra.  For many purposes no very strict
control of the temperature is necessary and if only the absorber need be cooled it is
quite easy to improvise cryostats in a laboratory workshop. For this purpose foam
plastic is sufficient for the insulation and most attention has to be paid to the
avoidance of condensation of moisture on the absorber and source.

When both source and absorber must be cooled or careful control of temperature
is important, or when liquid helium or lower temperatures are necessary, it is best to
obtain a commercial cryostat designed for this purpose. The avoidance of
unintentional vibration of source and absorber is often a problem in such systems.
A helium cryostat is a valuable accessory not only to permit the numerous
applications for which low temperatures are needed but also to permit the use of a
superconducting magnet. This is required to produce the very powerful magnetic
fields needed to give hyperfine effects from an external field  (See section 3.7).

## 2.8 TREATMENT OF THE DATA

The prime data comprise a sequence of several hundred six, or more, digit figures, representing the number of photons recorded in each channel of the analyser during collection of the spectrum. Typically there will be 512 or 1024 channels. The data are especially well suited to computer processing. If the data collection system is itself a small computer direct treatment of the data will be possible.

Several programs have been written for the analysis of the data. They suppose that the spectrum consists of n inverted peaks arising from the absorption and scattering events suffered by the Mossbauer
photons, each peak having a Lorentzian form. For an individual absorption line the computed content of channel i, represented by $N^{calc.}$ will be given by;

$$N_i^{calc.} = b_i - \frac{N_i^0}{1 + (2[i-i_0]/\Gamma)^2} \,.$$

where i is the channel number, $i^0$ is the channel for maximum absorption for this line and $N^0$ the number of photons recorded in this channel, $b_i$ is the number of photons recorded in the absence of Mössbauer absorption and $\Gamma$ is the line width.

The program corrects for the usually small parabolic variation of b with i due to the changing distance between source and detector during the cycle of motion.(See 2.4.3.2).

For n absorption lines one obtains:

$$N^{calc.} = b_i - \sum_{j=1}^{n} \frac{N_j^0}{1 + \{2(i - i_j^0)/\Gamma_j\}^2}$$

The program then varies $i^0$, $N^0$ and $\Gamma_j$ to reach a minimum of the function:

$$\chi^2 = \sum_{i=1}^{k} W_i (N_i^{obs.} - N_i^{calc.})^2 \,.$$

$W_i$ is a weighting factor of $1/N^{obs}$, allowing for the error in $N^{obs}$ which is determined by the Poisson distribution of the radioactive decay events. k is the number of channels in the data collection system. After each iteration, yielding a new value of $\chi^2$, the values of $d\chi^2/dq$, where q represents one of the variables $i^0$, $N^0$ or $\Gamma_j$, are calculated and the iteration proceeds until all $d\chi^2/dq \Rightarrow 0$.

In this way the best fit values of the parameters are obtained in terms of numbers of channels. The values are no longer necessarily integral.

The calibration spectrum is treated in the same way and using the known velocities for the peaks in this spectrum, including the zero velocity position, the values of the parameters in the first spectrum can be converted from units of channels to velocities ( $mm.s^{-1}$). The program will carry out this conversion. The combination of $N^0$ with $\Gamma_j$ gives the area under each peak, which is linearly related to the Mössbauer absorption for that kind of event.

A good spectrum should have a $\chi^2$ which if divided by the number of degrees of freedom, k - (3n +3), gives a quotient of not much more than one.

Programs are also available to treat the data from spectrometers using a sinusoidal drive system, when the velocity is not linearly related to the channel number. The program can also draw attention to any channels where the number of photons recorded deviates from the calculated best fit value by more than, say, twice the probable error on that number. This will identify any instrumental errors in the data collection system.

These programs are steered programs, that is to say that computing time is saved by the operator feeding in estimates of values of the parameters defining the spectrum. In this way the number of iterations the computer has to carry out is substantially reduced. These estimates may be made visually by inspection of the spectrum either directly or after making a plot of number of photons recorded against channel number. [Ref.2.2]

The program generally incorporates facilities for restraining some of the parameters. The commonest restraints are to fix the line widths at a previously determined value, useful in extracting unresolved quadrupole splittings, and setting equal the areas under a quadrupole split pair. Both these restraints need to be used with caution, especially the latter. Other parameters may need to be fixed in order to separate the contribution of a known spectrum from an unknown one in mixtures of compounds.

On remeasurement of a Mössbauer spectrum collecting a larger number of events in each channel and thus reducing the error due to the random distribution of decay events, it is often found that $\chi^2$ has increased. This is because $\chi^2$ is not a measure of the validity of the model but simply how closely the experimental data can fit the model chosen to calculate the $N^{calc}$. With complicated spectra, where several interpretations of the spectrum may be possible, another quantity, M, called MISFIT, is useful to enable one to decide which model is most likely.[Ref.2.3]. M = D/S where:-

$$D = \sum_{i=1}^{k} (N_i^{obs} - N_i^{calc})^2 \, /(N_i^{calc} - 1) \text{ and } S = \sum_{i=1}^{k} (N_i^b - N_i^{calc})^2 \, /(N_i^{calc} - 1)$$

$N_i^b$ is the background count in channel i.

The smaller the value of M the more appropriate the model. [Ref.2.3]

With simple spectra it may be remarked that if the line shape is assumed to be Gaussian, instead of Lorentzian, the $i^0$ obtained and therefore the computed isomer shifts and quadrupole splittings, are hardly changed.

With very poor quality spectra estimates of the Mössbauer parameters may be obtained using a Fourier transform procedure. This effectively removes background and line shape from the spectra. [Ref.2.4]

With certain types of spectra a much more elaborate analysis is needed. Complex spectra with overlapping lines and spectra obtained with thick absorbers demand a more complicated treatment. The line shape is no longer Lorentzian and to obtain reliable Mössbauer parameters a more elaborate computer program taking the transmission integrals into account is required. Fortunately several such programs are available. [Ref.2.5].

**References**
2.1 See for example:- Woodham, F.W.D. and Reader, S.M.(1991) *Meas. Sci. Technol.*, 2, 217 or Faigel,G.,Haustein P.E. and Siddons, D.P.(1986) *Nuclear*
                         *Instrum. Methods, Phys.Res.Sect.* B, 17, 363.
2.2 For discussion and references to simpler programs see Longworth,G. (1984) in *Mössbauer Spectroscopy Applied to Inorganic chemistry* Vol.1, Ed.Long,G.L., Plenum Press.
 2.3 Ruby,S.L.(1973) in *Mössbauer Effect Methodology* Vol.8, Eds. Gruverman, I.J. and Seidel, C.W. Plenum Press, N.Y.
 2.4 Stone, A.J. (1970) *Chem.Phys. Letts.*, 6, 331 and for recent application Vesely,V, (1997) *Nuclear Instrum. Methods, Phys.Res.Sect.* B, 18, 88.
2.5 See for example Jernberg,G. (1985) *Atomindex*, 16, 012558 and for discussion Shenoy, G.K., Freidt, J.M., Malatta, H. and Ruby, S.L. (1974) in *Mössbauer Effect Methodology*, Plenum Press. For another approach see Digar Ure, M.C. and Flinn,P.A.(1971) in *Mössbauer Effect Methodology* Vol.7, Ed. Gruverman I.J Plenum Press.

# 3

# Further consideration of principles

## 3.1 INTENSITIES AND ANGULAR DISTRIBUTIONS

Although different intensities are shown in Fig.1.8.a. for the absorption lines in the magnetically split spectrum, no explanation of why the different absorption transitions should have different probabilities has been given. In addition it has been tacitly assumed that the absorber, giving rise to the various spectra illustrated, was composed of randomly orientated crystallites.

A classical approach suggests some angular dependence of photon intensity if the radiating nucleus is oriented in an electric or magnetic field. An oscillating electric dipole emits radiation predominantly in directions normal to the axis of the dipole. The intensity of photon emission, the number of photons per unit time and unit solid angle, depends on $\sin 2\theta$, where $\theta$ is the angle between the direction in which the photon is emitted and the axis of the dipole.

A rigorous quantum mechanical treatment of the spontaneous emission of radiation is very complex and will not be attempted [Ref.3.1]. But the results are important and an indication of how they are obtaineed is useful.

### 3.1.1 Multipolar radiation

The emission of a photon leads to a change in the angular momentum of the emitting nucleus; the photon carrying one or more units of angular momentum, as determined by the conservation law. The number of units carried decides the multipolarity of the emission. Dipolar emission leads to a change of one unit, $\hbar$ of angular momentum. In quadripolar emission two units are lost by the emitting nucleus. In principle higher polarities are possible but, fortunately for our present purpose, each increase in multipolarity leads to a very substantial reduction in the probability of emission. Hence in Mössbauer spectroscopy we only encounter dipolar and quadripolar radiation. De-excitation of the nucleus takes place predominantly by the lowest polarity process consistent with the selection rules deduced in the next section.

Photon emission can be associated with the oscillation of both electric and magnetic multipoles, but the probability of emission from a magnetic multipole is thirty to two hundredfold less likely than for the corresponding electric multipole, supposing both were possible. As a result of these differences in emission probability most Mossbauer transitions are E(1), E(2) or M(1), where E(1) and M(1) denote electric and magnetic dipolar emission and E(2) electric quadripolar emission.

### 3.1.2  **Selection rules**

Symmetry considerations are very helpful in deciding what kind of photon emission process will be involved.

The parity of a function or operator indicates what happens to the sign of the function if all the coordinates change sign, that is to say one inverts in the origin, $x \Rightarrow -x$, $y \Rightarrow -y$, and $z \Rightarrow -z$. For positive parity the function does not change its sign, for negative parity it does. Parity behaves just like + and - , hence a positive and a negative parity combine to give negative overall, but two negative parities combine to give an overall positive parity. Thus, whereas x has negative parity $x^2$ has positive parity. Clearly if one integrates a function of negative parity over $\infty$  to $-\infty$  the result must be zero, the two hatched areas in Fig.3.1 just cancel.

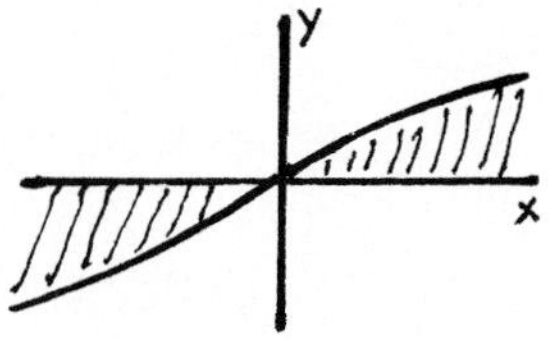

Now the probability of the de excitation of the Mössbauer nucleus by photon emission is given by Fermi's Golden Rule and is of the form  $K \int \Psi_i O \Psi_f \, d\tau$  where

O is the operator for the emission process and $\Psi_I$ and $\Psi_f$ the wave functions for the initial and final states of the nucleus. The parities of these wave functions for the Mössbauer nuclei are given in Table 2.1.

The operator for electric dipole radiation is of negative parity; it is of the form r and therefore has components depending on x, y and z, which are all of negative parity.  Consequently  one can conclude that

**Fig.3.1**

if $\Psi_i$ and $\Psi_f$ are of the same parity the probability of electric dipole emission becomes zero.

For electric quadrupole radiation the operator depends on terms of the type $x^2$, xy etc., and the parity is positive, so that it can be concluded that there will be zero probability of emission if there the parities of the two states are different.  For magnetic dipole radiation there is no x term in the operator and its parity is positive.

One can arrive at the selection rules shown in Table 3.1.   The quantum number specifying the total angular momentum of the nucleus, I, is tabulated in Table 2.1.

### **Table 3.1**

| Change in I | Change in parity | Acceptable multipole character |
|---|---|---|
| Even | No | E(2) |
| Odd | Yes | E(1) |
| Odd | No | M(1) / E(2) |
| 0 | Yes | E(1) |
| 0 | No | M(1) / E(2) |

Most of these combinations lead to a clear indication of the kind of emission process. Only in the case of magnetic dipole emission does one sometimes find considerable admixture of the alternatively permitted electric quadrupole emission.

The kind of emission is important because it is involved in determining the intensities for the different transitions in the Mössbauer emission or absorption processes.

The two processes follow the same rules and for convenience the treatment will be given in terms of the emission process.


## 3.2  INTENSITIES

The relative probabilities of the various modes of decay of the excited Mössbauer nucleus depend on the multipolarity of the emission, the combination of I values involved and the change in the component of I along the the principal axis of the electric field gradient or of the direction of the magnetic field, $\Delta m_I$. This reference direction is always taken as the z axis.    Quantities associated with the excited and ground states will be distinguished by the indices ' and " respectively.  Each transition will be associated with a change in an orbital angular momentum quantum number, L, which can assume the values I' + I", I' + I" -1, through to |I'- I"|.    The probability of a given transition decreases very rapidly as L increases and  generally one need only consider values of |I' - I"| and |I' - I"| +1.

The probability of photon emission by the state specified by |I' m'> to yield the state |I" m"> is given approximately by an expression of the form

$p = K f(L) [<I'.m'|(kr)^L Y_{L,m} (\theta,\varphi)|I".m">]^2$ where K involves the energy of the photon emitted as well as various constants, f(L) is a function of L, defined above, k being the wave vector for the photon emitted, r the nuclear radius and $Y_{L,m}$ the spherical harmonic corresponding to L and $\Delta m = m' - m"$. The quantity kr is very small so that $(kr)^L$ decreases exremely rapidly as L increases and only values of L of 1 and 2 are important.    These correspond to dipolar and quadrupolar emission respectively.

With  dipolar  emission  $\Delta m$  is  restricted  to  the  values  0, ±1  and  for  quadrupolar emission $\Delta m = 0, ±1, ±2$.

The above expression for the emission probability can be separated into two parts, one is independent of the direction of emission of the photon in relation to the z axis, the other depends on the value of m but is otherwise a purely angular term.

The angular independent part is given by the coupling coefficients for the |I' m'> and |I" m"> states. These are purely numerical terms and are called the Clebsch Gordan or, sometimes, the Wigner coefficients. The detailed derivation of these coefficients, which are common to all changes in angular momentum, is complex and laborious and will not be given here [Ref.3.2].

The observed intensity depends on the square of the Clebsch Gordan coefficient, $C^2$. Some of the most important relative values are tabulated in Tables 3.2 to 3.4.

The starred values are forbidden transitions for M(1), which only permits $\Delta m = 0, ±1$.

These values account for the 3.2.1 : 1.2.3 pattern of intensities in the magnetically split $^{57}Fe$ spectrum of a randomly orientated ferromagnetic material shown in Fig.1.8.c.

**Table 3.2**

Coefficients for I' = 3/2 and I" = 1/2

| | For M(1) multipolarity | | | For E(2) multipolarity |
|---|---|---|---|---|
| m' | m" | Δ m | $C^2$ | $C^2$ |
| 3/2 | 1/2 | 1 | 3 | 1 |
| 1/2 | 1/2 | 0 | 2 | 2 |
| -1/2 | 1/2 | -1 | 1 | 3 |
| -3/2 | 1/2 | -2 | 0 * | 4 |
| 3/2 | -1/2 | 2 | 0 * | 4 |
| 1/2 | -1/2 | 1 | 1 | 3 |
| -1/2 | -1/2 | 0 | 2 | 2 |
| -3/2 | -1/2 | -1 | 3 | 1 |

Fig.3.2 shows the approximately 1:2:3:4 :: 4:3:2:1 pattern of intensities for the magnetically split spectrum from powdered $IrF_6$, which arises from the mixed E(2)/M1 decay of [193m]Ir. The Δm = ±2 transitions are permitted. As will be show the angular dependence averages to one in both these cases.

**Table 3.3**

Coefficients for I' = 5/2 and I" = 3/2.

| | | | For M(1) | For E(2) | | | | For M(1) | For E(2) |
|---|---|---|---|---|---|---|---|---|---|
| m' | m" | Δm | $C^2$ | $C^2$ | m' | m" | Δm | $C^2$ | $C^2$ |
| 5/2 | 1/2 | 2 | 0 * | 40 | -1/2 | -1/2 | 0 | 6 | 6 |
| 3/2 | -1/2 | 2 | 0 * | 32 | -3/2 | -3/2 | 0 | 4 | 36 |
| 1/2 | -3/2 | 2 | 0 * | 12 | 1/2 | 3/2 | -1 | 1 | 27 |
| 5/2 | 3/2 | 1 | 10 | 30 | -1/2 | 1/2 | -1 | 3 | 25 |
| 3/2 | 1/2 | 1 | 6 | 2 | -3/2 | -1/2 | -1 | 6 | 2 |
| 1/2 | -1/2 | 1 | 3 | 25 | -5/2 | -3/2 | -1 | 10 | 30 |
| -1/2 | -3/2 | 1 | 1 | 27 | -1/2 | 3/2 | -2 | 0 * | 12 |
| 3/2 | 3/2 | 0 | 4 | 36 | -3/2 | 1/2 | -2 | 0 * | 32 |
| 1/2 | 1/2 | 0 | 6 | 6 | -5/2 | -1/2 | -2 | 0 * | 40 |

\* Forbidden transitions

It will be noticed that the above table can be folded at the m' = 1/2, m" = 1/2 line, the values of $C^2$ repeating in reverse order, giving the two sets of columns shown. Only the top half of the next table will be given.

**Table 3.4**

Coefficients for I' = 7/2 and I" = 5/2 and M(1) multipolarity.

| m' | m" | Δm | $C^2$ | m' | m" | Δm | $C^2$ |
|---|---|---|---|---|---|---|---|
| 7/2 | 5/2 | 1 | 21 | -3/2 | -5/2 | 1 | 1 |
| 5/2 | 3/2 | 1 | 15 | 5/2 | 5/2 | 0 | 6 |
| 3/2 | 1/2 | 1 | 10 | 3/2 | 3/2 | 0 | 10 |
| 1/2 | -1/2 | 1 | 6 | 1/2 | 1/2 | 0 | 12 |
| -1/2 | -3/2 | 1 | 3 | | | | |

For I' = 2 and I" = 0 with an E(2) emission all five lines have the same intensities,

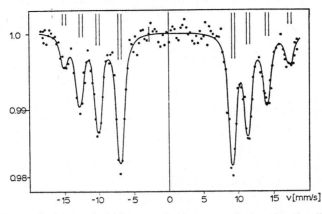

Yhe broad lines arise because the iridium source emits an unresolved quadrupole split pair of lines,

**Fig.3.2**

## 3.3 ANGULAR DEPENDENCE

The angular dependence of the intensity, although also difficult to derive, takes a simpler form. Supposing the angle between the direction of photon emission and the $z$ axis, as defined above, is $\theta$, (See Fig.1.6.) then the observed relative intensity is given by $C^2.f(\theta)$.

For both electric and magnetic dipolar emission $f(\theta)$ depends on $\Delta m$, but not on $I'$ or $I''$. Table 3.5 gives expressions for $f(\theta)$ for different values of $\Delta m$.

### Table 3.5

For dipolar multiplicity,

$$\Delta m = 0, \quad f(\theta) = 2\sin^2\theta,$$
$$\Delta m = \pm 1, \quad f(\theta) = (1 + \cos^2\theta)$$

For quadrupolar emission:

$$\Delta m = 0, \quad f(\theta) = 3/2\sin^2 2\theta.$$
$$\Delta m = \pm 1, \quad f(\theta) = \cos^2\theta + (\cos^2 2\theta)/4$$
$$\Delta m = \pm 2, \quad f(\theta) = \sin^2\theta + (\sin^2 2\theta)/4.$$

From these data several interesting conclusions can be drawn.

### 3.3.1 Intensities for quadrupole split spectra

The line intensities for purely quadrupole split spectra can be calculated from these data, remembering that the electric field gradient does not remove the $\pm$ degeneracy of the different m levels. For each emission mode, with a particular value of m, there will be a contribution to the total intensity of the line given by $C^2 f(\theta)$, multiplied by the number of ways that photons of that energy can arise with that value of $\Delta m$.

Taking as a first example a $3/2 \Leftrightarrow 1/2$ M(1) transition, the relative intensity of the $\pm 1/2 \Leftrightarrow \pm 1/2$ line includes two components with $\Delta m = 1$ and two with $\Delta m = 0$ and will be given by $2\text{x}2\text{x}2\sin^2\theta + 2\text{x}1\text{x}1(1 + \cos^2\theta) = 4 + 6\sin^2\theta$.

Because of the selection rules the $\pm 3/2 \Leftrightarrow \pm 1/2$ line only has two components and the relative intensity is given by $2\text{x}3\text{x}1(1 + \cos^2\theta) = 6 + 6\cos^2\theta$, or in the simplest form the relative intensities of the two lines are $2 + 3\sin^2\theta$ and $3(1 + \cos^2\theta)$.

Thus one finds for $\theta = 0°$ the ratio of intensities of the two lines is 2 : 6 and for $\theta = 90°$ the ratio becomes 5 : 3.   In principle this provides a means of identifying each line and thus determining the sign of the quadrupole splitting. But as will be considered later there are other factors that can alter the ratio of line intensities. The same result is obtained for a quadrupole split spectrum with an emission of E(2) multipolarity.

Supposing the particles in the absorber are randomly orientated, it will be necessary to obtain the average value of $f(\theta)$ as $\theta$ goes from zero to $\pi$ and the indeterminate $\varphi$ goes from zero to $2\pi$   (See Fig.1.6).

For $f(\theta) = 2\sin^2\theta$ the average value will be

$$\frac{1}{4\pi}\int_0^{2\pi}\int_0^{\pi} 2\sin^2\theta \sin\theta\, d\theta\, d\varphi \;=\; \int_0^{\pi}\sin^3\theta\, d\theta \;=\; \left[\, 1/3\cos^3\theta - \cos\theta\,\right]_0^{\pi} \;=\; 4/3.$$

Similarly the average value of $(1 + \cos^2\theta)$ is given by

$$\frac{1}{4\pi}\int_0^{2\pi}\int_0^{\pi}(1 + \cos^2\theta)\sin\theta\, d\theta\, d\varphi \;=\; \frac{1}{2}\int_0^{\pi}(2\sin\theta - \sin^3\theta)\, d\theta \;=\; 4/3.$$

Thus the relative values are 1 : 1. and the two lines should be of equal intensity.

The above calculations demonstrate the importance of working with really randomly orientated powders or with crystals in a known orientation. Fig.3.3 illustrates the intensities found for $\theta = 0°$, $90°$ and a random orientation, as in a powder, for a quadrupole split spectrum with a $3/2 \Leftrightarrow 1/2$ M(1) transition.

$\theta = 0°$                              $\theta = 90°$                              Powder

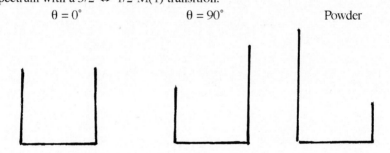

**Fig.3.3**

Using the data in Tables  3.3 and 3.5. one can obtain for a quadrupolar split $5/2 \Leftrightarrow 3/2$ transition with M(1) multipolarity the relative intensities shown in Table 3.6, and for a similarly split $7/2 \Leftrightarrow 5/2$ transition the values in Table 3.7.

| Table 3.6 | | | | Table 3.7 | | |
|---|---|---|---|---|---|---|
| m' | m" | $c^2 f(\theta)$ | | m' | m" | $c^2 f(\theta)$ |
| ±5/2 | ±3/2 | $10(1 + \cos^2\theta)$ | | ±7/2 | ±5/2 | $21(1 + \cos^2\theta)$ |
| ±3/2 | ±1/2 | $6(1 + \cos^2\theta)$ | | ±5/2 | ±3/2 | $15(1 + \cos^2\theta)$ |
| ±3/2 | ±3/2 | $8\sin^2\theta$ | | ±3/2 | ±1/2 | $10(1 + \cos^2\theta)$ |
| ±1/2 | ±3/2 | $(1 + \cos^2\theta)$ | | ±5/2 | ±5/2 | $12\sin^2\theta$ |
| ±1/2 | ±3/2 | $(1 + \cos^2\theta)$ | | ±3/2 | ±3/2 | $20\sin^2\theta$ |
| ±1/2 | ±1/2 | $6 + 9\sin^2\theta$ | | ±1/2 | ±1/2 | $12 + 18\sin^2\theta$ |
| | | | | ±1/2 | ±3/2 | $3(1 + \cos^2\theta)$ |
| | | | | ±3/2 | ±5/2 | $(1 + \cos^2\theta)$ |

### 3.3.2   **Intensities  in  a  magnetically  split  spectrum**

The intensities of the lines in a spectrum from a ferromagnetic or antiferromagnetic material with a $3/2 \Leftrightarrow 1/2$ transition can be obtained from the above data.

**Table  3.8**

| Absorption line | | Relative intensity | $\theta = 90°$ | $\theta = 0°$ |
|---|---|---|---|---|
| m' | m" | $C^2(\theta)$ | | |
| 3/2 | 1/2 | $3(1 + \cos^2\theta)$ | 3 | 6 |
| -3/2 | -1/2 | "    " | | |
| 1/2 | 1/2 | $2 \times 2 \sin^2\theta$ | 4 | 0 |
| -1/2 | -1/2 | " | | |
| 1/2 | -1/2 | $(1 + \cos^2\theta)$ | 1 | 2 |
| -1/2 | 1/2 | " | | |

Because, as was shown before, the average relative values of both $2\sin^2\theta$ and $(1 + \cos^2\theta)$ are 1 the six lines of the spectrum of a randomly orientated powder have intensities in the ratio 3 : 2 : 1 :: 1 : 2 : 3 as shown in Fig. 1.8.a. The spectra for observations at 90° and 0° in relation to a fixed magnetic field are shown schematically in Figs, 3.4.

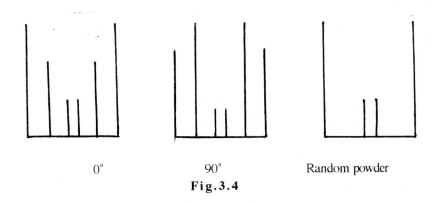

0°                              90°                              Random powder

**Fig.3.4**

## 3.4  **DEPARTURES  FROM  CALCULATED  VALUES**

Unfortunately there are several ways in which departures from these relative intensities can occur.

### 3.4.1  **The f factor anisotropic**

The Mössbauer factor for the emission, or absorption, process may be anisotropic. This is rather likely since it was shown in Section 1.2  that this factor is related to the distance between neighbouring atoms and the Mössbauer atom, which will depend on the direction of recoil in relation to the crystal axes.  This is known as the **Goldansky Karyagin effect**. Even in the case of the simple quadrupole split spectrum, from a $3/2 \Leftrightarrow 1/2$ transition, the 1 : 1 ratio for the lines may be invalidated.  A planar arrangement of ligands about the Mössbauer atom is especially conducive to this effect. Fig.3.5 shows a spectrum of the mineral Gillespite in which the $Fe^{2+}$ ions are bonded to four oxygen atoms in a planar configuration.

Perhaps the commonest cause of departures from the intensity pattern calculated for a powder absorber is due to texture: the crystallites in the absorber are not randomly orientated in relation to the photon beam. This is a particular hazard with absorber materials like the micaceous minerals.

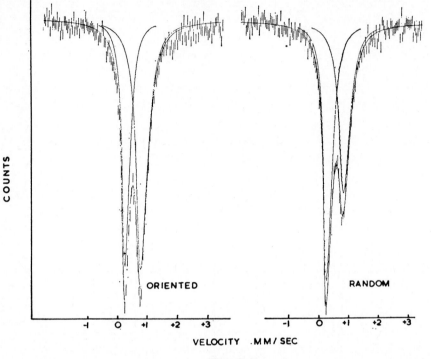

COUNTS

VELOCITY .MM/SEC

**Fig.3.5**

An undistorted spectrum can be obtained from a textured absorber in the following way. The thin planar absorber is placed so that a line normal to the surface of the absorber, at the point of incidence of the photon beam, makes an angle of 54.7° with the direction of the beam. A spectrum is accumulated for a time t. The absorber is then rotated by 90° about the normal as axis and another spectrum collected for the same time, t. The process is repeated following two further rotations of the absorber by 90°.
A composite spectrum is obtained by summing the data of the four measurements for each channel. This composite spectrum should be free of distortion due to imperfect randomisation of orientation of the crystallites in the absorber.

### 3.4.2  Incipient magnetic splitting

For some paramagnetic iron compounds the magnetic field, due to the orbital electrons of the iron, is not relaxing quickly enough, even at room temperature, to be fast compared to the Zeeman precession time of the excited Mössbauer state. In these circumstances incipient magnetic splitting begins, the first effect being to broaden the $\pm 3/2 \Leftrightarrow \pm 1/2$ line more than $\pm 1/2 \Leftrightarrow \pm 1/2$ line. This effect is especially noticeable with high spin iron(III) compounds of large molecular weight. Both the spin-lattice and relaxation mechanisms tend to be slow in such compounds. For this reason $\gamma$ irradiation of the absorber may sharpen the line by enhancing spin-spin relaxation by introducing spin free radicals into the lattice. Similarly spin-lattice relaxation increases with temperature so that heating the absorber may also lead to sharper lines in these circumstances.

### 3.4.3  Other factors

Besides the effects of texture, there are a number of other factors that can modify the relative intensities. In some cases the emission process is a mixture of M(1) and E(2) multipolarities. An example is $^{197m}Au$ which gives a eight line magnetically split spectrum, showing $\Delta m = \pm 2$ is permitted.

Notwithstanding the approximate nature of the above estimates, these intensities values are helpful in identifying the lines in quadrupole split spectra and hence deciding the sign of eq, for transitions other than $3/2 \Leftrightarrow 1/2$.

### 3.5  THE ELECTRIC FIELD GRADIENT

The electric field gradient, or EFG, arises from the asymmetry of the field due to all the charges surrounding the nucleus. It is useful to analyse the contributions to the EFG in terms of a highly polar model for the absorber.

Since the $V_{ZZ}$ due to a charge $q_i$ at $(r_i, \theta_i, \varphi_i)$ is given by $\sum_i q_i (3 \cos^2 \vartheta_i - 1) r^{-3}$

it will decrease rapidly as r increases (See 1.4.4) It is therefore useful to think in terms of two contributions to $V_{ZZ}$, a lattice contribution, $q_{lat.}$, due to the more remote charges in the lattice, beyond the Mössbauer atom and the molecular orbitals in which it is involved, and a valence shell contribution, $q_{val.}$, due to the electrons in these orbitals.

In fact the EFG the Mössbauer nucleus experiences is not the simple sum of these two terms because of polarisation of the inner, initially spherically symmetrical, electron cloud by the asymmetric electric field.

$$q = (1 - \gamma) q_{lat.} + (1 - R) q_{val.}$$

The quantities $\gamma$ and R are the **Sternheimer anti-shielding factors**. The first factor $\gamma$ is negative and substantially increases the effect of the lattice charges. Its value becomes more negative as the Z of the Mössbauer atom increases: for iron its value is about -9. R is positive and much smaller. It reduces the effect of any asymmetry of the field due to the electrons in the orbitals in which the Mössbauer atom participates. Its value is about 0.4 for iron. For a given element $\gamma$ and R are approximately constant and the two factors will subsequently be omitted, unless an absolute value of the EFG is involved.

The $q_{lat.}$ depends only on the charges on the ions in the lattice and the crystal structure; for a cubic lattice it will be zero and for a covalent molecular crystal lattice usually rather small. But it is not easy to calculate $q_{lat.}$, even for polar crystals of known structure, because the effective charges on the ions are difficult to evaluate.

### 3.5.1 Contribution to $q_{val}$ due to different orbitals

Conceptually it is useful to consider $q_{val}$ as composed of two terms, $q_{val.} = q_{fi} + q_{mo}$. The second term in this equation represents the contribution to $q_{val}$ from the free atom in the electronic configuration it assumes in the compound.

This term will be zero for the spherically symmetric $Sn^{4+}$ and low spin $Fe^{2+}$. The third term is the contribution from the electron distribution in the molecular orbitals formed with the ligands by the Mössbauer atom. Clearly this division is not entirely realistic.

To estimate the $q_{fi}$ arising from one electron in a p orbital it is necessary to calculate the quantum mechanical average value, or expectation value, of $(3\cos^2\theta -1)r^{-3}$ for a p orbital. Now the expectation value $<(3\cos^2\theta - 1)r^{-3}> = \int \Psi_p^* (3\cos^2\theta - 1)r^{-3} \Psi_p \, dv$ using normalised wave functions. If we can assume a hydrogen like p orbital the radial and angular parts are separable and $<(3\cos^2\theta-1)>$ is equal to $\int Y_{l,m}(3\cos^2\theta - 1)Y_{l,m} \, dv$ where the $Y_{l,m}$ are spherical harmonics, the angular part of the p wave functions. For a p orbital $l = 1$, $m = 0, \pm 1$. $Y_{1,0} = (3/4\pi)^{1/2}\cos\theta$, and the real part of $Y_{1,1} = (3/4\pi)^{1/2}\sin\theta\cos\varphi$ and of $Y_{1,-1} = (3/4\pi)\sin\theta\sin\varphi$. Hence for the $p_X$ orbital

$$<(3\cos^2\theta - 1)> = (3/4\pi)\int_0^{2\pi}\int_0^{\pi}\sin^2\theta\cos^2\theta(3\cos^2\theta - 1)\sin\theta \, d\theta \, d\varphi$$

$$= 3/2 \int_0^{\pi}\sin^3\theta(2 - 3\sin^2\theta) \, d\theta = \frac{3}{4}\left[2/5\cos\theta - 2/15\cos^2\theta\right]_0^{\pi} = -2/5$$

A similar calculation for the $p_y$ orbital gives the same result, -2/5, but for the $p_z$ orbital using $Y_{1,0}$ one gets:

$$<(3\cos^2\theta - 1)> = (3/4\pi)\int_0^{2\pi}\int_0^{\pi}\cos^2\theta\sin\theta(3\cos^2\theta)d\theta d\varphi$$

$$= 3/2 \left[3/5\cos\theta - 13/15\cos^3\theta\right]_0^{\pi} = 4/5.$$

The radial expectation value is the same in each case $<r^{-3}>$.

Analogous calculations for the d orbitals give $<(3\cos^2\theta- 1)>$ values of 2/7 for the $d_{xz}$ and $d_{yz}$ orbitals, -4/7 for the $d_{xy}$ and $d_{x^2-y^2}$ orbitals and +4/7 for the $d_{z^2}$. Similar calculations yielding the expectation values for $<(3\sin^2\theta\sin^2\varphi - 1)>$ and $<(3\sin^2\theta\sin^2\varphi - 1)>$ give the contributions to $V_{yy}$ and $V_{xx}$ respectively.

Calculation yields the values given in Table 3.9. for a charge e in each orbital. Note e is taken as the charge on the proton.

It will be seen that equal electronic occupation of all the p or d orbitals leads to a zero EFG. The same is true for the $d_{xy}$, $d_{xz}$ and $d_{yz}$ or the $d_{x^2-y^2}$ and $d_{z^2}$ subsets.

**Table 3.9.**

Relative values of $4\pi\varepsilon_0 q/(1-R)<r^{-3}>$ for unit charge in the orbital

| | $p_z$ | $p_x$ | $p_y$ | $d_{xy}$ | $d_{xz}$ | $d_{yz}$ | $d_{x^2-y^2}$ | $d_{z^2}$ |
|---|---|---|---|---|---|---|---|---|
| $V_{zz}/e<r^{-3}>$ | 4/5 | -2/5 | -2/5 | -4/7 | 2/7 | 2/7 | -4/7 | 4/7 |
| $V_{yy}/e<r^{-3}>$ | -2/5 | 4/5 | -2/5 | 2/7 | -4/7 | 2/7 | 2/7 | -2/7 |
| $V_{xx}/e<r^{-3}>$ | -2/5 | -2/5 | 4/5 | 2/7 | 2/7 | -4/7 | 2/7 | -2/7 |

$\varepsilon_0$ is the vacuum permittivity. The term $4\pi\varepsilon_0$ is needed to transform to S.I. units and will be omitted subsequently.

### 3.5.2 The ligand contributions

The quantitative interpretation of the quadrupole splitting due to the ligands was first investigated by nuclear quadrupole spectroscopists. This technique, developed some time before Mössbauer's discovery, also provides information about the electric field gradient at the nucleus, for nuclei for which $I \geq 1$, by measuring the splitting of levels in the nuclear ground state. It was suggested that a first approximation to the ligand effects could be obtained by associating a point charge with each ligand. The charge due to one ligand was supposed unaffected by the presence of other ligand charges.

This undoubtedly crude model has some interesting consequences. It implies that the ligand contributions are additive so that it should prove possible to establish a system of partial quadrupole splittings, due to the individual ligands. Consider situations for which the $q_{fi}$, as described above, is zero. Using the relations obtained in 1.4.2,

$$V_{zz} = q(3\cos^2\theta - 1)r^{-3}; \quad V_{yy} = q(3\sin^2\theta\sin^2\varphi - 1)r^{-3} \text{ and } V_{xx} = q(3\sin^2\theta\cos^2\varphi - 1)r^{-3}$$ and the data

collected for convenience in Table 3.10 and Fig.3.6 one can arrive at the results given in Tables 3.11.1 to 3.11.3.

In these tables the ligands A and B are located at the apices of the regular solids shown, the Mössbauer atom, denoted by M, is at the origin of coordinates. Each ligand is considered to bear a charge A or B. In each sketch, showing the arrangement of the ligands, the choice of the z axis leading to diagonalisation of the EFG is indicated.

**Fig. 3.6**            **Table 3.10.**

| | $\theta$ | $\varphi$ | $\sin\theta$ | $\cos\theta$ | $\sin\varphi$ | $\cos\varphi$ |
|---|---|---|---|---|---|---|
| 1. | 0 | 0 | 0 | 1 | 0 | 1 |
| 2. | $\pi$ | 0 | 0 | -1 | 0 | 1 |
| 3. | $\pi/2$ | 0 | 1 | 0 | 0 | 1 |
| 4. | $\pi/2$ | $\pi/2$ | 1 | 0 | 1 | 0 |
| 5. | $\pi/2$ | $\pi$ | 1 | 0 | 0 | -1 |
| 6. | $\pi/2$ | $3\pi/2$ | 1 | 0 | -1 | 0 |

The values of $V_{zz}$, $V_{yy}$ and $V_{xx}$ as well as $\eta$ are tabulated, but the constant multiplier $r^{-3}$ is omitted. In the case of the trigonal bipyramidal arrangements, different values are taken for A or B equatorial and A or B axial. This can take into account both differences in the charges and, or, differences in the lengths of the equatorial and axial bonds.

A number of comments can be made on the results:

(i) The z axis of the EFG is not always the principal axis of symmetry of the arrangement (See 3.11.1 No.2 above).

(ii) If the arrangement has at least a threefold axis of symmetry $\eta = 0$. (See 3.11.2 No.1 and 3.11.3 No.4 above).

(iii) The ligand contributions, $V_{zz}$ for $MAB_5$, cis $MA_2B_4$, trans $MA_2B_4$ and mer $MA_3B_3$ are in the ratio 2 : -2 : 4 : 3.

(iv) Unless an absolute value of A or B could be obtained a system of partial field gradients would have to be arbitrarily normalised, for example by setting $A = 0$ for a particular ligand.

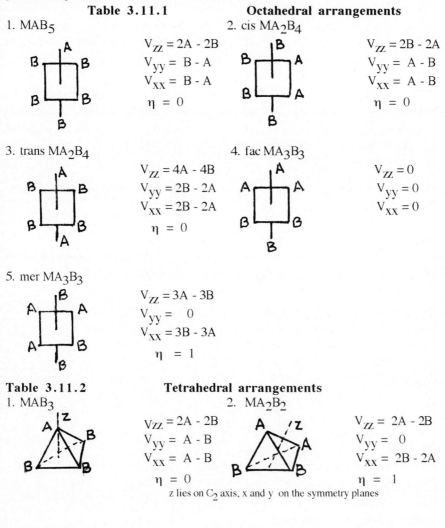

Table 3.11.1                    Octahedral arrangements

1. $MAB_5$

2. cis $MA_2B_4$

$V_{zz} = 2A - 2B$
$V_{yy} = B - A$
$V_{xx} = B - A$
$\eta = 0$

$V_{zz} = 2B - 2A$
$V_{yy} = A - B$
$V_{xx} = A - B$
$\eta = 0$

3. trans $MA_2B_4$

4. fac $MA_3B_3$

$V_{zz} = 4A - 4B$
$V_{yy} = 2B - 2A$
$V_{xx} = 2B - 2A$
$\eta = 0$

$V_{zz} = 0$
$V_{yy} = 0$
$V_{xx} = 0$

5. mer $MA_3B_3$

$V_{zz} = 3A - 3B$
$V_{yy} = 0$
$V_{xx} = 3B - 3A$
$\eta = 1$

**Table 3.11.2**                    **Tetrahedral arrangements**

1. $MAB_3$

2. $MA_2B_2$

$V_{zz} = 2A - 2B$
$V_{yy} = A - B$
$V_{xx} = A - B$
$\eta = 0$

$V_{zz} = 2A - 2B$
$V_{yy} = 0$
$V_{xx} = 2B - 2A$
$\eta = 1$

z lies on $C_2$ axis, x and y on the symmetry planes

**Table 3.11.3**     **Trigonal bipyramidal arrangements**

1. $MA_5$                    2. $MA_4B$, B equatorial

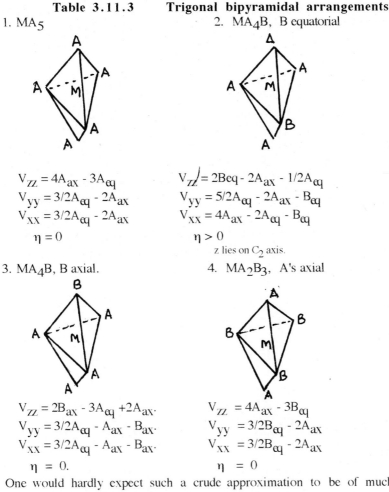

$$V_{zz} = 4A_{ax} - 3A_{eq}$$
$$V_{yy} = 3/2A_{eq} - 2A_{ax}$$
$$V_{xx} = 3/2A_{eq} - 2A_{ax}$$
$$\eta = 0$$

$$V_{zz} = 2B_{eq} - 2A_{ax} - 1/2A_{eq}$$
$$V_{yy} = 5/2A_{eq} - 2A_{ax} - B_{eq}$$
$$V_{xx} = 4A_{ax} - 2A_{eq} - B_{eq}$$
$$\eta > 0$$

z lies on $C_2$ axis.

3. $MA_4B$, B axial.              4. $MA_2B_3$, A's axial

$$V_{zz} = 2B_{ax} - 3A_{eq} + 2A_{ax'}$$
$$V_{yy} = 3/2A_{eq} - A_{ax} - B_{ax'}$$
$$V_{xx} = 3/2A_{eq} - A_{ax} - B_{ax'}$$
$$\eta = 0.$$

$$V_{zz} = 4A_{ax} - 3B_{eq}$$
$$V_{yy} = 3/2B_{eq} - 2A_{ax}$$
$$V_{xx} = 3/2B_{eq} - 2A_{ax}$$
$$\eta = 0$$

One would hardly expect such a crude approximation to be of much use. The existence of the trans effect in the chemistry of the second and third transition groups invalidates one of the assumptions on which this model is based. However, as will appear in later chapters, for certain groups of substances it works rather well.

### 3.5.3   **A molecular orbital approach**

A more sophisticated approach to the problem leads to substantially the same conclusions as the simple model, but it is valuable since it reveals the effects of a change of coordination number and identifies the kinds of substances for which the simple model works well and, equally importantly, what sort of properties invalidate this approach.

An accurate evaluation of the EFG can be obtained by calculating the expectation values for the three expressions for $V_{xx}$, $V_{yy}$ and $V_{zz}$ given in 1.4.3, for all the atomic and molecular orbitals in which the Mössbauer atom paticipates.

An approximation to such a fundamental approach will be explored. With N ligands, A,B, ---N, and a closed shell configuration, implying no unpaired electrons on the

Mössbauer atom, the EFG must arise from imbalance in electron density in the valence shell orbitals. Closed shell electrons, $d^{10}$, or full $d_{xy}$, $d_{yz}$ and $d_{xz}$ subsets, will make no contribution to the EFG. For the remaining valence shell electrons their wave function, $\Psi$ will be a product of 2N wave functions, $\psi_i^\alpha$ in pairs, $\alpha = +1/2$ and $-1/2$, allowing for two electrons in each orbital and $L = A, B, --- N$.

Then $V_{rs} = 2 \sum_{i=A}^{N} (1-R)<\psi|O_{rs}|\psi>$, $O_{rs}$ being the appropriate EFG operator for the component $V_{rs}$ of the EFG tensor. The r and s are combinations of x,y and z.

Thus $V_{zz} = 2 \sum_{i=A}^{N} (1-R)<\psi|(3\cos^2\theta = 1)r^{-3}e|\psi>$   (See 1.4.2)

Now for the $\sigma$ bonded structure the $\psi_L^\alpha$ comprising the wave function $\Psi$ can be transformed to a set of wave functions localised on the N bonds, $\Phi_L$, where $L = A,B, --- N$. so that

$V_{rs} = 2 \sum_{i=A}^{N} (1-R) <\Phi_L|O_{rs}|\Phi_L> = \sum_{i=A}^{N} V_{rs} [L]$. With such localised orbitals, rotationally symmetric about the bond axes, which become the local z axes, the various terms $V_{rs}[L]$ will be determined predominantly by the the properties of the ligand and different ligands will not affect each other substantially.

These localised orbitals will involve one or more sets of equivalent orbitals formed from the Mössbauer atom and each ligand. They can be approximated by linear combinations of equivalent orbitals on the Mössbauer atom, $\chi_M$ and ligand orbitals, $\chi_L$, $\Phi_L = c_1 \chi_M + c_2 \cdot \chi_L$

Hence: $<\Phi_L|O_{rs}|\Phi_L> = c_1^2 <\chi_M|O_{rs}|\chi_M> + 2c_1c_2<\chi_M|O_{rs}|\chi_L> + c_2^2<\chi_L|O_{rs}|\chi_L>$

Successive terms on the right hand side of this equation decrease rapidly because of the $r^{-3}$ term in the operator. Each is about an order of magnitude less than the previous term. For an approximate estimate only the first term will be taken into account.

Since the local axes are such that z lies along the metal ligand bond

$V_{zz} = 2(1-R)c_1^2 <\chi_M|(3\cos^2\theta - 1) r^{-3}|\chi_M> = 2e[L]$

### 3.5.3.1 *For a tetrahedral compound*

Only a single set of equivalent hybrid orbitals are concerned, $sp^3$,

and $\chi_M^{tet.}(z) = 1/2s + \dfrac{\sqrt{3}}{2} p_{z^2}$ where s and p are the wave functions combining to give the equivalent orbitals. The spherically symmetrical s orbital makes no contribution to the EFG. The angular part in the equation above has already been shown to amount to $-4/5$. (See Table 3.9)

Thus $[L]^{tet.} = -3/10 <r^{-3}>_p \sigma_L^{tet.}$ .

The parameter $\sigma_L^{tet.}$ is proportional to $c_1$ and is effectively a measure of the transferred electron density from the ligand to the metal.

### 3.5.3.2  *For an octahedral complex*

$$\chi_M^{oct.}(z) = \frac{1}{\sqrt{6}}s + \frac{1}{\sqrt{2}}p_z + \frac{1}{\sqrt{3}}d_{z^2}$$

A single set of equivalent metal hybrid orbitals is involved. Using the angular contributions taken from Table 3.9 one obtains:

$$[L]^{oct.} = [(-1/5)<r^{-3}>_p - (2/21)<r^{-3}>_d - (1/3\sqrt{\frac{2}{5}})<r^{-3}>_{sd}]\ \sigma_L^{oct.}$$

Now octahedral complexes are of two kinds, $sp^3d^2$ all of the same principal quantum number, from main group elements, and $d^2sp^3$ with the d orbital of principal quantum number one less than for the s and p orbitals, for transition elements.

  a) $sp^3d^2$.

  Numerical calculations of $<r^{-3}>_p$ , $<r^{-3}>_d$ and $<r^{-3}>_{sd}$ using for  example Herman Skillman wave functions, show that the last two quantities are, at most, a few percent of the first term. The Sternheimer antishielding factor modifies the effective values but to a reasonable approximation

$$[L]^{oct.}(sp^3d^2) = -1/5 <r^{-3}>_p\ \sigma_L^{oct.}\ .$$

  b) $d2sp\_$.

  For this situation $<r^{-3}>_d >> <r^{-3}>_p$ or $<r^{-3}>_{sd}$ ,

  so one obtains:    $[L]^{oct.}(d^2sp^3) = -2/21 <r^{-3}>_d\ \sigma_L^{oct.}\ .$

### 3.5.3.3.  *Trigonal  bipyramidal  complexes*.

  Here the situation is a little more complex. Two sets of hybrids, equivalent under $D_{3h}$, are involved and an arbitrary constant, $\theta$, must be introduced to allow for the division of the s and $d_{z^2}$ between the axial and equatorial sets.

Proceeding as before one obtains:

$$[L]^{ax} = \{(-1/5)<r^{-3}>_p - (1/7)\sin^2<r^{-3}>_d - \frac{1}{\sqrt{5}}\sin\theta\cos\theta <r^{-3}>_{sd}\}\ \sigma_L^{Ax.}$$

$$[L]^{eq} = \{(-4/15)<r^{-3}>_p + (1/21)\cos^2\theta<r^{-3}>_d - (1/3\ 5)\sin\theta\cos\theta<r^{-3}>_{sd}\}\ \sigma_L^{eq.}$$

  Fortunately the d and sd radial expectation values are very small compared to $<r^{-3}>_p$ so that one can approximate:

$$[L]^{ax.} = -\frac{1}{5}\langle r^{-3}\rangle_p\ \sigma_L^{ax.}$$

$$[L]^{eq.} = --\frac{1}{5}\langle r^{-3}\rangle_p\ \sigma_L^{eq.}\ .$$

These calculations provide some justification for a model associating a partial field gradient, or equivalently, partial quadrupole splitting, with each ligand, [L], the total EFG for the compound then being obtained by the expressions developed in Section 3.5.2.

The model should be applicable to compounds with filled orbitals, such as the compounds of Sn(IV) or the low spin complexes of Fe(II). ( See Chaps. 4 & 5.) It shows that the value of [L] should depend on the coordination number of the Mössbauer atom. Thus for a Mössbauer atom giving both $sp^3$ and $sp^3d^2$ complexes, if one supposes $\sigma^{tet.}$ and $\rho^{oct.}$ are not very different, then:

$$[L]^{oct.}/[L]^{tet.} \approx 0.67.$$

Any contribution of the $\pi$ bonding orbitals to the EFG cannot be treated in the same way. The complete set would involve 12 localised orbitals, two for each M-L bond. These cannot all be filled. Further one cannot preserve rotational symmetry about the bond axis. This analysis provides no answer to the question of how far the $\pi$ bonds affect the EFG. The model might well be unsuitable if such bonding is important.

Like the chemical shift there is no absolute reference standard for the [L] values. As seen in section 3.5.2 the experimental data always yield combinations of [A] and [B]   An arbitrary standard is used and the [L] value for some common ligand set equal to 0.

## 3.6 **Temperature dependence of Mössbauer Parameters**

Both the isomer shift and the quadrupole splitting are in general temperature dependent.

### 3.6.1 **Second order Döppler effect**

In a Mössbauer emission or absorption event the momentum of the affected atom is unchanged: it is a zero phonon event. But the mass of the emitting nucleus decreases by an amount equal to the mass equivalent of the the energy of the emitted photon, $\delta m$. Therefore if the initial and final velocities of the emitter are $v$ and $v'$ respectively

$$mv = (m - \delta m)v'.$$

If the corresponding kinetic energies are E and E'

$$E + E_{ex} = E' + E_\gamma$$

where $E_{ex}$ is the excitation energy of the nucleus.

Thus $E' = 1/2 \ (m - \delta m) \ v'^2$ and $E = mv^2$

$E' - E = 1/2 \ (m - \delta m)v'^2 - 1/2mv^2.$

$= 1/2 \ mv^2[m/(m-\delta m) -1] \approx mv^2\delta m /2m.$

But $\delta m = E_\gamma /c^2$ , $\therefore$ E' - E = 1/2 $E_\gamma \ v^2/c^2 = E_{ex} - E_\gamma$.,

or in terms of frequencies $\delta \upsilon/\upsilon = -v^2/2c^2$.

Now although the average value of $v$ over the lifetime of the excited state is zero, the average value of $v^2$, or $<v^2>$ is not. The observed $E_\gamma$ will be less than the excitation energy by an amount $E_\gamma <v^2>/2c^2$.  $<v2>$ is temperature dependent so that the observed chemical shift will also be temperature dependent;  the shift will decrease as the temperature of measurement increases. If we suppose that atoms in a solid behave as simple harmonic oscillators, then $M<v^2> = 1/2\ MU$, where M is the relative atomic mass and U the energy of the solid per unit mass.

Thus $\quad -\dfrac{1}{\upsilon}\left[\dfrac{\partial \upsilon}{\partial T}\right] = \dfrac{1}{Mc^2}\left[\dfrac{\partial U}{\partial T}\right] \quad == \ -C_P/2Mc^2 \quad$ where $C_p$ is the molar heat

capacity of the solid at constant pressure. At sufficiently high temperature $C_P \Rightarrow 3R$ so

that $\qquad\qquad \dfrac{1}{\upsilon}\dfrac{\partial \upsilon}{\partial T} = \dfrac{3kT}{2mc^2}.$

A more realistic analysis evaluates $<v^2>$ using a Debye model for the solid but the expressions become complicated. At very low temperatures differences in the second order Doppler effect for different solids will reflect differences in the zero point energies for the lattice vibrations in the solids.

An alternative, perhaps more elegant, approach to the **second order Döppler effect** employs the relativistic form of the equation for the Döppler effect:

$\upsilon = \dfrac{\upsilon_0\ (1-\upsilon\cos\ \theta/c)}{(1-v^2/c^2)^{1/2}} \quad$ where $v_o$ is the frequency emitted by the source, $v$ is the observed

frequency and v cos θ the relative velocity of the source in the direction of emission of the photon. Now during the lifetime of the excited state $<v\cos\theta>$ is zero

$$v \ = \ v_0/(1 - v^2/c^2)^{1/2} \quad \text{and} \quad \delta E \ = \ h(v_0 - v) \ \approx \ E_\gamma v^2/2c^2.$$

The two approaches are equivalent.

During absorption of a Mössbauer photon the lattice releases the same amount of energy, $E_\gamma<v^2>/2c^2$, as it absorbs in the emission event. If the source and absorber have the same composition and are at the same temperature there will be no isomer shift.

Because different compounds show different second order Döppler effects one must be very cautious of drawing conclusions from very small differences in chemical shifts. The observed shifts are most significant when measured at very low temperatures. By making a series of measurements at different temperatures and plotting:

$1/T$ against $c\delta E/E - 3kT/2mc^2$, the intercept on extrapolating to $1/T = 0$ gives the true chemical shift, supposed independent of T.

With complex spectra from mixtures of compounds it is sometimes possible to separate individual spectra by changing the temperature of measurement.

For iron compounds the change in the observed chemical shift for a 100° change in temperature ranges from about 0.02 to 0.1 mms$^{-1}$.

### 3.6.2 Temperature dependence of Quadrupole Splitting

Changes in the electric field gradient and the quadrupole splitting with temperature can arise from various causes. Excluding solid state phase changes, which are equivalent to a change of the absorber compound, the lattice expands on heating. In principle this leads to a change in the electric field gradient, but this effect is very small.

A much more important source of temperature dependence arises from changes in the electronic population of the orbitals of the Mössbauer atom.

Consider the case of a transition metal atom with six d electrons in a complex with a high spin configuration, Fig.3.6. In a regular octahedral ligand environment the $t_{2g}$ set of orbitals, containing four electrons, is degenerate. The fourth electron populates the constituent $d_{xy}$, $d_{xz}$ and $d_{yz}$ orbitals equally, so that the electric field gradient is zero. However, as the Jahn-Teller theorem predicts, a lower energy configuration can usually be reached by distortion of the stereochemical arrangement leading to removal of some or all the degeneracy. (Fig.3.6.b and c.) The fourth electron will then populate the resulting orbitals as determined by the Boltzmann distribution. The time taken in thermally activated electronic transitions (about $10^{-10}$ s) is much less than the quadrupole precession time.(about $10^{-8}$ s).

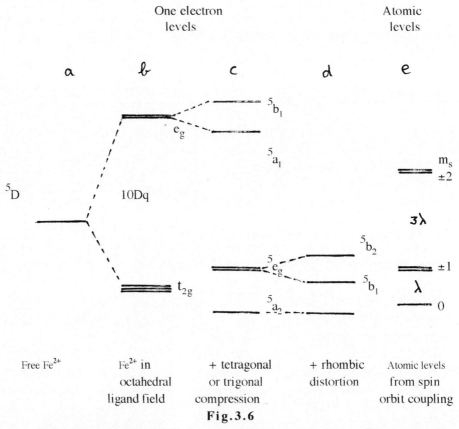

One electron levels

Atomic levels

|  a  |  b  |  c  |  d  |  e  |

Free $Fe^{2+}$       $Fe^{2+}$ in octahedral ligand field     + tetragonal or trigonal compression     + rhombic distortion     Atomic levels from spin orbit coupling

**Fig.3.6**

It should be noted that if the $d_{xy}$ orbital lies lowest in an iron(II) complex then the sign of $\Delta$ will be positive, but with $d_{xz}$, $d_{yz}$ lowest then $\Delta$ will be smaller and negative at low temperature.

The calculation that follows deals only with $q_{fi}$ and ignores any ligand contribution, $q_{mo}$, to the EFG (See 3.5.1.). The iron(II) compound is assumed to suffer tetragonal compression, along the z axis, so that $d_{xy}$ lies at the lower energy.

The Figure 3.6 d also includes the effects of rhombic distortion.

Only those orbitals derived from the $t_{2g}$ set need be considered, since those from the $e_g$ set lie at too high an energy. Considering only the situation of the sixth electron, since the other five give a spherically symmetrical field. The situation is shown in Fig.3.6 d, ignoring spin orbit coupling the distribution of the sixth electron at temperature T, will be: in $d_{xy}$ $1/S$; in $d_{xz}$ $(\exp.-\varepsilon'/kT)$ $/S$;

and in $d_{yz}$ $(\exp.-\varepsilon''/kT)$ $/S$. where $S = 1 + \exp.-\varepsilon'/kT + \exp.-\varepsilon''/kT$.

This leads to $\Delta^T = F. \Delta^0$, where F the reduction factor relates the quadrupole splitting at temperature T to the value at 0 K.

The reduction factor F is given by:-

$$\frac{[1 + \exp\text{-}2\varepsilon'/kT + \exp\text{-}2\varepsilon''/kT - \exp\text{-}\varepsilon'/kT - \exp\text{-}\varepsilon''/kT - \exp(\varepsilon'+\varepsilon'')/kT]}{S}$$

If $\varepsilon' = \varepsilon'' = \varepsilon$, ignoring any ligand contribution

$$\Delta^T = \Delta^0 \frac{[1 - \exp\text{-}\varepsilon/kT]}{[1 + 2\exp\text{-}\varepsilon/kT]}.$$

Thus one finds in such a case the quadrupole splitting increases as the temperature falls. (Fig. 3.7)

A similar treatment for the degenerate $d_{xz}$ , $d_{yz}$ pair lying lowest gives:

$$\Delta^T = \Delta^0 \frac{[1 - \exp\text{-}\varepsilon/kT]}{[2 + \exp\text{-}\varepsilon/kT]}.$$

Thus measurement of the quadrupole splitting as a function of temperature enables one to evaluate ligand field parameters, such as $\varepsilon'$ and $\varepsilon''$, which may not be accessible by optical spectroscopy.

At lower temperatures, usually below 150 K, spin orbit coupling removes some of the remaining spin degeneracy, as shown in Fig.3.6 e. Mixing of wave functions introduces some complexity, but expressions for the change in quadrupole splitting as a function of the differences in the energies of the split levels can be obtained by analogous calculations.

Fig.3.7 shows a plot of the reduction factor F as a function of temperature taking into account spin orbit coupling in tetragonally distorted octahedral compounds.

To obtain $V_{zz}$ etc., or $q_{val}$, each of the populations must be multiplied by the contribution that the orbital makes to the electric field gradient, as given in Table 3.9.

Different modes of distortion will lead to different sequences and separations of the orbitals, but can be treated in the same way. Many $Fe[L_6]$ complexes suffer trigonal distortion by elongation or compression along a threefold axis of the octahedron. The ensuing splitting of energy levels is similar to that due to tetragonal distortion. In both cases compression drives the orbital doublet to a higher energy and elongation lowers its energy, so that compression will yield an orbital singlet ground state and elongation an

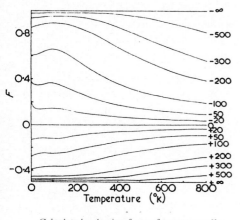

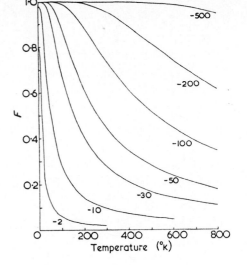

Calculated reduction factor for tetragonally distorted octahedral iron complex.
$Dq = 1000$ cm$^{-1}$, $\lambda = -80$ cm$^{-1}$.
Various values of $D_s$ ($D_s = \varepsilon/3$, Fig.3.6 )

**Fig.3.7**

Calculated reduction factors for tettragonally distorted tetrahedral iron complex.
$Dq = 1000$ cm$^{-1}$, $\lambda = -80$ cm$^{-1}$.
Various values of $D_s$.

**Fig.3.9**

orbital doublet ground state. However in the trigonal distortion case the wave functions involved are mixed functions.

$$\text{The } t_{2g}^{0} = |z^2\rangle, \quad t_{2g}^{+1} = [\sqrt{2/3}\ |x^2 + y^2\rangle - \sqrt{1/3}\ |xz\rangle]$$

$$\text{and } t_{2g}^{-1} = [\sqrt{2/3}\ |xy\rangle + \sqrt{1/3}\ |yz\rangle$$

One electron in the orbital singlet will produce about twice the contribution to $V_{zz}$ produced by an electron in the doublet. One would expect that, if the ligand contributions to $V_{zz}$ are not very large, the quadrupole splitting at low temperatures of a compound with trigonal compression should be substantially larger than one distorted by elongation. So far the two kinds of distortion lead to similar effects on the quadrupole splitting. The important difference is that with trigonal distortion the orbital singlet ground state will lead to a negative quadrupole splitting and a doublet ground state to a positive splitting.

The effect of temperature on the quadrupole splitting can be estimated using the same procedure as was used for the case of tetragonal distortion.

A corresponding treatment can also be used for moderately distorted tetrahedral complexes. Ligand field splittings are shown in Fig.3.8. The quadrupole splitting is generally much more sensitive to temperature for distorted tetrahedral than for distorted octahedral complexes. But spin orbit coupling effects are less important..

The reduction factor, F, as a function of temperature is shown in  Fig. 3.9.

From equations such as those derived above, data on the temperature dependence of the quadrupole splitting may enable  one to evaluate splittings such as $\varepsilon'$ and $\varepsilon''$, and to decide the sequence of energy levels of the orbitals.

Examples of such measurements will be found in Chap.5.

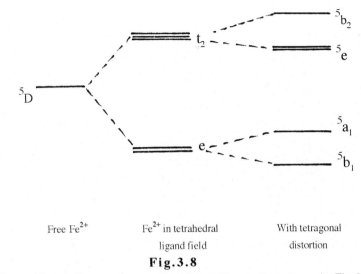

Free $Fe^{2+}$                    $Fe^{2+}$ in tetrahedral                    With tetragonal
                         ligand field                             distortion

**Fig.3.8**

Spin - orbit coupling produces a similar splitting to that shown in Fig.3.7

## 3.7 Sign of the Quadrupole Splitting

In cases other than those involving a 3/2 ⇔ 1/2 transition the pattern of line intensities will usually determine the sign of the EFG provided the resolution of the lines is adequate (Sec.3.2). Even with poor resolution the shape of the unresolved spectrum may be sufficient.

However for the rather commonly occurring 3/2 ⇔ 1/2 magnetic dipole case, the quadrupole split spectrum does not reveal whether the |3/2> or the |1/2> excited state lies at the higher energy. But the question may be answered by applying a strong magnetic field to the absorber. Usually in ferro- or antiferro-magnetic absorbers the magnetic field at the Mössbauer nucleus is very large indeed, 10 - 50 Tesla, and the quadrupolar interaction can be treated as a small perturbation of the larger magnetic interaction.

But such large fields are barely accessible in the laboratory even with superconducting electromagnets.

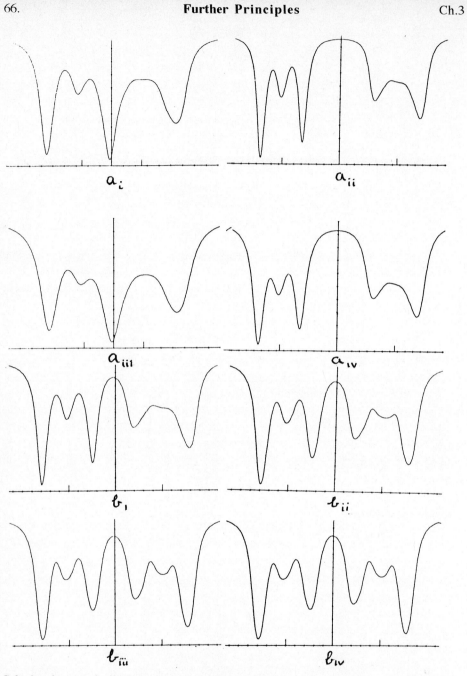

Calculated magnetically perturbed spectra for a diamagnetic iron complex. Applied field 4.5 T parallel to photon beam for all spectra. For $a_{ii}$ and $a_{iii}$ $\Delta = 1.0$, for $a_{ii}$ $\Delta = 3.0$ and $\Delta = 2.0$ for all others. $\Gamma = 0.25$ except for $a_{iii}$ $\Gamma = 0.35$. $a_i$ to $a_{iv}$ $\eta = 0$.

For $b_i$, $b_{ii}$, $b_{iii}$ and $b_{iv}$ $\eta = 0.4, 0.6, 0.8$ and $0.9$, respectively.

**Fig.3.10**

Suppose a diamagnetic absorber is subjected during measurement to a magnetic field parallel to the photon beam. As the field strength is increased, the quadrupole split line corresponding to transitions between the $|+1/2>$ excited state and the $|+1/2>$ ground state, which will eventually yield four lines as the magnetic field removes the $\pm$ degeneracy, will broaden and then split. But because only transitions with $\Delta m \leq 1$ are permitted, the other line, corresponding to transitions to the $|+3/2>$ state, will eventually yield two lines.

Computer calculations have been made (using the complete Hamiltonian for the magnetic and quadrupolar interactions) of the spectra to be expected for different quadrupole splittings,line widths, asymmetry parameters, and field strengths.These calculations usually assume the quadrupolar interaction is weaker than the magnetic interaction, so that it can be treated as a perturbation. However lengthier calculations avoiding this simplification can be made and may be essential.

Strong fields of several Tesla are needed for these experiments. The predicted spectra depend on whether the applied field is parallel or perpendicular to the direction of the $\gamma$ photon beam. Using a parallel magnetic field some examples of the predicted spectra for iron compounds are shown in Fig.3.10. (See also Ref.3.3)

As the field is increased the two lines of the zero field spectrum broaden and the line due to transitions to the $|+1/2>$ excited state begins to split. In the case of $^{57}$Fe spectra, for which the magnetic moments of the ground and excited states, both small, are positive and negative respectively, this line becomes a triplet and thatderived from the $|+3/2>$ excited state becoming a broad line doublet. For large values of the asymmetry parameter, $\eta \geq 0.7$, both lines become triplets and as $\eta$ approaches 1 the spectrum becomes a nearly symmetric pair of triplets. In these circumstances the sign of the EFG is no longer meaningful. Fortunately $\eta$ is less than 0.6 for a very large number of compounds and inspection of the magnetically split spectrum will often decidewhich line is which in the original quadrupole spectrum. If the triplet lies at more negative velocity than the doublet, $\Delta$ is positive and, since Q is positive for iron, eq and $V_{ZZ}$ are also positive.

An estimate of the asymmetry parameter from the spectrum is also often possible, although the probable error may be substantial.

As an example the magnetically perturbed spectrum of ferrocene, the first to be measured using this technique, is shown in Fig.3.11. The same procedure is applicable to paramagnetic absorbers provided the measurement is made at a temperature substantially above those at which any magnetic ordering takes place i.e. above the Curie or Néel temperature for the absorber.

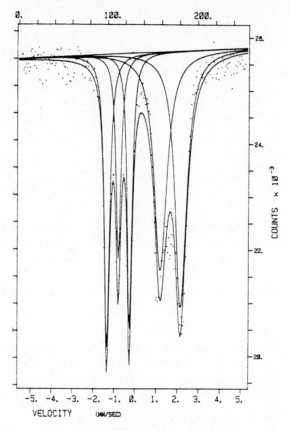

**Fig. 3.11**

## References

3.1    See for example Landau, R.H. (1990) in *Quantum Physics* , Chap.20,.
John  Wiley.

3.2    See  e.g.  Rose, M.E. (1957) in *Elementary Theory of Angular*
*Momentum.*  John Wiley..

3.3    See ref. for Fig.3.11

## Acknowledgements

Fig.3.2    Reproduced with permission from Wagner, F. and Zahn, U, (1970)
*Z. Physik*, **233**, 1.

Fig.3.5    Reproduced with permission from Clark, M.G., Bancroft, G.M. and
Stone, A.J. (1967) *J.Chem.Phys.*, **47**, 4250.

Fig.3.8    Reproduced with permission from Gibb, T.C. (1968)  *J.Chrm.Soc.*,
**1968A**, 1440.

Fig.3.10   Reproduced with permission from Gibb, T.C. (1968)
J.Chem.Soc.,**1968A**, 1441.

Fig.3.11   Reproduced with permission from Collins, R.L. and Travis, J.C.
(1967) in *Mössbauer Effect Methodology*, Vol.3, Ed. Gruverman, I.J.
Plenum Press.

Fig.3.12   Reproduced with permission from Collins,R.L. (1965) *J.Chem.Phys.*,
**42**, 1072.

# 4

# Mössbauer Spectroscopy in Tin Chemistry

## 4.1 MöSSBAUER SPECTRA OF MAIN GROUP ELEMENTS

The Mössbauer spectrum of tin compounds reflects the electronic environment of the tin atoms in the absorber. It is determined principally by the participation of the tin 5s, 5p and 5d orbitals in the interaction of the tin with its ligands.

The isomer shift arises predominantly from the occupation of the 5s orbital, like all the s orbitals this has a substantial electron density at the tin nucleus. Naturally all the other tin s orbitals make a greater contribution to $|\Psi(0)|2$ but the isomer shift is always a difference measurement. It reflects the changes in $|(0)|2$ in the absorber in relation to some chosen tin reference compound. Thus the isomer shift depends on the very small changes in electron density at the tin nuclei, due to changes in the extent of involvement of the 5s orbital in the interaction of the tin with its ligands.

However, the electron distribution in the 5s orbital is affected by screening due to electrons in the 5p and 5d orbitals and indeed, to some extent, by the occupation of the 5s orbital itself. The 5d orbitals are sufficiently remote that their effects can be ignored.

In principal it might be expected that: $\delta = an_s + bn_p + cn_sn_p + dn_s^2$ where $n_s$ and $n_p$ are the effective numbers of electrons in the tin 5s and 5p orbitals. The parameter a is positive but b, c and d, which allow for the screening effects, are all negative.

The numbers $n_s$ and $n_p$ should not be taken too literally. They refer rather to the coefficients weighting the participation of the tin 5s and 5p orbitals in the molecular orbitals formed by the tin and its ligands.

An expression with four independent parameters is rarely very useful so that, as a first approximation, one often supposes that: $\delta = an_s - bn_p$. This still means that $n_s$ and $n_p$ cannot be determined from the isomer shift alone.

The quadrupole splitting, dependent on the electric field gradient, arises predominantly from imbalance in the occupation of the 5p orbitals on the tin. The $<r^{-3}>$ term in the expression for the electric field gradient decreases sufficiently rapidly with increase in r so that the occupation of the 5d orbitals will only produce a much smaller contribution, as in many cases do more remote charges in the lattice. To a crude approximation one might expect that: $\Delta = k[np_z - 1/2(np_x + np_y)]$

## 4.2　NUCLEAR ASPECTS OF TIN MöSSBAUER SPECTROSCOPY

The Mössbauer transition in $^{119m}Sn$ involves an excited state with positive parity, and spin 3/2 and a ground state with the same parity, and spin 1/2. (v.Chap.2, table 2.1.). The emission arises from a pure magnetic dipole transition, M1.

The theoretical line width at half maximum absorption is such that a Ba $^{119m}SnO_3$ source generally shows a line width not less 0.7 mm.s$^{-1}$ ($2\Gamma_t = 0.63$). In favourable cases the isomer shift can be determined with a probable error of $\pm0.02$ mm.s$^{-1}$.

Quadrupole splittings are not resolved unless $\Delta > 0.7$ mm.s$^{-1}$. Below this value a computer analysis of the shape of the broadened line, knowing the line width from the source and a single line absorber, enables an estimate of $\Delta$ down to about 0.3 mm.s$^{-1}$ to be made. However several factors can render such estimates unreliable and such data should be used with caution. For resolved quadrupole splittings $\Delta$ can normally be determined with a probable error of about $\pm0.03$ mm.s$^{-1}$.

Isomer shifts are now generally referred to $BaSnO_3$ as zero, although $SnO_2$ has also been used. It has a small isomer shift relative to barium stannate, 0.03 mm.s$^{-1}$, not very much more than the error in the measurement. It is occasionally convenient to use $Mg_2Sn$ for reference because its isomer shift relative to barium stannate, 1.90 mm.s$^{-1}$, places it between the average tin(II) and tin(IV) values.

Unless otherwise stated all $\delta$ values quoted in this chapter will be with respect to barium stannate, with the absorber at 80 K, and the source at room temperature. All values given in this chapter for both isomer shifts and quadrupole splittings will be in mm.s$^{-1}$.

For the interpretation of isomer shifts it is important to know the sign and magnitude of $\Delta R/R$. The sign can be deduced with some confidence from the observation that the $\delta$ values for tin(II) compounds (2.0 to 4.5) are almost always greater than for tin(IV) compounds (-0.4 to 2.0). Since the tin(II) derivatives must have the higher 5s orbital occupation, the sign of $\Delta R/R$ must be positive. For most purposes the isomer shift for $\alpha$tin, 2.10, marks a convenient dividing point between Sn(II) and Sn(IV).

The magnitude of $\Delta R/R$ is less well established. It is difficult to choose compounds of tin for which the occupation of the 5s orbital on the tin is known. Even in a solid such as $K_2SnF_6$ the $a_1$ molecular orbital, involving the 5s electrons on the tin, is not entirely localised on the the fluorine atoms so that the tin is not exactly a $Sn^{4+}$ ion. Similarly in tin(II) compounds the tin is never exactly an $Sn^{2+}$ with a $5s^2$ configuration. Even the $|\Psi(0)|^2$ value for tin atoms in a solid argon matrix does not correspond to an unperturbed $5s^2 5p^2$ configuration. In a solid compound a pure, spherically symmetric, $5s^2$ configuration is unlikely to be found.

There are properties other than the isomer shift that are determined by $|\Psi(0)|2$. These can also be used to measure the sign and magnitude of $\Delta R/R$. The internal conversion coefficients in the different shells of electrons, associated with a photon emission, also depend on the s electron densities at the nucleus contributed by each shell.

Using thin sources of compounds of $^{119m}Sn$, soft electron spectroscopy enables one to compare the probabilities of conversion in, for example, the N and O shells of the tin. The latter can be expected to vary with the compound of tin in the same way as the isomer shift. The results confirm the sign of $\Delta R/R$. Theoretical calculations, using relativistic wave functions, give $|\Psi(0)|^2$ for the 4s electrons and hence the 5s density for the different compounds can be obtained

Similar information can be obtained by measuring the lifetimes for these decay modes. The different estimates indicate $\Delta R/R$ to be about $1.6 \times 10^{-4}$ (perhaps $\pm$ 15%).

## 4.3 GENERAL FEATURES OF TIN MOSSBAUER SPECTRA

Tin atoms in different oxidation states or crystal environments give rise to different Mössbauer absorption spectra. However the differences are sometimes so small that a compound with tin in two different environments may give an apparently single spectrum. But a spectrum composed of two component spectra always indicates two environments and, or, oxidation states for the tin atoms. Thus the spectra of two different crystal modifications of the same compound are generally different. Phase changes can often be identified in this way. A very simple example is found in the spectra recorded for elemental tin. The cubic diamond structured grey $\alpha$ tin gives a single line spectrum, of normal line width, with $\delta = 2.10$. The closer packed tetragonal and metallic white $\beta$ tin gives a rather broader line with $\delta = 2.54$. The distorted octahedral arrangement around each tin atom in this form leads to an unresolved quadrupole splitting estimated at about 0.03, which accounts for the broader line. The different crystal modifications of $SnWO_4$, $SnF_2$ and SnO also give readily distinguishable spectra.

The diamond structure of $\alpha$ tin suggests it should be considered as a Sn(IV) species, while the distorted six coordinate $\beta$ tin structure, suggesting some steric effect of a lone pair, might be considered a Sn(II) species. Now the isomer shift for $\alpha$ tin is less than for $\beta$ tin so that an isomer shift somewhere between these values might mark the division between tin(II) and tin(IV). From a practical point of view a value nearer to the shift for $\alpha$ tin seems more appropriate. However it will be shown later that anomalous values can be expected.

There are many tin compounds in which the tin occupies more than one kind of crystal site and these give two spectra. The compound of empirical formula $(CH_3)_4 Sn_3 (SO_3F)_8$ gives a quadrupole split absorption with $\delta = 1.91$ and $\Delta = 5.56$. The negative velocity line of the pair partly overlaps another line for which $\delta = -0.22$. (Fig.4.1) This is consistent with the formulation of the compound as $[(CH_3)_2SnSO_3F]_2Sn(SO_3F)_6$.

This spectrum presents a commonly occurring problem. To evaluate the $\delta$ and $\Delta$ parameters one must decide which pair of the three lines constitute the quadrupole split pair. Which of the two lines at negative velocities is to be paired with the positive

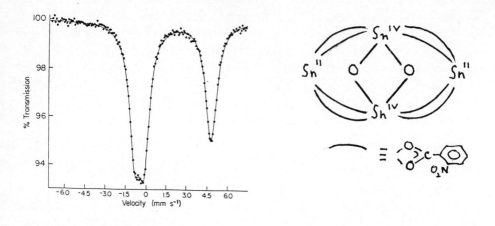

$$\text{Fig.4.1} \qquad\qquad\qquad\qquad \text{Fig.4.2}$$

velocity line?  Guidance can be obtained by considering the isomer shifts shown by other salts of the same anion. The barium salt gives $\delta = -0.31$ and indicates that the outer pair of lines arise from the quadrupole splitting.

The absence of two spectra cannot be taken as definitive evidence of a single tin environment. An example of such behaviour is the complex $Sn_2Br_4(tu)_5(H_2O)_2$, (tu = thiourea) in which the crystallographic structure shows there are two kinds of tin site, but a single spectrum is seen. The compound $[(CH_3)_2SnClterpy]^+[(CH_3)_2SnCl_3]^-$ (terpy = 2.2'.2" terpyridine) also only gives one quadrupole split pair, but it undoubtedly has two tin sites. There are many examples of such behaviour.

Aside from the possibility of accidental coincidence of the Mössbauer parameters for the two sites, there is also the fact that the rather large line width, $\approx 0.7$, limits the selectivity of tin Mössbauer spectroscopy for distinguishing sites with closely similar Mössbauer parameters.

Mössbauer spectra are particularly useful in the study of mixed valence compounds. The compound of empirical formula $SnCH_3(SO_3F)_2$ gives a spectrum composed of a line with $\delta = 3.84$ partly overlapping one line of a quadrupole split absorption with $\delta = 1.87$ and $\Delta = 5.38$, suggesting a formulation $Sn(II).Sn(IV)(CH_3)_2(SO_3F)_4$. Again data for the barium salt of the anion, which give $\delta = 1.76$ and $\Delta = 5.37$, allows identification of those lines which comprise the quadrupole split pair. The isomer shift for the Sn(II) is low indicating interaction of the cation with the $SO_3F$ groups on the anion.

The compound $[Sn_2(O_2C.C_6H_4NO_2)_4.O.THF]_2$ gives a spectrum with a quadrupole split pair with $\delta = 3.597$ and $\Delta = 1.823$ and a single line at $\delta = 0.068$ corresponding to Sn(II) and Sn(IV) atoms respectively. Crystallographic analysis reveals the bonding shown in the Figure 4.2.

## 4.4  ISOMER  SHIFTS  IN  TIN(IV)  COMPOUNDS

Isomer or chemical shift data on tin are seldom values corrected for the second order Doppler effect, but the correction may be within the error on the measurement. The values of the isomer shift for organometallic tin compounds normally fall in a narrow range between about 1.0 and 1.8

### 4.4.1  Halogen  compounds,  complex  halo  anions  and  $SnR_4$

Examination of the data for the isostructural $SnX_4$ and  the salts of the $SnX_nY_{6-n}$ anions shows that the isomer shift increases as the mean electro- negativity of the halogens decreases. This is compatible with the positive sign of $\Delta R/R$: as the Sn-X bond becomes more covalent the occupation of the $a_1$ molecular orbital, with a high tin 5s content, increases and thus $|\Psi(0)|^2$ increases.

It is interesting to explore this relation more quantitatively. Several systems of quantitative expression of electronegativity are currently in use. Fortunately the same conclusions can be reached whichever system is used. In Fig.4.3 the isomer shifts for these compounds are plotted against electronegativity values from a compilation using the Pauling system.

The plots show a linear relation between $\delta$ and $\chi$, the mean electronegativity. The data for the 4- and 6-coordinate species lie on separate lines of similar, but not identical, slope. The 6-coordinate species always show smaller isomer shifts than the 4-coordinate compounds. There are small differences, somewhat more than the probable error, in the isomer shifts for salts of the same halo anion with different cations. The plots of the data in Fig 4.3 use values for the tetraethyl ammonium salts wherever available. The data for the mixed fluoro complexes $SnCl_4F_2$ and $SnBr_4F_2$ lie notably above a least squares fitted line for the other data.

Omitting the two mixed fluoro compounds, one obtains the following linear relations:

For 4-coordinate compounds   $\delta = 4.86 - 1.28\chi$ .

*and*      for 6-coordinate compounds   $\delta = 4.59 - 1.27\chi$.

Isomer shifts calculated using the above equations agree with the experimental values to within an error of $\pm0.03$, which is not much in excess of the probable error on the experimental data. Generally the isomer shifts for the 6 coordinate species are about 0.25 less than for a 4 coordinate species at constant electronegativity. The increase in coordination number leads to longer bonds, a change in the effective charge on the tin atom and a reduction in 5s electron density at the nucleus. The formal $sp^3$ hybridisation changes to $sp^3d^2$, but screening of the 5s electron density by the 5d electrons seems unlikely to be sufficiemt to account for the difference. Perhaps the occupation of the tin 5p orbitals increases and produces more screening.

The $\delta - \chi$ relation can also be used to derive electronegativities from isomer shifts for compounds with other groups attached to the tin. Table 4.1 gives electronegativities for a number of groups calculated in this way. The procedure is only meaningful for compounds in which the tin is tetrahedrally bonded to the attached groups in the solid.

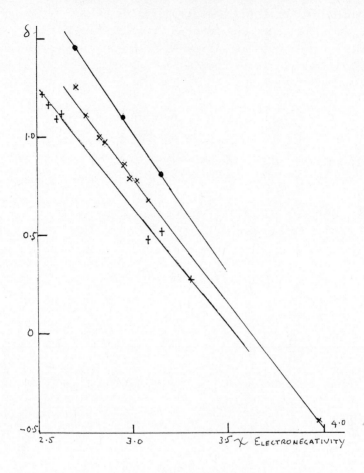

**Fig.4.3**

The $\delta$ - $\chi$ relation can also be used to derive electronegativities from isomer shifts for compounds with other groups attached to the tin. Table 4.1 gives electro- negativities for a number of groups calculated in this way. The procedure is only meaningful for compounds in which the tin is tetrahedrally bonded to the attached groups in the solid.

**Table 4.1**

| Radical | $CH_3$ | $C_2H_5$ | $nC_4H_9$ | $C_6H_5$ | $C_6F_5$ | $C_6Cl_5$ | H | $C_6H_{11}$ |
|---|---|---|---|---|---|---|---|---|
| Isomer Shift | 1.25 | 1.31 | 1.35 | 1.22 | 1.04 | 1.14 | 1.27 | 1.46 |
| $\chi$ calc. | 2.81 | 2.76 | 2.73 | 2.83 | 2.98 | 2.91 | 2.80 | 2.66 |

| Radical | 2Furyl | 3Furyl | 2Thienyl | 3Thienyl | $(CH_3)_2N$ | $(C_2H_5)_2N$ |
|---|---|---|---|---|---|---|
| Isomer Shift | 1.06 | 1.09 | 1.10 | 1.16 | 0.84 | 0.76 |
| $\chi$ calc. | 2.96 | 2.94 | 2.93 | 2.88 | 3.13 | 3.20 |

One should not perhaps attach much importance to the absolute values, but the relative values are interesting. On this scale the alkyl and aryl groups appear between iodine and bromine. Clearly these electronegativity values do not measure quite the same property as the Pauling or Mullikan scales. The discrepancy is most marked for the case of hydrogen.

Using the relation for the 6-coordinate species the isomer shifts found for

$$Sn(N_3)_6^{2-} \; 0.48; \; Sn(NCO)_6^{2-} \; -0.05 \text{ and } Sn(SO_3F)_6^{2-} \; -0.30$$

give electronegativities, on this scale, of 3.23, 3.65 and 3.85 respectively.

The mixed halo anions all give single line spectra, whereas one might expect two kinds of ligands to yield an electric field gradient and a quadrupole split spectrum. The large line widths found for some of the spectra, especially for those mixed halo anions containing iodine, suggest there may be some unresolved quadrupole splitting.

### 4.4.2 Organometallic compounds

The data for the halogen compounds above imply that the ligands make contributions to $|\Psi(0)|^2$ proportional to their electronegativities, so that a system of partial isomer shifts, one associated with each ligand, might be established. It will now be shown that the actual situation is more complicated.

There are spectral data for salts of several stannate anions of the type $nBuSnX_mY_{5-m}$. The isomer shifts prove to be linearly related to the mean halogen constituents. (v.Fig.4.2)   By introducing an appropriate electronegativity value for the nBu group one should find the points to lie on the same line as that given by the 6-coordinate mixed halostannates. However the electronegativity value for nBu needed to effect this transformation is about 1.25! This value should be contrasted with the value 2.73 indicated by the isomer shift for $Sn(nBu)_4$.

Another demonstration that the ligand contributions to the isomer shift are not additive can be found in the data for compounds of the kind $SnR_nX_{4-n}$. Additivity would imply a linear relation between n and $\delta$ for a sequence of compounds with the same R and X, provided the environment of the tin was tetrahedral for all the compounds. The latter condition is not easily fulfilled for a set ranging from n = 1 to 3; polymerisation in the solid often takes place by halogen bridging, with the tin increasing its coordination number. Some sets of data for four coordinate compounds are given in Table 4.2.

**Table 4.2**

Isomer shifts for:

|  | $SnR_4$ | $SnR_3X$ | $SnR_2X_2$ | $SnRX_3$ | $SnX_4$ |
|---|---|---|---|---|---|
| $R = C_6H_5, X = C_6F_5$ | 1.22 | 1.28 | 1.63 | 1.11 | 1.04 |
| $R = C_6H_5, X = H$ | 1.22 | 1.40 | 1.39 | 1.38 | 1.27 |

.

The simplest and perhaps most convincing observation is that the isomer shift for a compound $SnR_3X$, the X being more electronegative than R, generally proves to be larger than for the compound $SnR_4$.

### 4.4.3 **Effects of changes in hybridisation of tin orbitals**

An explanation of the results outlined above is best developed using a valence bond approach. In $SnR_3X$ the bonding hybrids are no longer equivalent as they are in $SnR_4$. As Bent's rule indicates the less covalently bonded and more electronegative X will have lower 5s character and more 5p character than the Sn - R bonds. Thus the Sn - R bonds acquire greater 5s character and the greater covalence of these bonds leads to an increase in $|\Psi(0)|^2$. At the same time the increase in 5p character in the Sn - X bond is at the expense of the Sn - R bonding orbitals, so that some reduction in the screening of 5s electron density at the tin nucleus by 5p electrons takes place.

The latter probably only accounts for a small part of the increase in $|\Psi(0)|^2$. Another small contribution to the isomer shift may arise from the increase in the charge on the tin atom due to the polarity of the Sn-X bond.

Clearly rehybridisation in the above way will be much less marked as X becomes Br or I. These changes in hybridisation have an important effect on the parameters of the Mössbauer spectra. Conversely the Mössbauer data provides a source of information about these changes. Because of such changes the effect of the electronegativity of the halogen, in a series of analogous organometallic compounds for example $[(CH_3)_2C_6H_5C.CH_2]_3SnX$, is far less marked than for the simple halides.(Table 4.3.a.) This sequence can be contrasted with the data for the phthalocyanine tin halides shown in Table 4.3.b. In the latter case one must conclude that there are also some changes in hybridisation of the tin bonds taking place as one moves from F to I, but less marked than in the case of the alkyl tin halides.

### Table 4.3

|     | X = | F    | Cl   | Br   | I    |
|-----|-----|------|------|------|------|
| (a) | $\delta$ = | 1.33 | 1.39 | 1.42 | 1.41 |
| (b) | X = | F    | Cl   | Br   | I    |
|     | $\delta$ = | 0.03 | 0.28 | 0.34 | 0.45 |

For the phthalocyanines the isomer shift is linearly related to the electronegativity of the halogen but the slope of the plot is much less than for the tin halides or hexahalostannates.

Changes in hybridisation of the tin orbitals will usually be associated with a distortion from a regular tetrahedral disposition of the bonds. Most frequently the greater the distortion the greater the isomer shift.

### 4.4.4 **Six Coordinate Organometallic Tin Compounds**

The formation of adducts, with an increase in the coordination number of the tin, usually leads to a small reduction in the isomer shift. However the reduction is less than for the halides and with very soft strong donors there may even be an increase in $\delta$. Hence the isomer shift is of little value in diagnosing coordination number. However in conjunction with the quadrupole splitting it can be useful as will be shown in Section 4.6.

Table 4.4 shows the changes in isomer shift with ligand for adducts of the type $SnCl_4L_2$ in order of increasing $\delta$

**Table 4.4**

| Ligand L | $CH_3CN$, | $(CH_3)_2SO$, | PhCN, | PyO, | Py, | $N(CH_3)_3$, | $Pph_3$ . |
|---|---|---|---|---|---|---|---|
| Isomer shift | 0.38 | 0.38 | 0.41 | 0.42 | 0.51 | 0.59 | 0.78 |

| Ligand L | $PPh_2(CH_3)$ | $P(C_2H_5)_3$ | $PPh(CH_3)_2$, | $P(nC_4H_9)_3$ |
|---|---|---|---|---|
| Isomer shift | 0.79 | 0.84 | 0.85 | 0.87 |

Only the soft strong donors lead to an increase in δ over the value, 0.80, found for $SnCl_4$. A similar sequence has been found for the adducts formed from $nBuSn\overline{Cl}_5$.

A more useful difference is found between the isomer shifts for cis and trans arrangements of the alkyl or aryl ligands in six coordinate complexes. The trans arrangement leads to substantially enhanced 5s character in the linear C-Sn-C system and thus produces a larger isomer shift.    Some examples are shown in Tables 4.5a & b which shows isomer shifts for compounds of known cis and trans configuration.

**Table 4.5 a,b & c**

(a)

| Cis $R_2SnX_2$ | δ | Trans $R_2SnX_2$ | δ |
|---|---|---|---|
| $Ph_2Sn(acac)_2$ | 0.74. | $(CH_3)_2Sn(acac)_2$ | 1.18 |
| $Ph_2Sn(NCS)_2bipy$ | 0.82. | $(C_2H_5)_2Sn(NCS)_2bipy$ | 1.43 |
| $Ph_2Sn(oxin)_2$ | 0.77. | $(PyH)_2Ph_2SnCl_4$ | 1.44 |

(b)

| Cis dithiocarbamate | δ | Trans Dithiocarbamate | δ |
|---|---|---|---|
| $Ph_2Sn(S_2CNPh_2)_2$ | 1.19. | $(CH_3)_2Sn[S_2CNPh_2]_2$ | 1.54 |
| $Ph_2Sn(S_2CN(C_2H_5)_2)_2$ | 1.17. | $(CH_3)_2Sn[S_2CN(C_2H_5)_2]_2$ | 1.72 |
| $Ph_2Sn(S_2CN(CH_2)_4)_2$ | 1.17. | $(CH_3)_2Sn[S_2CN(CH_2)_4]_2$ | 1.59 |
| $Ph_2Sn(S_2CN(CH_2Ph)_2)_2$ | 1.08. | $(C_4H_9)_2Sn[S_2CN(CH_2Ph)_2]_2$ | 1.69 |

In complexes where fac and mer forms are possible the fac form shows the greater isomer shift. The following complexes can be isolated in either form. (Table 4.5.c)   It will be noted that the difference in the quadrupole splittings for the two isomers is more marked than for the isomer shifts.

**Table 4.5c**

| Compound | δ | Δ |
|---|---|---|
| mer $nBuSnCl_3NiSalphen$ | 1.05 | 1.48. |
| fac $nBuSnCl_3NiSalphen$ | 1.21 | 2.13. |
| mer $nBuSnCl_3NiSalmphen$ | 1.08 | 1.53. |
| fac $nBuSnCl_3NiSalmphen$ | 1.15 | 2.18. |
| Salphen | | Salmphen |

This cis trans difference can be expected to be smaller the more covalent the bonding of the ligands, other than the alkyl or aryl groups, to the tin. The difference is also smaller if neither alkyl nor aryl groups are present.

Even in eight coordination the isomer shift does not increase very much if the ligand bonding is fairly polar. Values for a few eight coordinate compounds are given in Table 4.6.

### Table 4.6

| Compound | $Sn(NO_3)_4$ | $Sn(C_2O_4)_4^{4-}$ | $SnTrop_4$ | $Sn(O_2CCH_3)_4H_2O$ |
|---|---|---|---|---|
| Isomer shift | -0.04 | -0.05 | -0.02 | 0.08 |

Trop = Tropolone anion

### 4.4.5 Relation of the isomer shift to other physical data

It has already been mentioned that there are other properties of tin compounds closely related to the isomer shift. Data are often available for the spin-spin coupling constants arising from the interaction of the $^{119}Sn$ spin with that of its nearest neighbour $^{13}C$ or next nearest neighbour $^{1}H$. Such interactions give rise to the fine structure found in the nuclear magnetic resonance spectra of compounds such as $(CH_3)_3SnI$. The mechanism of this coupling depends on the extent of s character in the Sn-C and C-H bonds in the molecules. One might expect therefore a correlation between J $^{119}Sn$ - $^{13}C$ or J $^{119}Sn$ - $^{1}H$ and the isomer shift. Existing data do show some connection between these quantities, but the relation is not a simple one. The isomer shift depends only on the 5s electron density at the nucleus; the coupling involves other features of the bonding not reflected in the isomer shift.

There is also some correlation of the isomer shifts with the binding energies of the 3d electrons on the tin in the different compounds. This energy can be determined by the X-ray photo-electron spectroscopy. The energy of the electrons released by photoelectric excitation, using monochromatic photons from a synchrotron, is determined by soft electron spectroscopy. The difference in the photon and electron energies measures the binding energy.

### 4.4.6 Limits to isomer shifts for tin(IV) compounds

The lowest shifts are found for the compounds with the most polar bonding to the ligands, these will be the closest to an idealised $Sn^{4+}$ species. Nonetheless, even in $SnF_6^{2-}$ ($\delta = -0.44$) the electronic environment of the tin is a long way from this limit. Some other examples of low tin(IV) isomer shifts are shown in Table 4.7.

### Table 4.7

| Compound | $Sn(SO_3F)_6^{2-}$ | $Sn(SO_3CF_3)_6^{2-})$ | $SnP_2O_7$ | $SnF_6^{2-}$ |
|---|---|---|---|---|
| Isomer shift | -0.30 | -0.24 | -0.40 | -0.44 |

(Data for barium salts)

The highest values of $\delta$ can be expected for compounds with covalently bonded strong soft donors.

An interesting group of compounds are those containing a linear, or nearly linear, R - Sn - R unit combined with very polar ligands. The isomer shift for $(CH_3)_2Sn(SO_3F)_2$ is 1.89, an unusually high value. It is probable that the linear $CH_3$-Sn-$CH_3$ can be regarded as bonded by sp hybrid orbitals on the tin, the covalent character of the Sn - C bonds leading to a high 5s electron density on the tin. The interaction of the tin with the $SO_3F$ units being very polar, the electron density resides mainly on the ligands with very little p screening of the 5s electrons. In effect the $CH_3$-Sn-$CH_3$ behaves like the linear doubly charged cation.

Other compounds of the type $(CH_3)_2SnX_2$ show isomer shifts that correlate with the value of the Hammett function for $X^-$, the isomer shift increases as the basicity of $X^-$ falls. The highly polar character of the Sn - X interaction leads to the population of the tin $p_X$ and $p_y$ orbitals being largely located on the ligand X groups. Data for several such compounds are shown in Table 4.8

### Table 4.8
Isomer shifts for $(CH_2)_2SnX_2$

| X = | $SO_3CH_3$ | $PO_2F_2$ | $TaF_6$ | $SO_3Cl$ | $SO_3CF_3$ | $Sn(SO_3F)_6$ | $SbF_6$ | $Sb_2F_{11}$ |
|---|---|---|---|---|---|---|---|---|
| δ = | 1.52 | 1.53 | 1.69 | 1.75 | 1.79 | 2.01 | 2.04 | 2.08 |
| Δ = | 5.05 | 5.13 | 5.23 | 5.20 | 5.51 | 5.64 | 6.04 | 6.02 |

Further consideration will be given to the Sn(IV) - Sn(II) border in section 4.5 where some compounds with Sn - M bonds (M a transition metal) will be discussed.

## 4.5 ISOMER SHIFTS IN TIN(II) COMPOUNDS

Solid tin compounds never contain an idealised $Sn^{2+}$ ion, with a filled spherically symmetric 5s orbital. Indeed crystallographic evidence for many tin(II) compounds suggests the lone pair of electrons is stereochemically active and resides in a directed molecular orbital with some 5p character. Taking the ligands into account one can anticipate a wide range of isomer shifts never rising as high as the value for the hypothetical $5s^2$ tin(II), which is certainly in excess of 5.0, but extending down to the tin(IV) region.

One can appreciate this if one considers compounds of the $SnCl_3^-$ anion. The compound could contain tin with an unperturbed $5s^2$ configuration with the Sn-Cl bonds being formed from the three 5p orbitals and the Cl-Sn-Cl angles being 90°
.But it is much more likely that the lone pair will be stereochemically active and localised in an s-p hybrid orbital, directed away from the chlorines, around the principal axis of the molecule. The 5p-5s mixture will depend on the nature of the ligands, in the present case chlorine, and will be reflected in the Cl-Sn-Cl angle which will no longer be 90°.

Now consider the lone pair orbital interacting with the cation, or notional cation, in the compound. As the interaction grows stronger the tin takes on the character of tin(IV) and if one goes to the extreme case, for the cation $Cl^+$, one obtains $SnCl_4$, clearly a tin(IV) compound. There is no apparent reason why all intermediate situations should not arise so that isomer shifts could extend from the tin(II) to the tin(IV) region.

### 4.5.1 The simpler tin(II) compounds

Because of the lone pair the structural chemistry of tin(II) compounds is much more complex than for the tin(IV) species. Some commonly occurring situations are shown in Fig.4.4

Fig.4.4

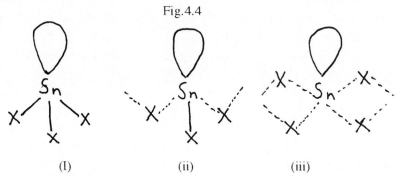

(I)                              (ii)                              (iii)

For a compound $SnL_2$ structure (ii) implies one short and two long Sn-L bonds, the latter involving the bridging ligands. In fact in many solids this description is very approximate and more distant ligand atoms may be interacting with the tin so that the coordination number of the tin is higher than three (or four for type (iii)). Clearly these structures will lead to many of the solids containing polymeric chain or sheet arrangements.

Examining structure (i) of $C_{3v}$ symmetry, which might also be an approximation to (ii), the 5s and $5p_z$ orbitals on the tin will be involved in two **molecular** orbitals formed with the ligands: both these orbitals possess the same symmetry as the tin environment. One, $a_1$, is located largely within the trigonal pyramid formed by the tin and the three ligand atoms and helps bind the four atoms together. The other, $a_2$, is a non-bonding orbital located around the symmetry axis on the side of the tin remote from the ligand atoms. This orbital accomodates the lone pair.

The admixture of tin 5s and 5p orbitals used in forming these molecular orbitals depends on the properties of the ligands and will be that leading to the lowest energy for the assembly, allowing for inter-ligand repulsion effects and optimisation of orbital overlap. If the ligands require a high proportion of 5p character in forming the $a_1$ orbital, the $a_2$ orbital, which must be orthogonal to the $a_1$ orbital, must have a high proportion of 5s character. This should be the case for an electronegative ligand such as chlorine. The composition of these molecular orbitals and the degree of covalency of the tin-ligand interaction, which affects the radial distribution of the orbitals, will determine $|\Psi(0)|^2$ and the isomer shift. The L-Sn-L angles will also depend on the composition of these orbitals.

Excepting $SnF_2$ the tin halides approximate to type (ii). $SnCl_2 \; \delta = 4.17$;

$SnBr_2$ $\delta$ = 3.98; $SnI_2$ $\delta$ = 3.87.

The isomer shift decreases as the bonding becomes more covalent and orbital overlap requirements lead to enhanced 5s character in the $a_1$ molecular orbital. Similar effects are found for the mixed halides SnXY.

$SnF_2$ ( $\delta$ = 3.44 ) is not included in the above halide group because its structure is so different. When fluorine is present it dominates the bridging between the tin atoms.

Crystallographic study of $Sn_3F_5Br$ has shown that it cotains a polymeric cationic $Sn_3F_5$ unit and bromide ions not attached to the tin. Although the crystal environments of all the tin atoms are not the same, they are similar enough that a spectrum with only one component is found.

In the sequence of salts of the $SnX_3$ anions shown in Table 4.9. the isomer shift increases as the halogen becomes less electronegative. These changes must reflect, at

**Table 4.9**

| Compound | $KSnF_3$ | $KSnCl_3$ | $KSnBr_3$ | $KSnI_3$ |
|---|---|---|---|---|
| Isomer Shift | 3.12 | 3.81 | 3.83 | 4.06 |

least in part, changes in the 5s-5p hybridisation preferred by the different halogens. The order is the reverse of that found for the $SnX_2$ series. Perhaps steric considerations lead to an increasing proportion of 5p character in the $a_1$ orbital.

Some other examples are shown in Table 4.10.

**Table 4.10**

| Compound. | $SnHPO_3$ | $KSn(O_2CH)_3$ | $KSn(O_2C.CH_3)_3$ |
|---|---|---|---|
| Isomer Shift. | 3.15 | 3.05 | 3.08 |

In the carboxylate anion the ligands are monodentate. The isomer shifts for other carboxylate complexes lie in the range 2.7 to 3.2.

An interesting and unusual mixed valence compound has been found as a oxidation product of $Sn(O_2C.CF_3)_2$.of composition $Sn_5O_2(O_2C.CF_3)_8$, the Mössbauer spectrum shows one tin(IV) atom, with parameters $\delta$ = 0.09, $\Delta$ = 1.12 and four tin(II) atoms with parameters $\delta$ = 3.65, $\Delta$ = 1.85. The tin(IV) occupies an approximately octahedral position with the two oxygens in trans configuration and with four tin(II) atoms attached by carboxylate groups to its remaining coordinating sites. Pairs of tin(II) atoms are attached to the oxygen and to each other by carboxylate bridges.

Formation of adducts, or increase in coordination number of the tin by attachment of more of the same kind of ligand, generally leads to a reduction in the isomer shift. As in the case of tin(IV) this is probably due to changes in bond length, effective charge on the tin atom, and increased screening of 5s electron density by p and d electrons. Some examples are given in Table 4.11

However it must be observed that the isomer shift for $BaSnF_4$ has been reported to be 3.38. But the structure of this tetragonal compound is very different, the tin being in an $SnF_5L$ environment, L representing the sterically active lone pair.

Even in $Sn(NCS)_2$, where the tin is nine coordinate, the isomer shift is as high as 3.49.

## Table 4.11

| Compound | SnF2 | $KSn_2F_5$ | $KSnF_3$ | $SnSO_4$ | $SnSO_4$ 2tu. |
|---|---|---|---|---|---|
| Isomer Shift | 3.44 | 3.24 | 3.04 | 3.95 | 3.25 |

| Compound | $SnCl_2$ | $SnCl_2Py$ | $SnCl_2$ 1.4 Dioxane |
|---|---|---|---|
| Isomer Shift | 4.17 | 4.01 | 3.73 |

| Compound | $Sn(O_2CH)_2$ | $KSn(O_2CH)_2$ | $Sn(O_2C.CH_3)_2$ | $KSn(O_2C.CH_3)_2$ |
|---|---|---|---|---|
| Isomer Shift | 3.33 | 3.05 | 3.23 | 3.08 |

The additional ligand usually leads to a square based pyramidal configuration around the tin resembling 4.4.(iii).

The adduct formed by $SnCl_2$ and the cyclic 16-crown-6 ether is unusual, the spectrum shows two tin(II) sites with parameters $\delta = 3.89$, $\Delta = 2.11$ and $\delta = 3.33$, $\Delta = 0.99$. The latter suggests the presence of $SnCl_3^-$ indicating that the compound should be formulated as $SnCl.[16\text{-}crown\text{-}6]^+SnClCl_3^-$.

The effects of different cations in salts of the tin(II) anionic complexes is rather larger than in the case of the tin(IV) anions, as can be seen in the series shown in Table 4.12.

## Table 4.12

| Compound | $CsSn_2F_5$ | $RbSn_2F_5$ | $KSn_2F_5$ | $NaSn_2F_5$ | $NH_4Sn_2F_5$ |
|---|---|---|---|---|---|
| Isomer Shift | 3.15 | 3.22 | 3.31 | 3.37 | 3.38. |

| Compound | SnOx | $Cs_2SnOx_2H_2O$ | $K_2SnOx_2H_2O$ | $Na_2SnOx_2H_2O$ |
|---|---|---|---|---|
| Isomer Shift | 3.70 | 3.17 | 3.38 | 3.67 |

$$(Ox = C_2O_4)$$

It seems that the polarising effect of the smaller cations enhances the isomer shift for the tin in the anion. Similar effects are found for the salts of the malonato anion.

The highest isomer shifts are found in very polar tin compounds where there is very little covalent interaction of tin with the ligands. Some examples are given in Table 4.13.

## Table 4.13

| Compound | $Sn(SbF_6)_2$ | $Sn(SO_3F)_2$ | $Sn(SbF_6)_2(AsF_3)_2$ | $Sn.Sn(SO_3CF_3)_6$ |
|---|---|---|---|---|
| Isomer shift of Sn(II) | 4.44 | 4.48 | 4.66 | 4.69 |

The stannate(II) salts are generally polymerised in the solid state. They show very low isomer shifts for tin(II). Thus $BaSn(OH)_4$  $\delta = 2.54$ and its dihydrate has $\delta = 2.44$. Generally the isomer shifts of this group lie in the interval 2.4 - 2.6.

The oxide SnO is interesting because the tetragonal and orthorhombic forms have quite different Mössbauer parameters, $\delta = 2.71$ with $\Delta = 1.45$ and $\delta = 2.60$ with $\Delta = 2.20$, respectively.

### 4.5.2  Organometallic tin(II) compounds

The majority of alkyl and aryl group compounds of the type $SnR_2$ are highly polymerised with extensive Sn-Sn bonding. They are effectively tin(IV) species with four shared bonds on each tin. The isomer shifts are less than 2. However, by substitution at the two and six positions on the phenyl ring stereochemical inhibition of tin-tin bonding occurs and monomeric $SnR_2$ compounds can be obtained. A selection of data are given in. Table 4.14

**Table  4.14**

| Substituents on ring | 2.6. $CF_3$ | 2.4.6.tBu | $(SnPh_2)_n$ |
|---|---|---|---|
| Isomer Shift | 3.37 | 3.28 | 1.56 |
| Quadrupole Splitting. | 1.93 | 1.90 | small |

Other very bulky R groups also give monomeric $SnR_2$. Both $[\{Si(CH_3)_2\}_2CH]_2$ Sn and $[\{Si(CH_3)_3\}_2N]_2$ Sn are monomeric in solution and in the gas phase. The latter is also monomeric in the solid with isomer shift 3.52. The former dimerises in the solid state to give Mössbauer parameters of $\delta = 2.16$ and $\Delta = 2.31$. There is an approximately tetrahedral arrangement of bonding around the tin atoms and it has been suggested that each tin functions both as a donor and acceptor.

Stannocene, $\eta Cp_2Sn$, is monomeric in the solid state with parameters $\delta = 3.72$ and $\Delta = 0.86$. Tin(II) forms several $\beta$ diketonate complexes whose isomer shifts lie in the range 3.0 to 3.6, clearly tin(II) species.

One could expect that $SnR_2$ species would act both as both Lewis acids and bases. Indeed $Cp_2Sn$ reacts with boron trifluoride in tetrahydrofuran, but the product is not a simple adduct. The Mössbauer parameters of the product are not very different from those of stannocene and the crystal structure has shown it to be

$$[(\eta^5Cp)_2Sn(\mu\eta^5CpSn)]^+BF_4.THF.$$

An unusual related mixed valence compound is: $[(CH_3)_3Sn(IV)C_5H_4]_2Sn(II)$. The parameters for the two kinds of tin in this compound are  Sn(IV) $\delta = 1.3$, $\Delta = 0$ and Sn(II) $\delta = 3.58$, $\Delta = 0.89$.

In the following compounds $Cp_2Sn$ does appear to act as a simple donor. None of them show quadrupole splitting. The isomer shifts are:

$Cp_2SnAlCl_3$  3.71;  $Cp_2SnBBr_3$  3.77;  $Cp_2SnAlBr_3$  3.83.

However, the crystal structures of these compounds have not been established.

There is one compound in which both simple donor and acceptor functions seem to be displayed: $F_3B\ SnCl_2\ N(CH_3)_3$ with parameters $\delta = 3.87$, $\Delta = 0$. But as the isomer shift indicates the bonding to the tin is probably rather weak in this compound.

Another group of compounds showing characteristic tin(II) parameters, with isomer shifts of about 3, are shownin Fig.4.5.

### 4.5.3  Compounds with tin-metal bonds

It is convenient to discuss these compounds in two groups, those derived from an $SnX_3^-$ unit and those from an $SnX_2$ moiety.

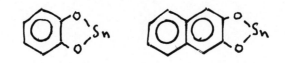

$$\delta = 2.95, \quad \Delta = 1.76. \qquad\qquad \delta = 3.08, \quad \Delta = 1.82.$$

**Fig.4.5**

### 4.5.3.1. *Compounds of the type $L_nMSnX_3$*

Such compounds display a wide variety of isomer shifts ranging from about 3, corresponding to tin(II), to as low as 1.5, which lies within the range found for tin(IV) compounds.

The concept of formal oxidation state for the tin in these compounds is no longer useful. Consider the compound $L_nMSnX_3$, in the form $L_nM^+.Sn\overline{X}_3$ which would be described as a tin(II) compound with the $Sn\overline{X}_3$ functioning as a donor ligand. Now suppose the compound udergoes an internal oxidation and reduction reaction, the oxidation state of the M is reduced by two units, the compound becomes $L_nMSnX_3$, formally a tin(IV) derivative. No change will have taken place that is detectable by classical analysis. The change is entirely a matter of an internal electronic rearrangement within the molecule. The two states described are simply the extremes of a continuous distribution of possible electronic states and according to the precise properties of L,M and X all intermediate situations could be expected to occur.

Donation by the $Sn\overline{X}_3$ group, leading to the formation of the Sn-M bond, will lead to a reduction in the s electron density on the tin and the product will have a lower isomer shift than a non-interacting $Sn\overline{X}_3$

. Data have been accumulated for a very large number of componds of this type. A selection of such data is given in Table 4.15.

In compounds 1-3 the $Sn\overline{Cl}_3$ unit is virtually free, that is to say very little s electron density is lost by interaction with the cationic entity. The first compound, with the highest isomer shift, must have the weakest interaction.

In compounds 4 and 5 the $Sn\overline{Cl}_3$ acts as a simple donor ligand; s electron density and indeed some p density is transferred to the cationic complex from the lone pair orbital. But rehybridisation more than replaces the p character of the Sn-M bond. More and more transfer of electron density takes place as one goes from compounds 6 to 13 and there is a continuous fall in isomer shift throughout this series. When $PPh_3$ is replaced by the

## Table 4.15

| Compound | $\delta$ | $\Delta$ | Compound | $-$ | $-$ |
|---|---|---|---|---|---|
| 1) Et₂OH.SnCl₃ | 3.54 | 1.00 | 11) dppeCpFe.SnCl₃ | 1.92 | 1.76 |
| 2) Et₄N.SnCl₃ | 3.47 | 1.00 | 12) PPh₃(CO)₄Mn.SnCl₃ | 1.70 | 1.78 |
| 3) CsSnCl₃ | 3.40 | 1.20 | 13) PPh₃(CO)CpFe.SnCl₃ | 1.86 | 1.89 |
| 4) (dppe)₂ClCo.SnCl₃ | 3.10 | 1.36 | 14) Cp(CO)₂Fe.SnCl₃ | 1.71 | 1.81 |
| 5) PPh₃CpNi.SnCl₃ | 3.22 | 1.20 | 15) Cp(CO)₃Cr.SnCl₃ | 1.67 | 1.54 |
| 6) (Pph₃)₃Cu.SnCl₃ | 2.46 | 1.88 | 16) PnBu₃(CO)₃CoSnCl₃ | 1.59 | 1.67 |
| 7) (Pph₃)₃Ag.SnCl₃ | 2.54 | 1.69 | 17) (CO)₄Co.SnCl₃ | 1.42 | 1.20 |
| 8) (Pph₃)₃Au.SnCl₃ | 2.44 | 1.58 | ********* | | |
| 9) (AsPh₃)₃Au.SnCl₃ | 2.26 | 1.67 | 18) Sn[Co(CO)₄]₄ | 1.96 | 0.0 |
| 10) (cod)₂Ir.SnCl₃ | 2.06 | 1.77 | 19) PhSnCl₃ | 1.27 | 1.79 |

Et = C₂H₅   dppe = Ph2P.CH2.CH2.PPh2   Cp = C₅H₅

cod = 1.5 cyclo octadiene, C₈H₁₂.

weaker donor AsPh₃ in compounds 8 and 9 there is still more loss of s electron character to the gold moiety.

For compounds 6-13 the concept of oxidation state of the tin is no longer meaningful. With further loss of s character and growth of p character in the Sn-M bond in compounds 14-17 the tin becomes effectively tin(IV). In compounds 16 and 17 it is seen that replacing the carbon monoxide on the transition metal by the stronger σ donor PnBu₃ less electron density is transferred from the tin and the isomer shift rises.

Further consideration of these data will be given when the significance of the quadrupole splitting data has been explored. (Section 4.6).

The more electronegative the X in the SnX₃ unit the greater the p character in the $a_1$ orbital and hence the greater the s character of the $a_2$ lone pair orbital, so that the isomer shift for analogous complexes of $\overline{SnX}_3$ would be expected to decrease as X becomes less electronegative. This prediction is only partly realised as can be seen in Table 4.16.

## Table 4.16

| Compound | $\delta$ | $\Delta$ | Compound | $\delta$ | $\Delta$ |
|---|---|---|---|---|---|
| Cp(CO)₂Fe.SnCl₃ | 1.71 | 1.81 | Cp(CO)₃Cr.SnCl₃ | 1.67 | 1.54 |
| Cp(CO)₂Fe.SnBr₃ | 1.76 | 1.61 | Cp(CO)₃Cr.SnBr₃ | 1.77 | 1.34 |
| Cp(CO)₂Fe.SnI₃ | 1.97 | 1.48 | Cp(CO)₃Cr.SnI₃ | 1.86 | 1.06 |
| Cp(CO)2Fe.SnPh₃ | 1.40 | 0.3 | | | |
| Cp(CO)2Fe.Sn(CH₃)₃ | 1.35 | 0 | Cp(CO)₃Cr.Sn(CH₃)₃ | 1.41 | 1.36 |

Iodine and bromine give higher shifts, as found with the $\overline{SnX}_3$ salts. Steric factors and perhaps some role for the 5d orbitals may be involved..

It will be observed that the isomer shift for compound 19 in the table 4.15. implies an effective electronegativity of 2.26 for the Co(CO)₄ group, making it less electronegative than, for example, the nBu group.

4.5.3.2. *Compounds derived fron* $SnR_2$

Monomeric compounds in which an $SnR2$ group donates to a transition metal moiety giving a three coordinate tin species are only found if the R group sterically inhibits dimerisation, a typical example is $R = [(CH_3)_3Si]_2CH-$ As seen in Table 4.17. these compounds give very low isomer shifts for tin(II).

### Table 4.17

| Compound | δ | Δ |
|---|---|---|
| $(CO)_5Cr.SnR_2$ | 2.21 | 4.43. |
| $(CO)_5Mo.SnR_2$ | 2.15 | 4.57. |
| trans $(CO)_5Cr.(SnR_2)_2$ | 2.21 | 4.04. |
| trans $(CO)_5Mo.(SnR_2)_2$ | 2.13 | 4.24. |

In many cases a dimer is formed with an $Sn \overset{M}{\underset{M}{\rightleftharpoons}} Sn$ ring,

An example is:

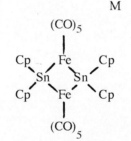

These contain effectively four covalent tin and have isomer shifts in the range 1.7 - 1.95, as shown in Table 4.18.

### Table 4.18

| Compound | δ | Δ | Compound | δ | Δ |
|---|---|---|---|---|---|
| 1.$[(CH_3)_2Sn.Fe(CO)_4]_2$ | 1.22 | 1.47 | 5.$[CpBrSn.Fe(CO)_4]_2$ | 1.82 | 1.60 |
| 2.$[Cp_2Sn.Fe(CO)_4]_2$ | 1.86 | 0 | 6.$[tBu_2Sn.Fe(CO)_4]_2$ | 1.83 | 1.16 |
| 3.$[Cp_2' Sn.Fe)CO)_4]_2$ | 1.95 | 0 | $Cp = C_5H_5$ | | |
| 4.$[CpClSn.Fe(CO)_4]_2$ | 1.79 | 1.58 | $Cp' = C_5(CH_3)_5$ | | |

A large number of compounds with Sn-M rings are known, but Mössbauer spectra have only been recorded for a small fraction of them. The isomer shifts all fall well into the tin(IV) region, for example:

$(CO)_4Fe$ — $Sn$ $\big\langle \begin{matrix} tBu \\ tBu \end{matrix}$      $\delta = 1.73, \ \Delta = 1.53.$
$(CO)_4Fe$

The compounds $Ph_3Sn.MgBr$, $PhCp_2Sn.MgBr$ and analogous zinc compounds have isomer shifts around 1.5 and may be ring dimers. The monomeric compounds and the dimers described above add another ligand to give monomeric compounds of the type $LR_2Sn.Fe(CO)_4$. The isomer shifts for these adducts lie at the top of the tin(IV) range. For $LtBu2Sn.Fe(CO)_4$ the Mössbauer parameters are for L = Tetrahydrofuran $\delta = 2.11,$

$\Delta = 1.44$; for L = Dimethylsulphoxide $\delta = 1.98$, $\Delta = 3.60$; and for L = Pyridine $\delta = 2.01$, $\Delta = 3.44$. The configuration of ligands around the tin is approximately tetrahedral.

The tin(II) diketonates, mentioned in section 4.5.2.,form numerous compounds of the type $k_2Sn.M(CO)_n$ where k represents a $\beta$ diketone anion. Some examples are given in Table 4.19.

Table 4.19

| | $\delta$ | $\Delta$ | | $\delta$ | $\Delta$ |
|---|---|---|---|---|---|
| $Acac_2Sn.$ | 3.12 | 1.89 | $dbm_2Sn$ | 3.07 | 1.98. |
| $Acac_2Sn.Cr(CO)_5$ | 1.81 | 2.28 | $dbm_2Sn.Cr(CO)_5$ | 2.02 | 2.07. |
| $Acac_2Sn.Mo(CO)_5$ | 1.82 | 2.30 | $dbm_2Sn.Mo(CO)_5$ | 1.93 | 2.51. |
| $Acac_2Sn.W(CO)_5$ | 1.80 | 2.35 | $dbm_2Sn.W(CO)_5$ | 1.90 | 2.39. |

Acac = Pentane 2.4 dionate; dbm = bis 4.phenyl butane 2.4 dionate

## 4.6. QUADRUPOLE SPLITTING IN TIN COMPOUNDS

Unlike the isomer shift, which is determined mainly by the occupation of the tin 5s orbitals, the quadrupole splitting is determined only by the imbalance of occupation of the 5p orbitals. Because of the $r^{-3}$ term in the expression for the electric field gradient, occupation of the 5d orbitals can only make a very small contribution.

Thus $\Delta = c[n_z - 1/2(n_x + n_y)]$, where the n refer to occupation of the $5p_z$, $5p_x$ and $5p_y$ orbitals on the tin, respectively.

For tetrahedral and octahedral arrangements of one kind of ligand about the tin no electric field gradient can arise $\Delta = 0$. For the trigonal bipyramidal arrangement, which occurs rather commonly with tin, the axial and equatorial bonds are not equivalent and an electric field gradient may be found even for $SnL_5$, for example $\Delta = 0.63$. for $SnCl_5^-$.

For the more symmetric structures any electric field gradient must be due to differences in electron donation to the tin 5p orbitals by the different ligands. However, because the line width in tin Mössbauer spectra precludes the accurate measurement of small quadrupole splittings, an apparently zero splitting may be found for some tetrahedral or octahedral compounds with more than one kind of ligand. Thus compounds of the type $SnR_nR_{4-n}$, where R and R' are different alkyl or aryl groups, generally show a single line spectrum. This is also the case for the octahedral mixed halides: but the large line widths indicate unresolved quadrupole splitting.

Large splittings can be expected if the ligands include both very weak and rather strong donors.

### 4.6.1  Applications to tin(IV) compounds

A useful empirical rule, combining isomer shift and quadrupole splitting data, states that the ratio $\Delta$ / $\delta$ is less than 1.9 for four coordinate tin and greater than 2.1 for five or six coordinate tin compounds. Like most empirical rules it must be used with caution. The circumstances in which it is valid will become apparent in the next section.

### 4.6.2  The partial quadrupole splitting approach

The basis of the partial quadrupole splitting analysis of Mössbauer data has been treated in Section 3.5.3. For accurate application of partial quadrupole splittings (p.q.s.) to the prediction of the splitting in compounds, other than those used to set up the system of p.q.s. values, very restictive conditions apply to the bonding. Each bond must be independent of the other bonds in the molecule. The existence of phenomena like the trans effect clearly show that this is not generally true.

Changes in hybridisation of the bonds, such as were suggested by the isomer shift data for the $SnR_nX_{4-n}$ compounds, must also limit the applicability of the p.q.s. approach. As was shown in Section 3.5.3, provided they are both tetrahedral molecules, $SnR_3X$ and $SnRX_3$ should have the same magnitude of quadrupole splitting but opposite sign if the p.q.s. approach is valid. To test this conclusion one must be careful to find compounds that give monomeric solids. This has been found to be the case for the pairs of compounds with $R = C_6H_5$ (Ph) and $X = Cl$ and with $R = CH_3$ (Me) and $X = I$.

The spectra show:

$Me_3SnI$    $\delta = 1.43$, $\Delta = 2.91$;         $MeSnI_3$    $\delta = 1.58$, $\Delta = 1.68$.

$Ph_3SnCl$   $\delta = 1.39$, $\Delta = 2.48$;         $PhSnCl_3$   $\delta = 1.19$, $\Delta = 1.73$.

From the isomer shift data one might expect that the discrepancy between the two splittings should decrease as the bonding becomes more covalent and the electronegativity of the X decreases. The following data supports this conclusion:

$Me_3Sn(C_6F_5)$   $\delta = 1.27$,  $\Delta = 1.31$   .$MeSn(C_6F_5)_3$   $\delta = 1.19$,  $\Delta = 1.14$

$Ph_3Sn(C_6Cl_5)$   $\delta = 1.27$,  $\Delta = 0.84$.  $PhSn(C_6Cl_5)_3$   $\delta = 1.11$,  $\Delta = 0.80$

Another semi-quantitative test is to measure the quadrupole splittings of both cis and trans forms of compounds of the type $SnX_4L_2$, the p.q.s approach predicts these to be in the ratio -1 to 2. Unfortunately there are not many isomeric pairs of tin compounds of this kind for which the splittings are large enough for a satisfactory test.

Indeed the first example of cis-trans isomers in tin chemistry arose from Mössbauer evidence. Preparations of $[(C_3H_7)_2SO_2]_2SnCl_2$ were found sometimes to give the parameters $\delta = 0.39$, $\Delta = 0.83$ and, with slightly different preparations, $\delta = 0.43$ and $\Delta = 1.57$. $SnBr_4(HCONMe_2)_2$, prepared by mixing solutions of the components in pentane gives a product with different Mössbauer parameters from that obtained when $SnBr_4$ is added to an excess of dimethylacetamide. The values are $\delta = 0.66$, $\Delta = 0.44$ and $\delta = 0.66$, $\Delta = 0.83$, suggesting cis and trans compounds, respectively. Similar preparations of $SnCl_4(CH_3CONMe_2)_2$ give products with parameters $\delta = 0.38$, $\Delta = 0.45$

and $\delta = 0.38$, $\Delta = 0.78$.  Similar differences are reported for a number of other adducts of $SnX_4$, but for all of them the splittings of the cis form could only be obtained by computer analysis of unresolved spectra and the values are subject to rather large probable errors.  The presumed trans form always gave the larger splitting but the trans/cis ratio was always less than 2.  In such compounds it is quite likely there is a change in bonding hybridisation on going from the cis to the trans form and this may account for the discrepancy.

    Another group of compounds that can be used for a similar test are those of the type $SnR_2Cl_2(L-L)$.  Rehybridisation effects might be rather smaller for these compounds. The ligand L-L is a bidentate ligand, 22'bipyridyl or a substituted bipyridyl.  There are three possible isomers, since the L-L must occupy a cis site; cis,cis,cis, trans R and trans Cl (cis R).  The p.q.s treatment gives (Section 3.5.3):

$$\text{All cis  } V_{ZZ} = [Cl] + [L-L] - 2[R]$$
$$\text{Trans R  } V_{ZZ} = 4[R] - 2[L-L] - 2[Cl]$$
$$\text{Trans Cl  } V_{ZZ} = 4[Cl] - 2[L-L] - 2[R]$$

(For all cis the z axis lies along Cl - L, for Trans R the z axis lies along R - R and for Trans Cl the z axis lies along Cl - Cl).

    Bearing in mind that the system of p.q.s values is normalised by setting $[Cl] = 0$ and that $[R] > [L-L]$, the $V_{ZZ}$ values for all cis to trans Cl to trans R should be in the ratios 1 to about 1  to about -2.  The R were $4X-C_6H_4$.  The trans R  form was made by substituting the bidentate ligand for dimethyl sulphoxide in the complex $R_2SnCl_2(dmso)_2$; this could be converted to the cis form (either all cis or trans Cl) by recrystallisation from a non-polar solvent, Table 4.20.

### Table 4.20
#### Results for 44'dimethyl22'bipyridyl complexes,

|           |     | $\delta$ | $\Delta$ |       | $\delta$ | $\Delta$ |                           |
|-----------|-----|------|------|-------|------|------|---------------------------|
| X = H     | cis | 0.91 | 2.07.| trans | 1.20 | 3.41.| Ratio of $\Delta$ values 1.65 |
| X = Cl    | cis | 0.84 | 1.99.| trans | 1.10 | 3.33.| Ratio of $\Delta$ values 1.67 |
| X = CH$_3$ | cis | 0.75 | 2.35.| trans | 1.20 | 3.41.| Ratio of $\Delta$ values 1.48 |
| X = CF$_3$ | cis | 0.86 | 1.86.| trans | 1.12 | 3.32.| Ratio of $\Delta$ values 1.78 |
| X = MeO   | cis | 1.01 | 2.27.| trans | 1.28 | 3.63.| Ratio of $\Delta$ values 1.60 |

Average ratio 1.63

With the 22'bipyridyl complexes, only X = MeO gave both forms. The cis form gave $\delta = 0.88$, $\Delta = 1.95$; and the trans $\delta = 1.26$, $\Delta = 3.54$.  Ratio of $\Delta$ values 1.81

    The ratio is invariably less than 2 but the trans always has a substantially higher quadrupole splitting.  There is crystallographic evidence that these compounds are seldom strictly octahedral around the tin and this probably accounts for the discrepancy (see Section 4.6.2.4).

    Nonetheless, despite these sources of inaccuracy, the p.q.s treatment is useful.  One can get better agreement between the experimental and calculated splittings by collecting p.q.s values for a group of similar compounds and applying them only to compounds of a similar type, but much of the value is lost in this way.  It is more useful to distinguish only between sets of p.q.s values applicable to tetrahedral, octahedral and trigonal bipyramidal tin compounds.

### 4.6.2.1 *Sign of the quadrupole splitting*

To establish a set of partial quadrupole splitting parameters it is necessary to determine not only the magnitude of the splitting but also its sign for a number of tin compounds containing the ligands of interest. Although the sign can be determined using absorbers comprised of orientated crystals, this is generally inconvenient since such absorbers are hard to prepare.

More usually the sign is deduced from spectra obtained when the absorber is subjected to a strong magnetic field. The basis of the procedure has been treated in Section 3.7.

The spectra obtained with tin compounds are rather different from those described for $^{57}Fe$. The quadrupole moment of $^{119m}Sn$ is negative so that a positive sign for $\Delta$, or $e^2qQ$, means that $V_{zz}$, $= eq$, is negative. Further the magnetic moment of the ground state $^{119}Sn$ is rather large and negative, while that of the excited state is smaller and positive (Table 2.1).

In the case of iron the magnetic field can be treated as a perturbation of the quadrupolar interaction, but for tin the magnetic and quadrupolar interactions are often more nearly equal. Simple inspection of the magnetically perturbed spectrum will not suffice to decide the sign, as can often be done for iron spectra. But comparison with numerically computed spectra made for the relevant magnitudes of $\Gamma$, the line width for the measurement and $B_{ext}$, the applied field, enables one to decide on the sign and obtain an approximate value for $\eta$, the asymmetry parameter. Some computed spectra for fields parallel to the source-absorber-detector axis are shown in Fig. 4.6.

The magnetic field leads to some mixing of nuclear states and the selection rule $\Delta m = 0$, $\pm 1$ is no longer rigidly obeyed and eight lines become possible. The magnetically perturbed tin spectra are not as clear cut as the related iron spectra. The $1/2 \Leftrightarrow 1/2$ line yields four lines with increasing field and the $1/2 \Leftrightarrow 3/2$ line yields two rather broad lines. If the latter pair of lines lies at more negative velocity the sign of $\Delta$ is negative.

Notwithstanding these complexities the sign of $\Delta$ can usually be identified and an estimate of the value of $\eta$ made, provided $\eta > 0.6$.

Since the sign of Q for $^{119m}Sn$ is negative a positive $\Delta$ implies a negative EFG. A negative EFG implies a higher p electron density along the z axis than in the xy plane, and vice versa. If the donor power of the ligands along the z axis exceeds that of those in the xy plane a negative EFG will ensue.

The sign of $\Delta$ is generally positive for (i) trans $R_2SnX_4$ , (ii) cis $R_2SnX_4$, (iii) $RSnX_5$, (iv) $RSnX_3$ (when tetrahedral) (v) $R_2SnX_3$ with R equatorial and for practically all Sn(II) compounds. Negative values are found for (i) $R_3SnX$, (ii) $R_3SnX_2$ with X axial, R can represent alkyl or aryl groups or the transition metal carbonyl moities; X a halogen or other rather weakly donating ligand. Irregularities may be found for compounds whose structures depart seriously from tetrahedral or octahedral symmetry. The reason

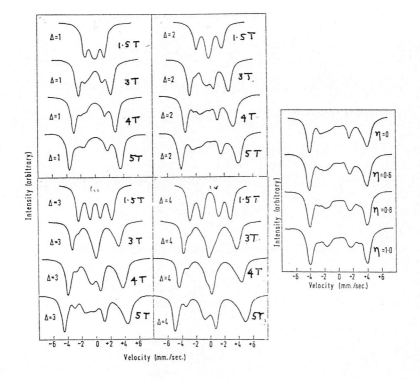

Calculayed magnetically perturbed spectra for different fields (a) showing effect of different quadrupole splittings and (b) different asymmetry parameters. Field applied parallel to the photon beam.

## Fig.4.6

for these generalisations becomes apparent when the p.q.s values given in the next section are considered.

In developing such tables compounds with polar lattices should be avoided to ensure there is no substantial lattice contribution to the quadrupole splitting. Where this is not possible large co-ions should be used to reduce cation-anion interactions (See section 4.5.1.).

4.6.2.2  *p,q,s,  values for tetrahedral and octahedral tin compounds*

### Table 21 a     p.q.s. values for 4 coordinate compounds

| Ligand | p.q.s | Ligand | p.q.s | Ligand | p.q.s |
|--------|-------|--------|-------|--------|-------|
| NCS | +0.21 | $HCO_2$ | -0.18 | H | -1.06 |
| Cl | 0 | $CF_3$ | -0.63 | $4.FC_6H_4$ | -1.12 |
| Br | -0.07 | $C_6F_5$ | -0.70 | $C_6H_5$ | -1.26 |
| $CH_3CO_2$ | -0.15 | $C_6Cl_5$ | -0.83 | Alkyls | -1.34 |
| I | -0.17 | $2.CF_3C_6H_4$ | -1.04 | | |

[Cl] is set at 0

### Metal carbonyl ligands

| Ligand | p.q.s | Ligand | p.q.s | Ligand | p.q.s |
|--------|-------|--------|-------|--------|-------|
| $Fe(CO)_3NO$ | -0.59 | $Co(CO)_3PPh_3$ | -0.83 | $Mn(CO)_4PPh_3$ | -1.01 |
| $Cr(CO)_3Cp$ | -0.72 | $Mo(CO)_3Cp$ | -0.83 | $Fe(CO)_2Cp$ | -1.08 |
| $Co(CO)_4$ | -0.76 | $W(CO)_3Cp$ | -0.87 | | |
| $Ir(C_8H_{12})_2$ | -0.82 | $Mn(CO)_5$ | -0.97 | | |

$Ph = C_6H_5$    $Cp = C_5H_5$.

### Table 4.21 b          p.q.s for 6 coordinate compounds

| Ligand | p.q.s | Ligand | p.q.s |
|--------|-------|--------|-------|
| Ni Salen | +0.30 | Pyridine oxide | -0.05 |
| Cu Salen | +0.24 | 1/2Oxine | -0.05 |
| $Ph_3PO$ | +0.16 | Pic | -0.07 |
| $[(CH_3)_2N]_3PO$ | +0.12 | 1/2 22'Bipyridyl | -0.08 |
| $(CH_3)_2SO$ | +0.10 | Pyridine | -0.10 |
| PCP | +0.10 | 1/2Trop | -0.11 |
| P2CP | +0.07 | 1/2Koj | -0.13 |
| NCS | +0.07 | I | -0.14 |
| Picolinato | +0.06 | 1/2Dipyam | -0.17 |
| $HCON(CH_3)_2$ | +0.04 | 1/2Dmbipy | -0.17 |
| As2CAs | -0.01 | $1/2S_2CH(CH_2PH)_2$ | -0.19 |
| Piperidine | -0.01 | 1/2 Pysox | -0.23 |
| $Ph_3AsO$ | -0.02 | $nBu_3P$ | -0.27 |
| $(CH_3)_3N$ | -0.02 | 1/2 Edt | -0.56 |
| BipydO | -0.03 | $C_6H_5$ | -0.83 |
| 1/2Acac | -0.03 | nBu | -0.93 |
| 1/2Phen | -0.04 | Alkyls | -1.06 |

### Metal carbonyl ligands

| Ligand | pqs | Ligand | pqs |
|--------|-----|--------|-----|
| $Mo(CO)_3Cp$ | -0.66 | $Mn(CO)_4PPh_3$ | -0.73 |
| $Mn(CO)_5$ | -0.71 | $Fe(CO)_2Cp$ | -0.74 |

PCP = $Ph_2OPCH_2POPh_2$; P2CP = $Ph_2OPCH_2CH_2POPh_2$; As2CAs = $Ph_2OAsCH_2AsOPh_2$; BipydO = 2.2'bipyridyl NN'dioxide; Acac = Pentane 2.4 dionate; Oxin = Quinolin 8 olate; Pic = 3 Picoline; Trop = Tropolonate; Koj = Kojate; Dipyam = Di 2.pyridyl amine; Dmbipy = 4.4' dimethyl 2.2' bipyridyl; Pysox = 2 pyridinethiol 1 oxide; edt = ethane dithiolate

A number of comments must be made on these data:

(i) The precise p.q.s values depend to some extent on the choice of compounds made to derive them.

(ii) Since the p.q.s values relate to $V_{zz}$ and the sign of Q is negative for $^{119m}Sn$, the sign of the calculated splitting must be changed to obtain $\Delta$.

(iii) It will be observed that the middle region of p.q.s values, -0.3 to -0.8, is very little populated. As a result quadrupole splittings tend to be dominated by the contributions from alkyl, aryl and metal carbonyl ligands, especially if they are in a trans configuration. Compounds containing only ligands of small, or large p.q.s will have small quadrupole splittings, often too small to be resolved, for example, $SnX_nY_{4-n}$ and mixed alkyl/aryls.

(iv) Unfortunately there are not very many ligands for which p.q.s values are available for both tetrahedral and octahedral compounds. However a plot of tetrahedral against octahedral values with the existing data yields a line of slope 0.73. The molecular orbital treatment of p.q.s values predicts a slope of 0.67 (See 3.5.3). Considering that more than half the data refer to compounds involving metal carbonyl ligands, for which it is known that the tin - metal bond has an enhanced 5s content, the agreement between experimental and theoretical slopes is acceptable. The theoretical basis of p.q.s values assumes all four or six bonds are equivalent hybrids.

(v) As the p.q.s value decreases and becomes more negative, donation by the ligand into the 5p orbitals of the tin increases. Thus if one changes the sign of the p.q.s the sequence reflects increasing donor strength of the ligands with respect to tin.

(vi) The large p.q.s values for alkyl and aryl groups accounts for the dominant influence of these groups on most quadrupole splittings and for the very large splittings in trans R six coordinate complexes.

Supposing $[X]^{tet.} << [R]^{tet.}$, and similarly for the octahedral values, then $\Delta$ for $R_2SnX_2 \approx -2.7$: assuming it is tetrahedral, $\Delta$ for cis $R_2SnX_4 \approx -2$ and $\Delta$ for trans $R_2SnX_4 \approx 4.1$.

### 4.6.2.3 *Applications of p.q.s. values*

A simple illustration of the use of p.q.s. values is to identfy the structures of $(C_6H_5)_2Sn(Acac)_2$ and $(CH_3)_2Sn(Acac)_2$. Using p.q.s. values in the above tables one calculates that the quadrupole splittings for the cis and trans forms of the former compound are 1.60 and 3.20 and for the latter compound 2.06 and 4.12. The observed values are 2.14 for the phenyl and 4.12 for the methyl compound, clearly the former has a cis and the latter a trans configuration. The rather large discrepancy between calculated and observed values is most likely due to a rather distorted configuration. ( See below ).

### 4.6.2.4 *Applicability of p.q.s calculations*

In most cases quadrupole splittings calculated using p.q.s values agree with the experimental values to within 0.30. This is about an order of magnitude greater than the probable error on the experimental data. It is clear that p.q.s values give one a rather approximate estimate of $\Delta$. Indeed the most valuable aspect of the estimates is to identify compounds that, for one reason or another, deviate grossly from the model. A discrepancy between calculated and experimental values of 0.40, or more, may indicate that the supposed structure for the compound is wrong, that the bonding departs markedly from equivalent $sp^3$, or $sp^3d^2$, hybrids, or that the structure is substantially distorted from an idealised tetrahedral or octahedral configuration round the tin.

It must also be remembered that where measurements are made on salts there may be an appreciable $q_{lat.}$ term that is not taken into account in the p.q.s treatment.

Compounds of the type $RSnCl_3(L-L)$, where L-L is a bidentate ligand can assume one of two possible structures:

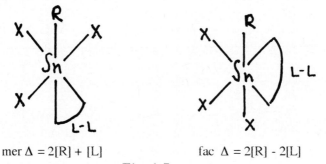

$$\text{mer } \Delta = 2[R] + [L] \qquad\qquad \text{fac } \Delta = 2[R] - 2[L]$$

**Fig.4.7**

The bidentate ligands used were the nickel and copper salen complexes, coordination to the tin takes place through the oxygen atoms, X = Cl, R = Ph. (See 4.4.4)   The quadrupole splitting for the $SnCl_3$ adducts of these ligands gave the p.q.s values of: NiSalen = +0.30; CuSalen = +0.25 and NiSalmphen = +0.26.

Although the values calculated for the mer form differ from the experimental value by several times the experimental error it seems clear these complexes have the mer configuration. Data are given in Table 4.22.

### Table 4.22

| Compound | $\Delta$ mer | $\Delta$ fac | $\Delta$ Expt. |
|---|---|---|---|
| PhSnCl_3NiSalen | +1.36 | +2.26 | 1.38 |
| PhSnCl_3CuSalen | +1.41 | +2.15 | 1.49 |
| CH_3SnCl_3NiSalen | +1.82 | +2.72 | 1.63 |
| nBuSnCl_3CuSalen | +1.61 | +2.35 | 1.67 |

However, with some complexes recrystallisation from chloroform gave a product with different Mössbauer parameters from those obtained on recrystallisation from acetonitrile. The former product changed to the latter on long storage as a solid. Data are given in Table 4.23.

**Table 4.23**

| Compound | $\Delta$ calc. | $\Delta$ Expt. | Compound | $\Delta$ calc. | $\Delta$ Expt |
|---|---|---|---|---|---|
| mer nBuSnCl$_3$NiSalphe | +1.63 | 1.48 | mer nBuSnCl$_3$NiSalmphen | +1.64 | 1.53 |
| fac nBuSnCl$_3$NiSalphen | +2.31 | 2.18 | fac nBuSnCl$_3$NiSalmphen | +2.30 | 2.18 |

Salmphen = NN'o 4 Methylphenylene bis Salicylidineiminate

With these two complexes it appears that both isomers can be isolated, the fac form being the less stable.

For most complexes of the type R$_2$SnX$_2$L$_2$ the values predicted for the different possible structures are too close for a decision on structure to be made.

The compounds for which the p.q.s predictions are seriously wrong fall into a few well characterised groups. It has already been noted that compounds where the different bonds involve substantially different sp hybrids cannot be expected to fit this model. Such compounds include SnX$_4$L$_2$ and compounds that combine very electronegative and strong donor ligands, including metal carbonyl ligands.

Another rather numerous group comprises compounds that contain an R$_2$Sn unit In this case the p.q.s prediction may differ from the observed quadrupole splitting in both magnitude and sign. Many of these compounds are substantially distorted from a regular tetrahedral or octahedral configuration.

### 4.6.2.5 *Effects of distortion*

Consider the contribution to the EFG made by two ligands, carrying unit charge, with an L-Sn-L angle of $2\theta$ (Fig.4.8). The EFG components they produce in relation to the axes 11, 22 and 33, as shown, are $V_{11} = 2(3\sin^2\theta - 1)$, $V_{22} = 2(3\cos^2\theta - 1)$ and $V_{33} = -2$. Hence $|V_{22}|$ will be largest and will define the z axis of the EFG; if $(3\sin^2 - 1) > 1$, that is if $\theta > 54°$ 70. $|V_{11}|$ will be largest. If $(3\cos^2\theta - 1) > 1$ or $\theta < 35°25$ and finally $|V_{33}|$ will only define z if $54°70 > \theta > 35°25$. The overall effect on the magnitude of the EFG is shown in Fig.4.9. At $\theta = 45°$ $V_{11} = +0.5$, $V_{22} = +0.5$ and $V_{33}$ or $V_{zz} = -1$. Thus $\eta = 0$ at $\theta = 45°$. At $\theta = 35°25$ or $54°70$ $\eta$ rises to 1, as shown in Fig.4.9. It can be seen that changes in $\theta$ can alter the overall EFG substantially, even to the extent of a change in its sign. If the EFG is dominated by the large p.q.s of R one would expect $\Delta = -4[R](1 - 3/4 \sin^2\varphi)$, where $\varphi = 2\theta$ ie the total R - Sn - R angle.

This expression agrees with quite a lot of the experimental data. For distorted R$_3$SnX compounds, preserving C$_{3v}$ symmetry, a similar treatment gives:

$$\Delta = 2[X]^{tet} - 3[R]^{tet}(1 - 3\cos^2\theta).$$

The above analysis is on a point charge representation of the ligands. It is also possible to develop a molecular orbital approach to the problem. In fact the latter treatment, although theoretically more acceptable, does not give any better agreement with the experimental data.

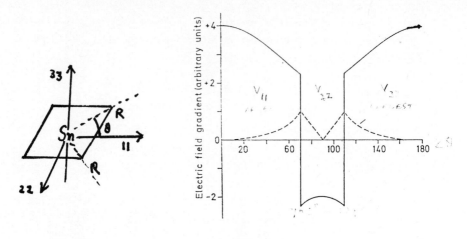

Fig.4.8                                        Fig.4.9

The quadrupole splittings found for the numerous compounds with distorted C-Sn-C angles are more conveniently treated using the point charge approach.

### 4.6.2.5 *Five coordinate tin compounds*

Tin is frequently five coordinate in its solid compounds.

Salts of the $SnCl_5^-$ anion show a quadrupole splitting of +0.63. In a trigonal bipyramidal configuration this requires that the equatorial and axial bonds are different. Indeed they differ in bond length for the same ligand. If therefore one is to extend the p.q.s approach to these compounds one will need to associate two p.q.s values with each ligand, $[L]^e$ for equatorial binding and $[L]^a$ when axially bound.

This considerably complicates the development of a set of p.q.s values for 5 coordinate tin. In addition the crystallographic data for such compounds are limited. Using the same procedure as in section 3.5.2 one expects:

$$\Delta \, SnCl_5^- = -4[Cl]^a + 3[Cl]^e.$$

Several compounds of the type $R_3SnL_2$ have been made and their Mössbauer spectra recorded. Fig.4.10 shows possible structures and predicted $\Delta$.

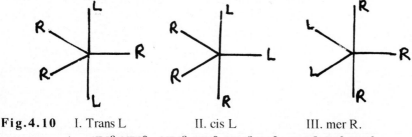

Fig.4.10    I. Trans L            II. cis L            III. mer R.

$\Delta = 4[L]^a\text{-}3[R]^e$    $2[R]^a\text{-}2[R]^e+2[L]^a\text{-}[L]^e$    $4[R]^a\text{-}[R]^e\text{-}2[L]^e$

If the quadrupole splittings of these compounds are plotted against the values for the analogous $R_2SnL_4$ compounds, the data cluster around three straight lines. Structural data

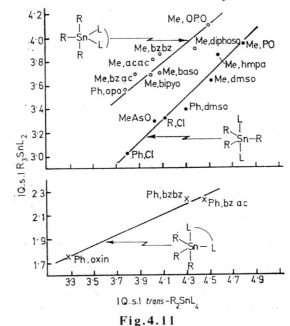

Fig.4.11

for some of these compounds indicate that the three lines distinguish the three forms shown in Fig.4.11.

With a collection of data for different R and L and setting $[Cl]^a = 0$, one can obtain a set of p.q.s values. (v.Table 4.24)   Data for $(CH_3)_3SnCl_2$ which has structure I, yields $[CH_3]^e$. Care must be taken to avoid highly distorted compounds. For example $(CH_3)_2SnBr_3$ has a positive quadrupole splitting, but the p.q.s treatment suggests a negative value. As shown in the previous section this could be explained by a C-Sn-C angle of about $123°$.

**Table 4.24**

| Ligand | Cl/Br | Alkyl | Ph | Acac | Bzac | Bzbz | P2CP | Bipyo |
|---|---|---|---|---|---|---|---|---|
| $[L]^a$ | 0 | -0.94 | -0.89 | -0.03 | -0.07 | -0.02 | +0.05 | -0.04 |
| $[L]^e$ | +0.2 | -1.13 | -0.98 | +0.20 | +0.15 | +0.22 | +0.24 | +0.15 |

( Bzac = Benzoyl acetone anion, Bzbz = Dibenzoyl methane anion, Bipyo = NN' bipyridyl oxide, others as in earlier tables.)

The absolute value of the p.q.s for the equatorial position is always greater than for the axial position but  with the more covalently bonding ligands $[L]^e$ is more negative than $[L]^a$.

### 4.6.3 Combined use of δ and Δ

The p.q.s data will often provide an indication of the structure of a tin compound if its Mössbauer parameters are known. However in many cases a definitive answer is not possible.  Predicted splittings for different configurations may be too close together, or even the same, or  distortion may lead to anomalous results.  However the combination of isomer shifts and quadrupole splittings provides interesting information about the ligand interactions.

Mössbauer data are available for a very large number of compounds of this kind. A plot of δ against Δ for several such compounds is shown in Fig.4.12. As the cationic component interacts more strongly with the $SnCl_3^-$ the isomer shift falls, indicating abstraction of 5s electron density from the tin. The quadrupole splitting rises concomitantly indicating an increase in the 5p density along the M-Sn axis. This trend continues, with varying proportions of the two changes, as the tin becomes more like tin(IV). Eventually, however, the quadrupole splitting falls sharply as the bonds approach $sp_3$ hybrids. The 5p imbalance disappears completely for $ClSnCl_3$.

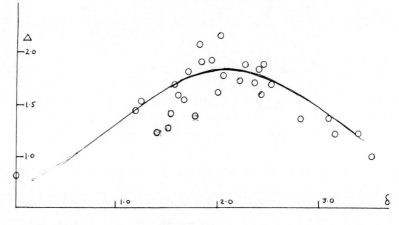

**Fig.4.12**

### 4.6.3.1 *Derivatives of SnCl_3*

An interesting and unusual feature arises with some of the simple salts of $SnBr_3^-$ and $SnCl_3^-$. $CsSnCl_3$ forms monoclinic crystals giving δ = 3.62, Δ = 0.9. These are colourless. But it is also possible to obtain a yellow cubic form with parameters δ = 3.70, Δ = 0. $CsSnBr_3$ gives black cubic crystals with parameters δ = 3.96, Δ = 0, which exhibit low temperature semiconduction. Solid solutions $CsSnCl_{3-x}Br_x$ show colours depending on composition, becoming less intense as x decreases. The isomer shifts decrease with x and line broadening, suggesting unresolved quadrupole splitting, develops.

It appears that the 4d orbitals on the bromide ions in the cubic lattice overlap to give a conduction band which is partly occupied by the 5s electrons of the tin. The position and occupation of this band, and colour of the salts, change with temperature and x. In the cubic environment there is no longer a directed lone pair orbital and the tin effectively suffers a delocalised interaction with the bromide using the three p orbitals on the tin.

### 4.6.3.2 *A more quantitative approach to LSnCl_3.*

A rewarding way of giving quantitative expression to the changes in the $SnCl_3$ moiety with change in L makes use of the relation of the isomer shift and the quadrupole splitting to the occupation of the tin 5s and 5p orbitals. In approximation one can write

**4.25 Table**

| No. | Compound | M-S $\delta$ / $\Delta$ | $\delta'$ | $n_p$ | $\Delta'$ | $n_s$ | Net donation |
|---|---|---|---|---|---|---|---|
| 1 | (PPh$_3$)$_3$CuSnCl$_3$ | 2.46/1.88 | 0.88 | 0.22 | 1.08 | -0.37 | 0.15 |
| 2 | (PPh$_3$)$_3$AgSnCl$_3$ | 2.54/1.69 | 0.69 | 0.17 | 1.00 | -0.35 | 0.18 |
| 3 | (PPh$_3$)$_3$AuSnCl$_3$ | 2.44/1.58 | 0.58 | 0.145 | 1.10 | -0.38 | 0.24 |
| 4 | cod$_2$RhSnCl$_3$ | 2.16/1.72 | 0.72 | 0.18 | 1.38 | -o.48 | 0.30 |
| 5 | cod$_2$IrSnCl$_3$ | 2.06/1.77 | 0.77 | 0.19 | 1.48 | -0.52 | 0.33 |
| 6 | (PPh$_3$)$_2$nbdRhSnCl$_3$ | 2.38/1.70 | 0.70 | 0.175 | 1.16 | -0.405 | 0.23 |
| 7 | (Pph$_3$)$_2$nbdIrSnCl$_3$ | 2.16/1.82 | 0.82 | 0.205 | 1.38 | -0.48 | 0.27 |
| 8 | (CO)$_3$CpCrSnCl$_3$ | 1.67/1.54 | 0.54 | 0.135 | 1.87 | -0.66 | 0.52 |
| 9 | (CO)$_3$CpMoSnCl$_3$ | 1.69/1.71 | 0.71 | 0.18 | 1.85 | -0.65 | 0.47 |
| 10 | (CO)$_3$CpWSnCl$_3$ | 1.67/1.77 | 0.77 | 0.19 | 1.87 | -0.66 | 0.49 |
| 11 | (CO)$_5$MnSnCl$_3$ | 1.62/1.57 | 0.57 | 0.14 | 1.92 | -0.68 | 0.54 |
| 12 | Ph$_3$P(CO)$_4$MnSnCl$_3$ | 1.70/1.78 | 0.78 | 0.19 | 1.84 | -0.61 | 0.42 |
| 13 | (CO)$_2$CpFeSnCl$_3$ | 1.72/1.80 | 0.80 | 0.20 | 1.82 | -0.64 | 0.44 |
| 14 | Ph$_3$P(CO)CpFeSnCl$_3$ | 1.88/1.88 | 0.88 | 0.22 | 1.66 | -0.58 | 0.36 |
| 15 | P(OPh)$_3$(CO)CpFeSnCl$_3$ | 1.79/1.82 | 0.82 | 0.21 | 1.75 | -0.61 | 0.40 |
| 16 | (CO)$_4$CoSnCl$_3$ | 1.42/1.20 | 0.20 | 0.05 | 2.12 | -0.75 | 0.70 |
| 17 | nBu$_3$P(CO)$_3$CoSnCl$_3$ | 1.59/1.67 | 0.67 | 0.17 | 1.95 | -0.69 | 0.52 |
| 18 | (Ph$_3$P)$_2$nbdRhSnCl$_3$ | 2.41/1.83 | 0.83 | 0.21 | 1.13 | -0.39 | 0.18 |
| 19 | (Ph$_3$As)$_2$nbdRhSnCl$_3$ | 2.28/1.89 | 0.89 | 0.22 | 1.26 | -0.44 | 0.22 |
| 20 | (Ph$_3$Sb)$_2$nbdRhSnCl$_3$ | 2.15/2.04 | 1.04 | 0.26 | 1.39 | -0.48 | 0.26 |
| 21 | (nbd)$_2$RhSnCl$_3$ | 2.23/1.72 | 0.72 | 0.13 | 1.31 | -0.46 | 0.33 |
| 22 | (C$_2$H$_5$)$_2$PCOCpFeSnCl$_3$ | 1.95/1.91 | 0.91 | 0.23 | 1.59 | -0.56 | 0.33 |
| 23 | (CO)$_4$FeSnCl$_3$ | 1.55/1.40 | 0.40 | 0.10 | 1.99 | -0.71 | 0.61 |
| 24 | (CO)$_4$Fe(SnCl$_3$)$_2$ | 1.53/1.26 | 0.26 | 0.065 | 2.01 | -0.71 | 0.64 |
| 25 | (Et$_4$N)$_2$Cl$_2$CORhSnCl$_3$ | 1.77/1.93 | 0.93 | 0.23 | 1.77 | -0.62 | 0.39 |
| 26 | (Et$_4$N)$_2$ClCORh(SnCl$_3$)$_2$ | 1.74/2.15 | 1.15 | 0.29 | 1.80 | -0.63 | 0.34 |
| 27 | (Et$_4$N)$_2$CORh(SnCl$_3$)$_3$ | 1.68/2.25 | 1.25 | 0.31 | 1.86 | -0.65 | 0.34 |
| 28 | (Ph$_3$P)$_3$ClPtSnCl$_3$ | 1.84/2.07 | 1.07 | 0.27 | 1.70 | -0.59 | 0.32 |
| 30 | (Ph$_3$P)$_2$Cl$_2$Pt(SnCl$_3$)$_2$ | 1.61/2.01 | 1.01 | 0.25 | 1.93 | -0.675 | 0.32 |
| 31 | (Ph$_3$P)CpNiSnCl$_3$ | 1.85/1.90 | 0.90 | 0.225 | 1.69 | -0.58 | 0.36 |
| 32 | (Ph$_3$P)$_2$CpNiSnCl$_3$ | 3.18/1.20 | 0.20 | 0.05 | 0.36 | -0.13 | 0.08 |
| 33 | CsSnCl$_3$ | 3.40/1.22 | 0.22 | 0.055 | 0.14 | -0.046 | 0 |
| 34 | SnCl$_4$ | 0.80/0.00 | 0.00 | 0.00 | 2.74 | -0.98 | 0.98 |

nbd = norbornadiene    cod = cyclo-octa 1.5 diene.

$\delta = (n_s - bn_p)/a$ and $\Delta = cn_p$. If one chooses the compound with the highest isomer shift of this type as a reference compound and puts $\delta' = \delta_{ref.} - \delta$, and similarly $\Delta' = \Delta - \Delta_{ref..}$ where $\delta$ and $\Delta$ relate to some other derivative $LSnCl_3$, then $n_s$ and $n_p$ calculated from the above equations, using the primed parameters, will measure the change in the populations of the 5s and 5p orbitals between the selected $LSnCl_3$ and the reference compound, a supposedly free $\overline{SnCl}_3$.

The constants a,b and c have been estimated, partly from experimental data and partly from theoretical calculations, to be $a = 2.8$, $b = 0.15$ and $c = 0.25$. (See section 4.8.2).

The best choice of reference compound appears to be $[(C_2H_5)_2OH]$ $SnCl_3$ with parameters $\delta = 3.54$, $\Delta = 1.00$. Table 4.25. collects values of differences in $n_s$ and $n_p$, in relation to the reference compound, of several $\overline{SnCl}_3$ derivatives calculated in this way. It will be seen that $n_s$ is negative, indicating the tin loses 5s electron density by donation to the L, but acquires 5p density by rehybridisation of the Sn-Cl bonds and, or, donation from the L.

Although not too much emphasis should be placed on the precise numerical values of $n_s$ and $n_p$, since the values of a,b and c are rather crude estimates, the data yield useful information about the M-Sn bonding. The following conclusions seem justified:

(i)The data for compounds 1-3, 4 & 5, 6 & 7 suggest that the net donation by the tin, loss of 5s less gain of 5p electrons, increases as the acceptor metal changes from the first to the second and then to the third family of transition elements. The effect is not very big and, indeed, is not apparent for the series 8-10.

(ii) Replacing a CO ligand by a stronger donor, such as $Ph_3P$, reduces the donation by the tin (11 & 12, 13-15 and 16 & 17).

(iii) Correspondingly stronger donors on the M lead to reduced donation by the tin (18-21, 14,15 and 22).

(iv) With more than one $SnCl_3$ group the donation by the tin remains practically constant (23 & 24 and 25-27).

(v) Increased oxidation state of the metal acceptor leads to increased donation. (28 & 29), but other factors are involved in such comparison.

(vi) The nickel complexes are interesting. These results suggest that compound 32 contains a practically free $\overline{SnCl}_3$ ion, while compound 31 does not. But the latter complex dissolves to give conducting solutions in acetonitrile!

### 4.6.3.3 *The lone pair*

In many tin(II) compounds the lone pair is stereochemically active and occupies an orbital with some p character, but it must not be assumed that this is always the case. In the intercalation compound $Nb_3S_6Sn$ the tin is undoubtedly tin(II), the isomer shift for the compound is 3.68, but there is no evidence of quadrupole splitting. The tin has an effectively cubic environment.

## 4.7 **THE RECOIL FREE FRACTION.**

It was stated at the beginning of this chapter that the data used would all be for measurements at 80 K. Indeed many of the compounds mentioned give no observable absorption at room temperature. Nonetheless, room temperature spectra can be recorded for many tin compounds; for example, $Ph_3SnCl$, $Ph_4Sn$, and $(CH_3)_2Sn(Acac)_2$.

This draws attention to a Mössbauer parameter, obtainable from the spectra, that has not yet been considered - the **Mössbauer or recoil free fraction**, f. It was shown in Section 1.2. that, with approximations, f depends on $<x^2>$, the mean square distance of neighbouring atoms from the Mössbauer atom in the lattice. The recoil free fraction depends on the strength of coupling of the Mössbauer atom with its neighbours, the rest of the molecule and the rest of the lattice. Information about this coupling, in principle, can be obtained from Raman and other spectra of the solid, but these data are not always available, nor is their interpretation straightforward.

With further approximations it was shown in 1.2. that f is related to the Debye temperature, $\theta$, of the solid and at temperatures such that $T \geq \theta / 2$, $f = exp.-6RT/k\theta^2$. Measurements of f itself are complicated and time consuming, but the above expression gives $dln f/dT = -6R /k\theta^2$ (R is the recoil energy). Now f is linearly related to the area under the Mössbauer absorption line, A, so that a plot of $ln A_T/A_{80}$ versus T should be a straight line. Again referring to 1.2. one can transform the above expression for the slope to: $dln [A_T/A_{80}]/ dT = 3kE_\gamma /h^2c^2M \omega^2$ where M is the effective mass opposing the the recoil of the Mössbauer nucleus and $\omega$ is the angular frequency of a lattice mode of the solid. Thus the slope of the plot should depend on the mass of that part of the solid coupling to the Mössbauer atom. For a monomeric solid it might correspond to the molecular mass, but it will be much greater for a polymeric solid. The frequency, $\omega$, might also be expected to depend on the coordination number of the tin and degree of polymerisation in the solid.

Thus the slope of the plot should be smaller for a polymeric material than for a monomeric solid. Preliminary experimental data suggested that monomeric solids gave a slope more negative than $-1.7 \ 10^{-2}$ $K^{-1}$. Unfortunately more extensive data has shown that the situation is more complex. A tightly packed monomeric solid may give quite a low slope: for example $SnCl_4(Pyrazine)_2$ gives $-0.74 \ 10^{-2}$, while some polymeric compounds give rather high slopes, for instance $(CH_3)_3 SnF$ gives $-1.75 \ 10^{-2}$.

Tin(II) compound generally give very high slopes, for example $SnCp_2$ $-3.13 \ 10^{-2}$ and $Sn(Acac)_2$ $-2.8 \ 10^{-2}$.

The plots are not always linear and overlap of inter and intra molecular vibrational modes may be involved. It maybe that the physical preparation of the absorber can influence the results.

## 4.8 **SPECTRA IN A MATRIX**

Mössbauer spectroscopy is more or less restricted to solid materials although small effects can be observed in very viscous liquids, for example molten $SnF_2$. However very satisfactory spectra can be obtained with tin species held in a solid matrix. Such systems include frozen solutions and tin species incorporated in a frozen gas. The matrix itself may be a volatile material such as a solid rare gas, or it may be a more or less refractory solid. In the latter case one can really only inject accelerated tin atoms. But extensive work has been done in this way using the injected tin atom as a probe to study refractory solids.

### 4.8.1. Frozen solutions

Care must be taken in the preparation of such systems. Rapid freezing is essential to produce a glass and avoid segegation of crystallites of the solute.

The hydrolysis of $(CH_3)_2Sn(ClO_4)_2$ and $(CH_3)_3Cl$ in aqueous solutions have been investigated in this way, exploring the Möössbauer parameters as a function of the pH of the solution.

1. $(CH_3)_2(ClO_4)_2$

| pH | 0.95 | 9.70 |
|---|---|---|
| Parameters | $\delta = 1.34, \Delta = 4.64$ | $\delta = 0.98, \Delta = 2.33$ |

2. $(CH_3)_3Cl$

| pH | 1.34 | 9.60 |
|---|---|---|
| Parameters | $\delta = 1.47, \Delta = 3.84$ | $\delta = 1.24, \Delta = 2.76$ |

Solutions of $LiSnPh_3$ and of $KSnPh_3$ in tetrahydrofuran gave two spectra:

$LiSnPh_3$  $\delta = 1.30, \Delta = 0, \approx 40\%, \quad \delta = 2.15, \Delta = 1.87, \approx 60\%$

$KSnPh_3$  $\delta = 1.42, \Delta = 0, \approx 40\%, \quad \delta = 2.16, \Delta = 1.82, \approx 60\%$

It was suggested that the single line arose from M - $SnPh_3$ ion pairs, the tin becoming effectively tin(IV) and the quadrupole split pair from $SnPh_3$ . More extensive study of these systems is needed.

### 4.8.2 Frozen gas Matrices

A valuable method of exploring the Mossbauer spectra of some rather reactive species is to incorporate the tin species in a solid matrix formed by condensation of a inert gas. Thus tin atoms can be incorporated in an inert gas matrix by condensation from a gas beam containing a small partial pressure of tin vapour. Satisfactory absorption spectra can be obtained at low concentrations of tin by using separated $^{119}Sn$.

Such a preparation yields two spectra, the dominant one a single line spectrum, $\delta = 3.21$, arising from atomic tin and a quadrupole split spectrum, $\delta = 3.05$, $\Delta = 3.90$, due to tin atoms in adjacent sites in the argon or xenon matrix.

The tin atoms are in the $5s^2 5p^2$ state, the p orbitals being degenerate, but the isomer shift does not imply a shift of 3.21 for an unperturbed filled $5s^2$ configuration. Some expansion of the 5s orbital in the inert gas lattice reduces the shift and a theoretically calculated correction must be applied. This information, combined with a corrected value for $K_2SnF_6$, allowing for the partial population of the 5s orbital by electron density from the fluoride ions, provides isomer shifts for the $5s^2$ and 5s configurations and leads to:

$$\delta = (n_S - 0.15n_p)/2.8$$

The quadrupole split spectrum arises from pairs of tin atoms on adjacent sites in the inert gas matrix. This is not a simple $Sn_2$ dimer, the tin atoms being too far apart for substantial covalent interaction. But the interaction is sufficient to split the 5p orbitals, the $p_z$ rising in energy, so that the tin atoms are effectively $5s^2 5p_{x,y}$ leading to a p imbalance of 1. After some theoretical correction for covalent interaction the observed splitting of 3.9 leads to the relation  p imbalance = 0.25 $\Delta$

Amongst other systems studied in this way may be memtioned tin in an $NH_3$ matrix and SnO in a nitrogen matrix.

## 4.9 **Resumé.**

The rather substantial line width for $^{119m}Sn$ limits the resolution and tin atoms in different sites, but the same oxidation state, are not always distinguished.

The absence of low-lying electronically excited states of tin renders the quadrupole splittings insensitive to change of temperature. Any variations of $\delta$ or $\Delta$ with temperature are due to the second order Döppler effect for the former, and lattice expansion for the latter quantity. However phase changes are generally apparent.

With tin(IV) compounds useful information about the ligand tin bonding can be obtained and changes in hybridisation are revealed.

With tin(II) compounds the spectra provide valuable information about the nature of the lone pair orbital and the character of Sn-M bonds.

The spectra are insensitive to the occupation of the 5d orbitals and there are very few instances of compelling evidence for $\pi$ bonding in the ligand tin bond.

In some cases the structures of tin complexes can be deduced.

## **Acknowledgements**

Fig.4.3  Reproduced with permission from Mallela, S.P., Tomic, S.T., Lee, K., Sams, J.R. and Aubke, F (1986) *Inorg.Chem.*, **25**, 2945.

Fig.4.6  Reproduced with permission from Gibb, T.C. (1970) *J.Chem.Soc.* **1970A**, 2503.

Fig.4.9  Reproduced with permission ftom Parish, R.V. and Johnson, C.E. (1971) *J.Chem.Soc.* **1971A,** 1905.

Fig.4.10   Reproduced with permission from Bancroft, G.M., Kumar Das, V.G., Sham,T.K.and Clark, M.G. (1976) *J.Chem.Soc.*,**1976A**, 643.

# 5

# Mössbauer Spectroscopy of Iron

## 5.1 MÖSSBAUER SPECTROSCOPY OF TRANSITION ELEMENTS

Unlike the main group elements, Mössbauer spectroscopy involving transition elements is dominated by the occupation of the partially filled (n-1)d orbitals. Although the molecular orbitals bonding the transition metal to its ligands involve some participation of the metal ns and np orbitals, the isomer shift will be determined to a considerable extent by the screening by the (n-1)d orbitals. The ns occupation also makes a contribution to the shift, but unless the ligands are very strong donors and, or, the bonding notably covalent, as in the lowest oxidation state compounds, this is unlikely to be the dominant factor determining the isomer shift. Similarly the occupation of np orbitals due to the ligand bonding is relatively unimportant compared with any imbalance in the (n-1)d orbitals. The value of $<r^{-3}>$ for the np orbitals is very much less than for the (n-1)d orbitals.

Many transition metal compounds have unpaired (n-1)d electrons and are paramagnetic. This leads to a rich variety of magnetic phenomena. Interaction between lattices of the ions may lead to cooperative effects such as ferro- or antiferro- magnetism. If the compound contains more than one transition metal atom per molecule, localised interaction may occur. In both cases the Mössbauer spectrum will be affected.

These elements display a number of oxidation states, and in the mixed valence compounds, delocalisation of an electron may occur with changes in Mössbauer parameters.

The ions of these metals frequently have low-lying electronically excited states. These may arise from Jahn-Teller distortions of molecular geometry or spin-orbit coupling, or in other ways. Excitation of such states modifies the Mössbauer parameters of the compound. Thus the quadrupole splittings are more temperature dependent than is the case for compounds of main group elements.

## 5.2 NUCLEAR CONSIDERATIONS WITH $^{57}$Fe

The theoretical line width for $^{57}$Fe is small, 0.194, and sources giving widths close to 0.2 are readily obtained (see Table 2.1). As in the previous chapter all isomer shifts and quadrupole splittings are given in units of mm.s$^{-1}$, with soft iron as reference substance. For inorganic compounds the Mössbauer fraction is usually greater than 0.2 and the cross section $\sigma_0$, at 225.7 X $10^{-3}$ cm$^2$, is unusually large so that the magnitude of the absorption is generally good. Spectra can be measured on hard crystalline solids at room temperature. The natural abundance of $^{57}$Fe is rather low, $\approx$ 2%, and a fifty-fold enhancement can be obtained, if necessary, by using an absorber made from separated $^{57}$Fe.

The isomer shift increases as the oxidation state of the iron decreases in high spin compounds. This suggests that the sign of $\Delta R/R$ is negative. A decrease in oxidation state of the iron means an increase in the number of 3d electrons on the iron atoms and therefore increased screening of the s electron density at the iron nuclei.

It is sometimes convenient to use powdered sodium nitroprusside ($Na_2Fe(CN)_6NO..2H_2O$) as the reference standard. The two nitroprusside lines lie at - 0.599 and 1.113 in relation to the soft iron centre. To convert $\delta$ values relative to sodium nitroprusside to the soft iron reference scale deduct 0.257.

As in the case of tin, the magnitude of $\Delta R/R$ for iron is not accurately known. The reason is essentially the same as for tin. Using improved relativistic wave functions to construct self consistent field molecular orbitals for clusters, such as $FeF_6^{3-}$ and $FeF_6^{4-}$ that are present in compounds for which the isomer shifts are available, a value of about - 6.8 x $10^{-4}$ has been calculated. Other methods of estimation give comparable values. It is often convenient to have the derived quantity, $\alpha$, the proportionality constant relating the isomer shift to the electron density at the nucleus. For $^{57}$Fe this is about -0.25 in units of $a_0$.mm.s$^{-1}$, where $a_0$ is the most probable electron radius for the ground state of hydrogen.

The favourable line width in iron spectra leads to the probable error in normal measurements of either $\delta$ or $\Delta$ being better than $\approx$ 0.01.

The magnetic moments of the ground and excited states of $^{57}$Fe, +0.091 and -0.155 nm. respectively, are of opposite sign to the tin values and substantially less in magnitude.

The nuclear characteristics of $^{57}$Fe are particularly favourable to Mössbauer spectroscopy. Comparing $^{57}$Fe with $^{119}$Sn the theoretical line width for the iron spectra is about one third that of tin. The isomer shift, $\delta$, is proportional to $(ZA^{2/3}/E_\gamma)\Delta R/R \, |\Psi(0)|^2$ (see Section 1.4.1). The value of $\Delta R/R$ for iron is about ten times the value for tin; the photon energy for the iron is rather more than half the value for tin, but the $ZA^{2/3}$ term is about one third that for tin. Altogether for a given electron density at the nucleus, the isomer shift for the iron should be about five times greater than that for tin. However because of the greater atomic number of tin, the tin compounds will generally have a larger $|\Psi(0)|^2$. In fact the span of isomer shift values for tin compounds is rather larger than for iron compounds. The more important factor is the narrower line width for iron.

The quadrupole moment of the excited state of $^{57}$Fe is more than twice as large as that of the $^{119}$Sn. But this advantage is offset by the EFG values being generally rather smaller for the iron compounds, so the ranges of splittings found for the two groups of compounds are not very different.

The cross section for Mössbauer absorption, $\sigma_0$, for iron is nearly twice as big as that for tin, but the natural abundance of $^{119}$Sn is about four times that of $^{57}$Fe.

For absorbers of natural isotopic composition and of average Mössbauer fraction, the tin compound will show the greater total absorption, that is, area under the absorption line. The narrower line width for the iron however leads to a greater "dip" or percentage fall in counting rate at the velocity of maximum absorption, for a given total absorption. This facilitates resolution of lines so that as before the narrower line width for iron is the most important factor.

## 5.3  GENERAL FEATURES OF IRON MöSSBAUER SPECTRA.

The chemistry of iron is rich and varied. It gives rise to numerous phenomena that can be investigated by Mössbauer spectroscopy: indeed more than half of the thousands of papers published on this subject refer to $^{57}$Fe spectra.

The Mössbauer spectrum of an iron-containing material usually identifies the oxidation and spin state of the iron atoms that are present. For many materials the oxidation state of the iron atoms can easily be deduced from other data, but this is not the case in some minerals and iron proteins or enzymes. The spectrum may also provide an estimate of the proportions of $Fe^{2+}$ and $Fe^{3+}$ that are present.

The good resolution of $^{57}$Fe spectra frequently permits the identification of different iron sites, because these generally differ in electric field gradient at the iron nuclei and therefore yield different quadrupole splittings. Measurement of the quadrupole splitting over a range of low temperatures provides quantitative information on low lying electronically excited states in, for example, $Fe^{2+}$ compounds. The partially filled 3d orbitals on iron will lead to many compounds with unpaired electrons and interesting magnetic properties.

A feature of iron chemistry is the formation of high, low, and occasionally, intermediate spin compounds; which are often distinguishable by their Mössbauer spectra. Magnetic measurements made many years ago showed that with some compounds the spin state changed with temperature. Mössbauer spectroscopy is an excellent technique for the study of these spin cross-over systems and has revealed the complexity of such behaviour.

The ready distinction between high spin $Fe^{2+}$ and $Fe^{3+}$ by Mössbauer spectroscopy finds application in the study of mixed valence compounds. Many years ago Verwey suggested that the sharp changes in the electrical, magnetic, and other properties of magnetite, that take place around 120 K, were associated with electron delocalisation between the $Fe^{2+}$ and $Fe^{3+}$ on the octahedral B sites. Mössbauer spectroscopy confirms such delocalisation takes place and provides evidence for delocalisation between pairs, or small numbers, of iron atoms in various mixed valence compounds and minerals. Similar processes have been identified by Mössbauer spectroscopy in the $Fe_4S_{=4}$ units that are found in iron enzymes.

Many of these processes have proved to be rather complex and, to obtain a satisfactory understanding, measurements need to be made over a wide range of temperatures and in the presence of external magnetic fields. The Mössbauer data in isolation are generally of limited use, but taken in conjunction with other methods of investigation will often lead to a much clearer picture of the processes involved.

Finally Mössbauer spectroscopy has the valuable property of being a non-destructive technique of investigation. This was of considerable importance in its application to the examination of the first moon rock samples.

## 5.4  HIGH SPIN IRON COMPOUNDS

### 5.4.1  Isomer Shifts

In principle values of $\delta$ are susceptible to calculation, but $\Delta R/R$ is not known very accurately and the calculations are very lengthy for all but the simplest compounds. Here a more empirical approach will be adopted.

The isomer shift will be determined by several factors. The most important are the following:

i/ The extent of screening of the s electrons by the 3d electrons. This indicates that $\delta$ will be sensitive to the state of oxidation of the iron.

ii/ The extent to which the iron acquires 4s electron density from its ligands.

iii/ Nephelauxetic effects changing the radial distribution of the 3d, and possibly 4s, electrons.

#### 5.4.1.1  *Effect of oxidation state*.

As seen in Fig.5.1 a & b, the isomer shifts for iron in its different oxidation states in high spin compounds are sufficiently separated, so that a Mössbauer spectrum usually identifies the state of the iron in the absorber. The higher the oxidation state of the iron the smaller the number of 3d electrons on the iron atoms and hence the greater $|\Psi(0)|^2$ and therefore the smaller the isomer shift.

The identification of the oxidation state of iron in this way is very useful. In numerous minerals the oxidation state of iron cannot be reliably inferred from the results of classical analysis or visual absorption spectroscopy. Oxidation often accompanies the solution of the mineral. The Mössbauer spectrum will usually decide whether $Fe^{2+}$, $Fe^{3+}$ or both are present. The spectra are more sensitive to the detection of $Fe^{2+}$ in the presence of excess of $Fe^{3+}$ than the inverse situation. As an example, the mineral tripuhyte formerly represented as $Fe_2Sb_2O_7$ is easily shown to contain predominantly $Fe^{3+}$ and should be formulated as $FeSbO_4$. The quantitative estimation of the two kinds of iron will be discussed in a subsequent section. (Section 8.3).

Isomer shift ranges. Room temperature values relative to soft iron.

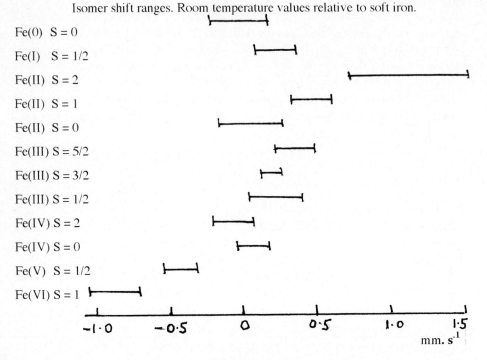

Fe(0)  S = 0

Fe(I)  S = 1/2

Fe(II) S = 2

Fe(II) S = 1

Fe(II) S = 0

Fe(III) S = 5/2

Fe(III) S = 3/2

Fe(III) S = 1/2

Fe(IV) S = 2

Fe(IV) S = 0

Fe(V)  S = 1/2

Fe(VI) S = 1

$-1.0$      $-0.5$      $0$      $0.5$      $1.0$      $1.5$

mm. s$^{-1}$

**Fig.5.1  a**

Range of isomer shifts for iron with oxygen coordination

Coordination number of iron

number of iron

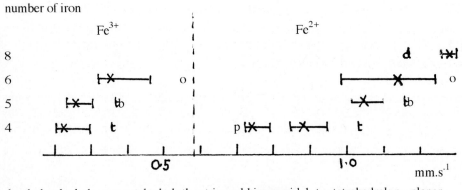

$Fe^{3+}$                                     $Fe^{2+}$

8                                                    d

6              X                    o                    X          o

5            X        tb                          X        tb

4          X        t              p X    X          t

$0.5$                              $1.0$

mm.s$^{-1}$

d = dodecahedral;  o = octahedral; tb = trigonal bipyramidal; t = tetrahedral; p - planar.

X  Average value for minerals.

**Fig.5.1  b**

### 5.4.1.2 *Factors influencing δ in high spin iron compounds*

The iron atoms in a solid will interact with the surrounding ligands which will donate some electron density into the iron 3d, 4s and 4p orbitals. The donation into the 3d and 4s orbital will have opposite effects, lowering and raising the $|\Psi(0)|^2$ respectively. Any nephelauxetic effects arising from covalence will reduce the screening by the 3d electrons. Further any back donation, $Fe \rightarrow L$, will also raise $|\Psi(0)|^2$.

It can be expected that the Mössbauer spectra of iron compounds will be more sensitive to π bonding involving the 3d orbitals than the tin spectra are to any analogous effects involving the 5d orbitals.

In a crude approximation one might expect $\delta = -k(\sigma + \pi)$, where σ and π are the σ donation and π acceptance power of the ligand.

For the simple $Fe^{2+}$ halides the isomer shift falls roughly linearly with the electronegativity of the halogen, suggesting the donation to the 4s orbital and increasing covalence are the more important effects. (Fig.5.2) Their adducts also often follow the same pattern.(Table.5.1)

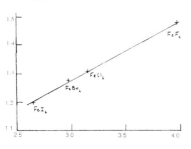

Isomer shift, δ
in mm.s$^{-1}$

χ, Halogen electronegativity

**Fig.5.2**

In the case of $Fe^{2+}$ compounds the effect is rather small and there are only limited data for compounds of similar structure. However, the sequence of isomer shifts for these compounds is not always: I < Br < Cl < F. For example, for the tetraethyl ammonium salts of the haloanions one finds: $FeCl_4^-$ δ = 0.29 and for $FeBr_4^-$ δ = 0.36 at 78 K, while for $Fe_2Cl_9^{3-}$ δ = 0.32 and for $Fe_2Br_9^{3-}$ δ = 0.36 as the pyridinium salts (78 K). The effect of changing the cation is quite small for these complex anions.

In view of the number of factors influencing δ enumerated above and noting they do not all lead to the same sign of the change in δ, it is not surprising that there is no uniformity of behaviour.

**Table 5.1**

| Compound | Isomer shift | Compound | Isomer shift |
|---|---|---|---|
| Fe quin$_2$Cl$_2$ | 0.87 | Fe Pic$_2$Cl$_2$ | 1.05 |
| Fe quin $_2$Br$_2$ | 0.83 | Fe Pic$_2$Br$_2$ | 0.99 |
| Fe quin $_2$I$_2$ | 0.78 | Fe Pic$_2$I$_2$ | 0.93 |

Data at 295 K. quin = Quinoline; γPic = 4MePyridine.

For the more polar six coordinate $Fe^{2+}$ compounds the effect of the ligand appears to follow its position in the nephelauxetic series, but for the low spin and more covalent complexes it rather follows the spectrochemical series. For $Fe^{3+}$ the pattern is similar but the isomer shifts cover a much narrower range.

### 5.4.1.3 *Effect of coordination number*.

A rather more useful difference in isomer shift is found between 4 and 6 coordinate complexes of the same ligand. If one supposes that donation to the 4s iron orbital is similar for the two complexes, the difference must arise from the increased screening of the s electrons by the enhanced occupation of the 3d orbitals in the 6-coordinate complexes.

Thus one finds $FeCl_6^{3-}$ δ = 0.51 and $FeCl_4^{-}$ δ = 0.29. and **for the $Fe^{3+}$**

environments in minerals, $FeO_8^{14-}$ ( distorted dodecahedral) δ = 1.3;

$FeO_6^{10-}$ (octahedral) δ = 1.05 to 1.25 and $FeO_4^{6-}$ (tetrahedral) δ = 0.90 to 1.15.

Occasional overlap of these groups occurs and some distorted tetrahedral iron(II) complexes show quite high isomer shifts, for example δ = 1.22 for the iron(II) l.sparteine complex shown in Fig.5.3.

The isomer shift is not very sensitive to moderate distortions from the idealised symmetry, but it does change if a radical change in symmetry takes place. Thus the rare examples of planar $FeO_4^{6-}$ have δ = 0.75.

The differences in δ with coordination number help in assigning $Fe^{2+}$ to cationic sites in minerals. Similar, but smaller, differences are found for $Fe^{3+}$; Octahedral $FeO_6^{9-}$ gives isomer shifts of between 0.6 and 0.75; tetrahedral $FeO_4^{5-}$ δ = 0.5 to 0.7. There is more overlap and the difference is less useful. Turning to $Fe^{3+}$ complexes such differences

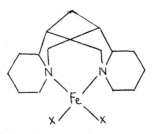

**Fig.5,3**

complexes such differences almost disappear, thus they hardly exist for the $Fe^{3+}$ complexes of ethylene diamine tetraacetic acid.

### 5.4.2 Quadrupole splitting in high spin iron compounds

In high spin $Fe^{3+}$ compounds there is a half filled set of 3d orbitals, and therefore there is no valence contribution to the electric field gradient. Any EFG must be due either to the presence of more than one kind of ligand or to a low symmetry environment, as in five coordination. Excluding the latter complexes this generally leads to rather modest quadrupole splittings.

Since $<r^{-3}>4p$ is much less than $<r^{-3}>3d$, the EFG will arise largely from the imbalance in the 3d orbitals. Despite the fact that 3d orbitals are concerned, the situation resembles that of the Sn(IV) compounds and, since the bonding in the iron compounds is fairly polar, one might expect that a system of p.q.s. values might be constructed. However the range of splittings found in $Fe^{3+}$ compounds is less than for tin, and insufficient data exist to pursue this possibility very far.

In high spin $Fe^{2+}$ compounds there is always a substantial free ion contribution to the EFG. This can be estimated if the symmetry and electronic configuration of the iron in the compound are known, in units of $e^2Q<r^{-3}>3d/4\pi\varepsilon_0$, using the angular components given in Table 3.5. This contribution is of opposite sign to the ligand contribution. As a result in a series of compounds containing quasi-octahedral $FeO_6$ units, those which are furthest from octahedral symmetry will show the lowest values of $\Delta$.

In a truly octahedral environment for the $Fe^{2+}$ there is a sharp change because the $t_{2g}$ orbitals become degenerate so that there is no EFG and $\Delta = 0$. Such a situation is only rarely encountered, $RbFeF_3$ has a perovskite structure and shows a sharp single line spectrum. A similar result is found for $Fe^{2+}$ doped into cubic magnesium oxide.

The quadrupole splittings in the tetrahedral iron(II) complexes, like the isomer shifts, are generally smaller than for the related octahedral species. They are also very temperature dependent.

### 5.4.2.1. *Low lying electronic states.*

For the majority of quasi-octahedral $Fe^{2+}$ compounds a Jahn-Teller distortion takes place and the $t_{2g}$ orbitals split into two or three levels (See Fig.3.6).

The iron(II) halides, $FeX_2$, X = Cl, Br, or I, have a trigonally distorted octahedral environment for the iron. Their quadrupole splittings are positive, so that the ground state is the orbital doublet. The bonding is sufficiently polar so that $q_{lig.}$ is unlikely to dominate the EFG. (See Section 3.6.2.) An analysis of the quadrupole splitting data, assuming no rhombic distortion, gives D values of 150, 175, and 183 $cm^{-1}$ for the chloride, bromide and iodide respectively, with a spin orbit coupling of $\lambda = -103$ $cm^{-1}$. $FeF_2$ has a much more distorted environment for the iron. The quadrupole spitting is less sensitive to temperature. An analysis of the data indicates excited states at the much higher values of 740 and 930 $cm^{-1}$.

Taking into account both the trigonal distortion and spin orbit coupling the Hamiltonian operator giving the energy levels is

$$: \mathbf{H} = D_s(L_z - 2) - [L_zS_z + 1/2 ( L_+S_- + L_-S_+)]$$

The $L_{\pm}$ and $S_{\pm}$ are orbital and spin angular momentum shift operators. A good approximation is to use as a basis set only the wave functions of the $t_{2g}$ set.

Both iron(II) fluorosulphate and iron(II) p.toluene sulphonate have trigonally distorted octahedra around the iron atoms. The quadrupole splittings are positive indicating an orbital doublet ground state due to trigonal elongation. The magnitude of the splittings confirms this structure. Their magnetically perturbed spectrum shows $\eta \approx 0$, so that there can be no sustantial rhombic distortion. The $\Delta$ - T data were fitted using $D_s = -96$ $cm^{-1}$ and $\lambda = -90$ $cm^{-1}$ for the fluorosulphate and $D_s = -93$ $cm^{-1}$ and $\lambda = -70$ $cm^{-1}$ for the p.toluene sulphonate.

### 5.4.2.2 *Effects of Pressure.*

Compression of a solid iron compound reduces the bond lengths and increases the overlap of the metal and ligand orbitals. Such changes lead to an alteration of the isomer shift and quadrupole splitting. Electron transfer between metal and ligand orbitals is facilitated so that compression of a iron(III) compound may give rise to the appearance of an iron(II) spectrum. The energies of the iron orbitals are modified by increased pressure and this may lead to a change in the spin state. Finally compression may lead to a phase change giving rise to a new Mössbauer spectrum.

All these changes are fully reversible, although there may be hysteresis effects.

These effects are of two kinds: firstly, a progressive change in the Mössbauer parameters with increase in pressure, and secondly, the appearance of new spectra arising from new chemical species produced, reversibly, by the change in pressure.

The spectrum of iron(III) phosphate under pressure, shown in Fig. 5.4, has a component due to an iron(II) species. The parameters of the iron(II) spectrum change with the applied pressure. The isomer shift falls from 1.405 at 25 kBar to 1.354 at 200 kBar while the quadrupole splitting increases from 2.12 to 2.21 for the same change in pressure. Up to 50% of the iron can be transformed to the iron(II) species.

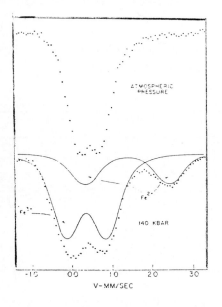

The isomer shift in high spin iron(II) compounds also falls during compression. This implies an increase in $|\Psi(0)|^2$. It seems possible that this change is not so much due to an increase in s orbital occupation as to a reduction in 3d shielding The isomer shift may fall by more than 10% as the pressure increases, at the same time the quadrupole splitting decreases.

Similar changes are observed for low spin iron(II) compounds. The isomer shift for $K_4Fe(CN)_6$ changes from -0.015 at atmospheric pressure to -0.20 under 200 kBar. A discontinuity, with an increase in isomer shift, takes place at about 50 kBar, reflecting a phase change. Further increase in pressure again leads to a fall in isomer shift.

In the $Fe(Phen)_2X_2$ complexes the effect of pressure on the spin state of the

**Fig.5.4**

iron is displayed. (Fig.5.5) For X = Cl,Br or I these complexes are high spin at room temperature and atmospheric pressure, but a low spin spectrum appears upon compression. For X = Cl the isomer shift decreases from 1.00 at 10 kBar to 0.74 at 170 kBar. At the same time the low spin spectrum grows from about 10% of the iron at 20 kBar to nearly 50% at 130 kBar. These changes reflect changes in the energies of the 3d orbitals on the iron. Their extent depends on the back bonding, Fe→X, in the complex and parallels the position of X in the nephelauxetic series. For X = NCO-, NCS- and $N_3$ which also normally give high spin complexes, a more complex pattern is found. A growing amount of a low spin spectrum is produced up to about 35 kBar, at still higher pressure the proportion of the low spin spectrum decreases. With X = CN-, a

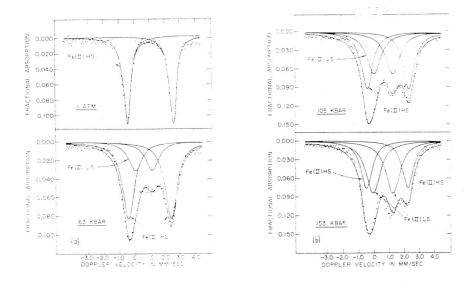

Effect of pressure on $Phen_2FeCl_2$

**Fig.5.5**

normally low spin species, application of pressure leads to the appearance of some high spin spectrum. These studies  are an example of Mössbauer spectroscopy yielding information that is difficult to obtain in any  other way.

## 5.5 LOW SPIN IRON COMPOUNDS

The effect on the Mössbauer parameters of changing from iron in oxidation state three to two in the low spin compounds can be contrasted with that for the high spin compounds. Octahedral low spin iron(II) compounds have a filled $t_{2g}$ subset and there is no valence contribution to the EFG, in which respect they resemble the high spin iron(III) compounds.  Any quadrupole splitting can only arise from a lattice term, usually small, and a ligand term, which requires more than one kind of ligand to be present.

Low spin iron(III) compounds on the other hand will have an appreciable valence contribution.

The bonding in the low spin compounds is notably more covalent than in the high spin compounds and the ready distinction between di- and tri-valent compounds found for the high spin compounds is no longer possible.(See Fig.5.2).

In general the low spin iron(II) compounds show rather larger isomer shifts than the analogous iron(III) compounds, but as seen in Fig.5.2 the ranges of δ for the two oxidation states overlap. However the quadrupole splitting will often decide the oxidation state.  Low spin $Fe(II)L_6$ have zero or very small quadrupole splitting, arising from only a lattice contribution. Low spin $Fe(III)L_6$ complexes have a valence contribution so that quadrupole splittings range from 0.28 to as much as 1.1.

For these low spin compounds the σ donor power of the ligand, tending to increase occupation of the 4s orbital of the iron, and the π acceptance by the ligand, reducing 3d occupation, assume greater importance. Several salts of the anions of the type $Fe(CN)_6X$ have been examined. The change in isomer shift with different X is shown in Table 5.2.

<h4 style="text-align:center">Table 5.2</h4>

| X = | $H_2O$ | $AsPh_3$ | $NH_3$ | $NO_2^-$ | Py | $SO_3^-$ |
|---|---|---|---|---|---|---|
| δ = | 0.04 | 0.03 | 0.02 | 0.01 | 0.03 | 0.01 |
| X = | $PPh_3$ | DMSO | $CN^-$ | CO | $NO^+$ | |
| δ = | 0.02 | 0.03 | -0.1 | -0.14 | -0.26 | |

Since $δ = -k(σ + π)$ if one changes the sign of δ the values obtained will place the ligands, X, in order of their increasing donor plus acceptor power. As can be seen the sequence corresponds roughly to the spectrochemical series.

### 5.5.1 Partial isomer shifts. p.i.s.

The bonding in these complexes is generally sufficiently polar and the double bond character sufficiently low that the isomer shifts might prove to be additive functions of contributions from the individual ligands. This requires that little rehybridisation accompanies the replacement of one ligand by another, that isomer shifts are insensitive to modest changes in bond angles, bond lengths and symmetry around the iron, and that π backbonding, metal to ligand, be not too important.

There is a very extensive body of data on the Mössbauer parameters of these compounds but it is not all well suited to testing this possibility. All the isomer shift data used to derive a set of **p.i.s** values must relate to a common reference compound and should be from spectra recorded at one temperature. Ideally all the data should have been corrected for the second order Döppler effect (See 3.6.1). Rather few such data are available. To minimise the SOD correction all the measurements should have been made at a constant low temperature. All these conditions are seldom fulfilled. However since the partial isomer shift can easily be shown to have only approximate validity this is not important.

The clearest evidence that p.i.s. values are not constant is provided by the data for some cis and trans complexes. Thus the isomer shift for cis $FeCl_2(ArNC)_4$ is 0.28, but the value for the trans isomer is 0.36. Similarly for $Fe(CO)_2I_2(PMe_3)_2$ with the CO groups cis the shift is 0.06 and with them trans 0.19. This is compatible with other physico-chemical data on cis and trans complexes.

A set of data that has been used to derive p.i.s. values is given in Table 5.3. Such a set must contain at least one complex of the type $FeL_6$ and another also containing the L ligand

Now the difference in p.i.s. values for Cl and H can be obtained from several pairs of compounds. Using data for 4 and 9 one obtains 0.19; from 5 and 8 one obtains 0.20; from 25 and 26 one gets 0.0; 12 and 16 yield 0.14 , and 2 and 19, 0.13.

### Table 5.3

| Compound | $\delta$ | Compound | $\delta$ |
|---|---|---|---|
| 1. $Fe(CH_3NC)_4(HSO_3)_2$ | -0.11 | 15. $FeHN_2(depe)_2^+$ | 0.07 |
| 2. $Fe(CH_3NC)Cl(dmpe)_2$ | 0.11 | 16. $FeHN_2(dmpe)_2^+$ | 0.10 |
| 3. $FeCl_2(dmpe)_2$ | 0.37 | 17. $FeHH_2(depe)_2^+$ | 0.08 |
| 4. $FeCl_2(depe)_2$ | 0.39 | 18. $FeHH_2(dmpe)_2^+$ | 0.06 |
| 5. $FeCl_2(depb)_2$ | 0.39 | 19. $FeH(CH_3NC)(dmpe)_2^+$ | -0.02 |
| 6. $FeBr_2(depe)_2$ | 0.47 | 20. $FeH(Bu^iNC)(dmpe)_2^+$ | -0.04 |
| 7. $FeBr_2(depb)_2$ | 0.39 | 21. $FeBr(CCPh)(depe)_2$ | 0.22 |
| 8. $FeH_2(depb)_2$ | -0.02 | 22. $FeCl(CCPh)(depe)_2$ | 0.21 |
| 9. $FeHCl(depe)_2$ | 0.20 | 23. $FeCl(CCPh)(dmpe)_2$ | 0.16 |
| 10. $FeH(C_2H_5)(dmpe)_2$ | 0.10 | 24. $FeH(CO)(depe)_2^+$ | -0.12 |
| 11. $FeBrN_2(dmpe)_2^+$ | 0.25 | 25. $FeH(CO)(dmpe)_2^+$ | -0.09 |
| 12. $FeClN_2(dmpe)_2^+$ | 0.24 | 26. $FeCl(CO)(dmpe)_2^+$ | -0.09 |
| 13. $FeH(CH_3CN)(dmpe)_2^+$ | 0.11 | 27. $FeBr(CO)(dmpe)_2^+$ | 0.05 |
| 14. $FeH(CH_3CN)(depe)_2^+$ | 0.10 | | |

The cationic species were measured as their salts with $BPh_4^-$.

depe = diethylphosphino ethane, dmpe = the analogous dimethyl compound, depb = o.bis diethylphosphino benzene.

Compound 8 has two hydrido ligands, compounds 17 and 18 have a hydrido ligand and a coordinated hydrogen.

These data show show that p.i.s. data will only give very approximate estimates of unmeasured isomer shifts. Nevertheless p.i.s. data for different ligands can give useful information about their bonding, especially in conjunction with p.q.s. data.

A series of p.i.s. values calculated from the data in Table 5.3 is shown in Table 5.4. The experimental data are for measurements at 298 K and a soft iron foil as the reference zero. The values shown in brackets are those obtained with a different set of compounds.

Clearly the p.i.s. values are dependent on the compounds chosen for the data set used in their derivation. Values obtained with data from a set of related compounds, for example all of the type FeXYP_2, where P_2 is a chelating di-phosphino ligand, will give good predictions of isomer shifts for other compounds of the same type.

Examination of a large body of data shows that those ligands where back donation, M→ L, is substantial usually show variable p.i.s. values. This is especially noticeable for CO. Indeed the main difference between the compounds in the two data sets used to derive the above two groups of p.i.s. values lies in the proportion of carbonyl derivatives.

## Table 5.4

| Ligand | p.i.s. | | Ligand | p.i.s. |
|---|---|---|---|---|
| 1. $Br^-$ | 0.13 (0.26) | 9. | 1/2 dmpe | 0.04 (-0.02) |
| 2. $Cl^-$ | 0.10 (0.22) | 10. | $P(OEt)_3$ | 0.02 (---). |
| 3. $C_2H_4$ | --- (0.12) | 11 | $N_2$ | 0.01 (0.09). |
| 4. $AsPh_3$ | 0.08 (---) | 12. | C=CPh- | --- (0.03). |
| 5. $H_2O$ | 0.08 (---) | 13. | $CN^-$ | -0.01 (0.01). |
| 6. 1/2 depb | 0.06 (-0.02) | 14. | MeNC | -0.02 (-0.02). |
| 7. 1/2 depe | 0.06 (-0.02) | 15. | $H^-$ | -0.10 (-0.07). |
| 8. $NH_3$ | 0.05 (0.04). | 16. | CO | -0.15 (-0.13). |
| | | 17. | $NO^+$ | -0.23 (---). |

The high value of $(\sigma + \pi)$ for CO must be largely due to its strong $\pi$ acceptance. The still higher value for NO+ suggests both high $\pi$ acceptance and $\sigma$ donation. In contrast the value for $H^-$ must arise entirely from its $\sigma$ donation. The rather low value for $N_2$ shows that this ligand is only a modest $\pi$ acceptor and presumably a poor $\sigma$ donor. $C_2H_4$ shows a still lower value of $(\sigma + \pi)$. . The two sets suggest that the p.i.s. of $H^-$ is variable. It may be observed that a direct value from the isomer shift for $Mg_2FeH_6$ gives a value, $\approx 0.0$, about midway between two tabulated values.

The difference in isomer shifts for some cis and trans isomers described above indicates that when a strongly $\pi$ accepting ligand is situated trans to a largely $\sigma$ donating ligand, such as a halide ion, its $\pi$ accepting ability is enhanced.

### 5.5.2 Partial quadrupole splittings, p.q.s

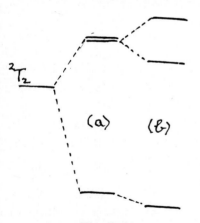

(a)     (b)

In the low spin iron(III) complexes there are both valence and ligand contributions to the EFG. . Usually there are low lying electronically excited states so that the quadrupole splittings are notably temperature dependent. Fig.5.6 shows how spin orbit coupling and distortion from octahedral symmetry lead to lowering of degeneracy and the appearance of two low lying excited levels. As in the case of high spin iron(II) compounds the temperature dependence of the quadrupole splitting may permit the evaluation of the energies of these excited levels. The ferricyanide salts also display a rather substantial, temperature independent, lattice contribution to the EFG.

**Fig.5.6**        The splitting in low spin iron(II) compounds arises almost entirely from the ligand contribution to the EFG. There are extensive data on these compounds and they provide an opportunity to examine the validity of a p.q.s approach. The situation is more favourable than in the treatment of isomer shifts because the splittings span a wider range, extending from about -1 to +2.

In 3.5.2.1 it was shown that the quadrupole splittings for $MAB_5$ : cis $MA_2B_4$ : trans $MA_2B_4 = 1 : -1 : 2.$, on the basis of additive p.q.s. contributions. The splittings for some cis and trans pairs and $MA_2B_4$ complexes are shown in Table 5.5.

### Table 5.5

cis   $Fe(CN)_2(EtNC)_4$   0.29;       cis   $FeCl_2(ArNC)_4$   0.78

trans   $Fe(CN)_2(EtNC)_4$   0.59;      trans   $FeCl_2(ArNC)_4$   1.55

$Fe(CN)(EtNC)_5$   0.17           $FeCl(ArNC)_5$   0.73.

The change in sign of the splitting from the cis to the trans forms has been verified in $FeCl_2(ArNC)_4$ as can be seen below..

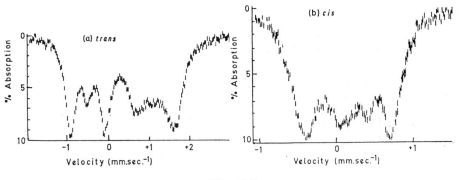

### Fig.5.7

The agreement with the predictions is not always as good as in the above cases, and when one proceeds to draw up a table of p.q.s. values one finds similar difficulties, although not as severe, as for the p.i.s. values. To minimise any lattice contributions one should avoid cationic and anionic complexes in the set chosen to derive the p.q.s. values. It is important to know the sign as well as the magnitude of the quadrupole splittings of the compounds chosen.

Since the data only yield differences between p.q.s. values (See 3.5.2.1), an arbitrary normalisation must be made and it is usual to set p.q.s. $Cl^- = -0.3$. Two sets of p.q.s. values are shown in Table 5.6. Column A gives values calculated from quadrupole splitting data on salts of $Fe(CN)_5X$ anions, taking the p.q.s. of $CN^- = -0.84$. Column B gives values based on a wide variety of iron(II) complexes. Where the two sets overlap the agreement is acceptable. As in the case of the p.i.s., the stronger $\pi$ bonding ligands tend to give variable results.

One would expect that p.q.s. $= -k'(\sigma - \pi) - l$ , where l is a correction for any contribution to the EFG from the lattice. This implies that as the donation by the ligand increases, its p.q.s. will become more negative, while enhanced $\pi$ accepting ability should have the opposite effect.

Quadrupole splittings calculated from p.q.s. values are not very accurate, probable errors amount to rather more than $\approx 0.1$. However this may still be sufficient for the experimental quadrupole splitting to decide which of various possible isomers of a complex has been measured. The sign of the quadrupole splitting can also be deduced using p.q.s. data.

### 5.5.3 Combination of p.i.s. and p.q.s. data

Since the p.i.s. $= -k(\sigma + \pi)$ while the p.q.s.$=- k'(\sigma - \pi)$ and for a number of ligands the $\pi$ contributions are not very big, a plot of the p.i.s. against p.q.s. values is roughly linear with a positive slope. Those ligands that lie substantially off this line, having p.q.s.values greater than the linear relation indicates for their p.i.s. values, are those whose $\pi$ contributions are substantial, notably CO and $NO^+$. A plot of this kind is shown in Fig.5.8.

The tabulated values for p.i.s. and p.q.s. show that $NO^+$ has the most negative p.i.s. and the most positive p.q.s values. This is consistent with it possessing both strong $\sigma$

### Table 5.6

| Ligand | p.q.s. A | p.q.s. B | Ligand | p.q.s. A | p.q.s. B |
|---|---|---|---|---|---|
| 1.  NO+ | +0.01 | --- | 26. Guanine | -0.47 | --- |
| 2.  $C_2H_4S$ | -0.22 | --- | 27. Pyrrolidine | -0.47 | --- |
| 3.  $Br^-$ | --- | -0.28 | 28.  $NH_3$ | -0.49 | -0.52 |
| 4.  $I^-$ | --- | -0.19 | 29.  $NH_2NH_2$ | -0.49 | --- |
| 5.  $Cl^-$ | --- | -0.30 | 30. Caffeine | -0.49 | --- |
| 6.  $H^-$ | -0.31 | --- | 31.  $SbPh_3$ | --- | -0.50 |
| 7.  $N_2$ | --- | -0.37 | 32.1Me.imidazole | -0.51 | --- |
| 8.  $Me_2SO$ | -0.37 | --- | 33.  $NCS^-$ | --- | -0.51 |
| 9.  Purine | -0.38 | --- | 34.  $AsPh_3$ | --- | -0.51 |
| 10.  MePz | -0.39 | --- | 35.  $PPh_3$ | --- | -0.53 |
| 11.  Morpholine | -0.39 | --- | 36.  $P(OPh)_3$ | --- | -0.55 |
| 12.  $NO_2^-$ | -0.40 | -0.42 | 37.  $PPh_2Me$ | --- | -0.58 |
| 13.l.Histidine | -0.40 | --- | 38.  1/2 depb | --- | -0.59 |
| 14.β Picoline | -0.42 | --- | 39.  1/2 dppm | --- | -0.61 |
| 15.  $SnCl_3^-$ | --- | -0.43 | 40.  1/2 dppe | --- | -0.62 |
| 16.  MeCN | -0.41 | -0.43 | 41.  1/2 depe | --- | -0.63 |
| 17.  $NHMe_2$ | -0.43 | --- | 42.  $P(OEt)_3$ | --- | -0.64 |
| 18.  $H_2O$ | -0.44 | -0.44 | 43.  $PMe_3$ | --- | -0.66 |
| 19.  Py | -0.44 | --- | 44.  1/2 dmpe | --- | -0.70 |
| 20.  1/2 opdp | --- | -0.47 | 45.  CO * | -0.66 | -0.74 |
| 21.  Pz | -0.44 | --- | 46.  MeNC * | --- | -0.74 |
| 22.  Ethanolamine | -0.44 | --- | 47.  ArNC | --- | -0.75 |
| 23.  $NH_2Me$ | -0.45 | --- | 48.  $CN^-$ | --- | -0.84 |
| 24.  Imidazole | -0.45 | --- | 49.  $H^-$ | --- | -1.04l |
| 25.  Adenine | -0.46 | --- | * Value very variable. | | |

Me = methyl; Py = pridine; Ph = phenyl; dppm = $Ph_2PCH_2PPh_2$; $R_2PCH_2CH_2PR_2$, R = Me for dmpe; R = Et for depe; R = Ph for dppe. opdp = o.bis diphenylphosphino benzene; depb = o.bis diethylphosphino benzene; Pz = pyrazine..

and $\pi$ characteristics. The strong $\sigma$ donor H⁻ has the most negative p.q.s. and a weakly positive p.i.s. value. The values for nitrogen agree with the ligand being a weak $\sigma$ donor and a rather stronger $\pi$ acceptor.

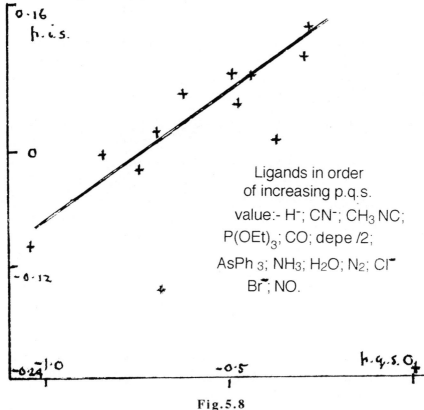

Ligands in order of increasing p.q.s. value:- H⁻; CN⁻; CH₃ NC; P(OEt)₃; CO; depe /2; AsPh₃; NH₃; H₂O; N₂; Cl⁻ Br⁻; NO.

**Fig.5.8**

### 5.5.4 Noteworthy low spin iron(II) compounds

Some iron(II) compounds deserve more detailed attention.

#### 5.5.4.1 *The nitroprussides*

Sodium nitroprusside $Na_2Fe(CN)_5NO.2H_2O$, is a diamagnetic compound, the NO effectively coordinating as $NO^+$ and the iron becoming iron(II). All the electrons in the 3d orbitals are paired. An approximate molecular orbital energy diagram is shown in Fig.5.9.

The compound forms orthorhombic crystals, space group Pnnm with four molecules in each unit cell. The iron sites are all equivalent and the Fe-N-O is linear. Hence the local symmetry of the iron is $C_{4V}$ and $\eta$ can be expected to be zero. The degenerate $d_{xz}$ and $d_{yz}$ orbitals are involved in substantial Fe→NO back bonding. A similar, but smaller, Fe→CN interaction affects the $d_{xy}$ orbital. The delocalisation of the 3d electrons arising in this way, together with a rather higher 4s population, are responsible for the

Metal orbital                    Ligand orbital

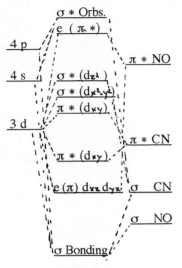

**Fig.5.9**

very high $|\Psi(0)|^2$ and consequently very low $\delta$

The Mossbauer parameters found for different salts of the nitroprusside anion are sensitive to the cation present, thus for the sodium salt $\delta = -0.257$ and $\Delta = 1.71$ while for the tetraethyl ammonium salt $\delta = -0.19$ and $\Delta = 1.97$. Hence there is some lattice contribution to the EFG.

This is confirmed by a difference between the parameters for the solid salts and their frozen solutions.

Single crystal absorber studies, measuring the line intensities as a function of the orientation of the crystal in relation to the $\gamma$ photon beam, have shown the sign of $\Delta$ is positive and, as the symmetry demands, $\eta$ is zero. They also yielded the principal axes of the EFG in relation to the crystal axes.

Since the $<r^{-3}>$ for the 4p orbitals is rather small, as is their population, if one ignores the small lattice contribution, the EFG for the nitroprussides will be determined by the imbalance of the 3d populations. In suitable units one can put

$\Delta = K[\ 4/7\ n_{xy} - 2/7(n_{xz} + n_{yz})]$ where the n are the effective populations of the orbitals indicated. Now electron spin resonance data for $K_4Fe(CN)_6$ give a covalent orbital reduction factor of 0.87, so that $n_{xy}$ should be about 1.74.

Now $n_{xz} = n_{yz}$ and taking K as about 4.8 one obtains $n_{xz} = n_{yz} = 1.38$. Thus the total population of this set of orbitals is only 4.5 accounting for the very low $\delta$ .

### 5.5.4.2 *Cyclopentadienyl and arene sandwich compounds.*

The Mössbauer spectrum of the classic sandwich compound, $Fe(C_5H_5)_2$, ferrocene, is of considerable interest. It was the first compound for which the sign of $\Delta$ was determined by the magnetic perturbation method (See Fig.3.12). The sign was shown to be positive, $\delta = 0.53$ and $\Delta = +2.37$.

A simple crystal field approach to the bonding in ferrocene, regarding it as a sandwich with $Fe^{2+}$ between two elliptical singly charged rings, predicted an electronic configuration $(d_{z^2})^2 , (d_{xz}, d_{yz})^4$, which would require a negative value for $\Delta$. A molecular orbital analysis of the bonding leads to the correct sign for $\Delta$.

A partial molecular orbital diagram for ferrocene is shown in Fig.5.10.

Fortunately it proves to be unimportant whether the compound adopts the staggered, $D_{5d}$, or the eclipsed, $D_{5h}$, configuration. The $p_z$ orbitals on the rings combine to give one non-degenerate and two doubly degenerate orbitals. On the two rings these combine to give g and u, orbitals, according to their symmetry with respect to inversion at the iron nucleus. Only the $4p_x$ and $4p_y$ have the right symmetry to combine with the ligand iron nucleus. Only the $4p_x$ and $4p_y$ have the right symmetry to combine with the ligand u orbitals. As the diagram shows, the binding is largely due to the $e_{1g}$ and $e_{2g}$ molecular orbitals. The eighteen electrons, 10 from the rings and 8 from the iron, are accomodated in the six lowest energy molecular orbitals.

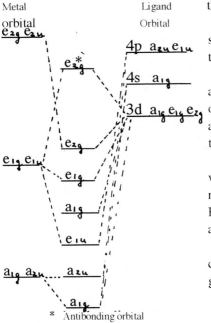

Different calculations give slightly different sequences to the energies of the orbitals, but they all lead to the same conclusion about $\Delta$.

Table 5.7 shows the results of such a calculation. The lowest lying orbital is at the bottom of the table. It can be seen that whether or not any contribution from the 4p orbitals is included, the sign of $\Delta$ must be positive.

Table 5.7 records a selection of data for various compounds in this group. First it will be noted that their isomer shifts are almost constant. Hence changes in the total 3d population must be accompanied by changes in 4s population.

The molecular orbital diagram for the di-arene complexes should be substantially similar to that given for ferrocene.

## Table 5.7

| Molecular Symmetry | Orbital Composition | Population p or d on iron | Contribution to $\Delta$ |
|---|---|---|---|
| $e_{2g}$ | $0.898(d_{xy}, d_{x^2+y^2}) + 0.52(Cp)$ | 3.224 | +1.84 * |
| $e_{1g}$ | $0.454(d_{xz}, d_{yz}) + 0.89(Cp)$ | 0.411 | -0.24 * |
| $a_{1g}$ | $(d_{z^2})$ | 2 | -1.14 * |
| $e_{1u}$ | $0.591(p_x, p_y) + 0.807(Cp)$ | 1.397 | +0.56 # |
| $a_{1u}$ | $0.471(p_z) + 0.882(Cp)$ | 0.444 | -0.35 # |
| $a_{2u}$ | $0.633(s) + 0.774(Cp)$ | -- | — |

\* In units of $e{<}r^{-3}{>}_{3d}/4\pi\varepsilon_0$;    \# In units of $e{<}r^{-3}{>}_{4p}/4\pi\varepsilon_0$.

Dropping the u and g to include less symmetric rings, Table 5.7 shows that the $e_1$ orbital is associated with ring$\rightarrow$ Fe donation and $e_2$ with Fe$\rightarrow$ ring donation. Changes in the $e_2$ orbital have twice the effect on the quadrupole splitting that changes in $e_1$ have. Hence $\Delta = k(2ne_2 - ne_1)$, where the n refer to the populations of the related 3d orbitals. Thus a change in the quadrupole splitting need not involve appreciable change in the total 3d population on the iron.

## Table 5.8

| A. Compound | $\delta$ | $\Delta$ | B. Compound | $\delta$ | $\Delta$ |
|---|---|---|---|---|---|
| 1 $FeCp_2$ | 0.53 | 2.37 | 7 $FeCpC_6H_6^+$ | 0.52 | 1.68 |
| 2 $FeCp.C_5H_4CH_3$ | 0.53 | 2.39 | 8 $FeCp.C_6(CH_3)_6^+$ | 0.53 | 2.18 |
| 3 $Fe(C_5(CH_3)_5)_2$ | 0.53 | 2.50 | 9 $FeCp.(C_6H_5CN)^+$ | 0.63 | 1.51 |
| 4 $Fe(C_5H_4CN)_2$ | 0.53 | 2.29 | 10 $FeC_5H_4NH_2.C_6H_6^+$ | 0.58 | 1.58 |
| 5 $Fe(C_5H_4COCH_3)_2$ | 0.52 | 2.15 | 11 $FeC_5(CH_3)_5.C_6(CH_3)_6^+$ | +0.55 | 1.40 |
| 6 $FeCp.C_5H_4PPh_2$ | 0.52 | 2.30 | | | |

| C. Compound | $\delta$ | $\Delta$ |
|---|---|---|
| 12 $Fe(C_6H_6)_2^{2+}$ | 0.53 | 1.90 |
| 13 $Fe(C_6H_5CH_3)_2^{2+}$ | 0.63 | 1.93 |
| 14 $Fe(C_6H_3(CH_3)_3)_2^{2+}$ | 0.56 | 2.01 |
| 15 $Fe(C_6(CH_3)_6)_2^{2+}$ | 0.64 | 2.10 |

Table 5.8 shows that electron donating substituents in the cyclopentadienyl or benzene rings lead to higher values of $\Delta$ without much change in $\delta$. However with one of each kind of ring, introduction of such substituents leads to a reduction in $\Delta$. The interaction of the iron with one kind of ring affects its interaction with the other. The asymmetry of the iron interactions is reflected in calculations of the charge distribution in $([\eta C_5H_5]Fe[\eta C_6H_6])^+$ which show that 65% Of the charge resides on the benzene ring and 11% on the cyclopentadienyl ring.

Table 5.7 shows that if one electron is removed from the highest occupied orbital in ferrocene the quadrupole splitting will be greatly reduced. In fact the Mössbauer parameters for ferrocinium are $\delta = 0.43$ and $\Delta = 0.2$ (estimate).

Biferrocenyl, $[\eta C_5H_5FeC_5H_4]_2$, has $\delta = 0.52$ and $\Delta = 2.36$.

## 5.6 SIMPLE APPLICATIONS

A very direct application of [57]Fe Mössbauer spectra is to explore the number and nature of the sites occupied by the iron atoms in the absorber compound. Crystallographically different iron sites imply different iron environments and these often give rise to measureably different spectra. Similarly, different crystal phases may also yield different spectra and, from measurements of the spectra at different temperatures, information about phase changes may be obtained.

Sharp first order phase changes can be distinguished from second order changes. The latter will usually display two component spectra over some range of temperatures.

Finally the spectra may show that the iron species in the lattice of the absorber are quite different from those indicated by the composition of the absorber compound.

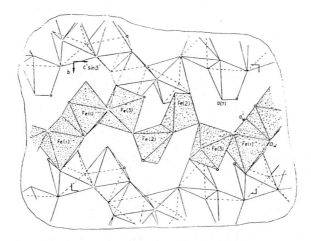

Fig.5.11 a

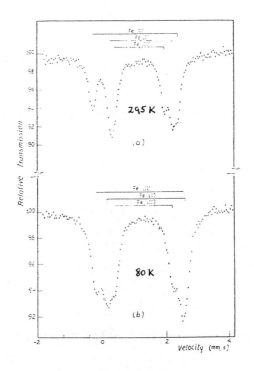

Fig.5.11 b

### 5.6.1 Different crystal sites

The crystal structure of the hydrated iron(II) phosphate, $Fe_3(PO_4)_2H_2O$, reveals a complex structure with three iron sites as shown in Fig.5.11a. In two sites the iron is in a distorted octahedron of oxygen atoms and in the third the iron is five coordinate in a very distorted tetragonal pyramid of oxygen atoms. The Mössbauer spectrum of the compound, shown in Fig.5.11b. is composed of three quadrupole split pairs with parameters Fe(1) $\delta = 1.36$; $\Delta = 2.48$;   Fe(2) $\delta = 1.16$; $\Delta = 2.72$ and Fe(3) $\delta = 1.29$; and  $\Delta = 1.80$ at 80 K. The isomer shifts indicate that Fe(2) occupies the pyramidal sites. There are equal numbers of the three kinds of sites and the three components to the spectrum appear in roughly equal proportions.

The octa hydrated iron(II) phosphate, vivianite, also possesses a crystal structure with two iron sites, present in proportions 1 : 2.   Its Mössbauer spectrum is composed of two barely resolved quadrupole doublets, with $\Delta$ values of 2.59 and 3.16 and practically the same isomer shifts at 80 K.  Computer analysis of the spectrum shows the ratio of the areas under the two pairs of peaks to be approximately 2, the less intense doublet having the smaller quadrupole splitting.

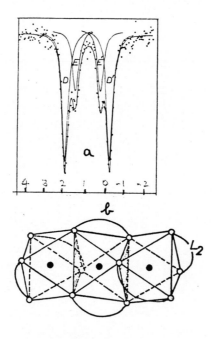

Fig.5.12  a  &  b

Anhydrous iron(II) salicylate also gives two doublets with intensities in the ratio 1 : 2, as shown in Fig.5.12 a. These data suggest two kinds of octahedral $Fe[O_6]$ units The crystal structure is not available but the compound may well possess the nickel acetylacetonate trimer structure shown in Fig. 5.12 b.

The crystal structure of $Fe_2PO_4Cl$ shows three kinds of iron site. The structure is complex and unusual, it is assembled from the units shown in Fig. 5.13 a & b.

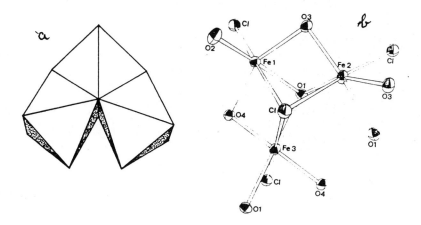

**Fig.5.13 a & b**

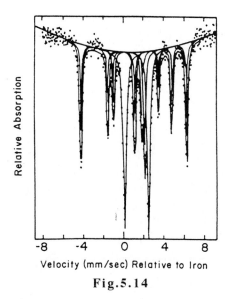

**Fig.5.14**

Two $Fe[O_4Cl_2]$ octahedra share faces with a third and there is a common edge to all three. (5.13 b)   At 40 K the spectrum comprises magnetically split and quadrupole split components in intensity ratio 2 : 1, as can be seen in Fig. 5.14.

The Fe(I) in Fig.5.13 b must be the atom that does not suffer magnetic splitting.

In several cases however crystallographically distinct sites yield indistinguishable Mössbauer spectra. Iron(II) oxalate, $FeC_2O_4 \ 2H_2O$, can be obtained in a monoclinic and an orthorhombic form form, but at 100 K both forms give a single quadrupole split spectrum, $\delta = 1.25$ and $\Delta = 2.00$. A single quadrupole split absorption is also found at 300 K. But at 200 K line broadening can be analysed in terms of splittings of 1.68 for the monoclinic and 1.71 for the orthorhombic form. However such a computer analysis of the data can only be justified on the basis of the crystallographic data.

Iron(II) acetate has a complex crystal structure with three types of iron site, in proportions of 1 : 2 : 2. At 78 K its spectrum shows only a quadrupole split pair with parameters $\delta = 1.35$ and $\Delta = 2.68$, but at 295 the best fit to the spectrum involves two pairs of lines with parameters $\delta = 1.17$ with $\Delta = 2.34$ and $\delta = 1.23$ with $\Delta = 2.17$.

### 5.6.2 **More profound changes in coordination**

An early example of different iron environments in a simple iron compound is the hydrated iron(II)formate. This compound gives a spectrum comprised of two well resolved quadrupole split absorptions, with equal intensities and parameters $\delta = 1.23$ with $\Delta = 0.60$ and $\delta = 1.26$ with $\Delta = 2.96$. The isomer shifts suggest that both arise from octahedral $Fe[O_6]$ species. Crystal data show one is due to iron coordinated to six oxygens from carboxylate groups, while the other has four water oxygens and two carboxylate oxygens in a trans configuration.

Measuring the spectrum with an applied magnetic field shows that $V_{zz}$ for the first mentioned site is negative while that for the second site is positive. This implies that site one is tetragonally distorted by elongation along the z axis with the ground state an orbital doublet. Site two is compressed along the z axis and the ground state is an orbital singlet. These deductions are only valid if one can assume the ligand contribution to $V_{zz}$ is not very big. The temperature dependence of the quadrupole splittings show that for site one the first excited state lies at 250 K above the ground state and for site two at 850 K. (see Section 3.6.2.).

Anhydrous iron(II) formate only gives one doublet in its spectrum, but the acid formate, $Fe(HCOOH)_2(OOCH)_2$ gives a pair with parameters $\delta = 1.14$ with $\Delta = 1.20$ and $\delta = 1.25$ with $\Delta = 1.93$ at 298 K.

Changes in structure in a series of compounds are usually clearly revealed by the Mössbauer spectra. The compounds FeterpyX$_2$ with X = Br or I each give one quadrupole split spectrum (terpy = 2.2'.2" tripyridyl). With X = Cl however two well resolved doublets with parameters $\delta = 0.21$ with $\Delta = 0.98$ and $\delta = 0.91$ with $\Delta = 1.69$ are found at room temperature and $\delta = 0.27$ with $\Delta = 0.99$ and $\delta = 1.06$ with $\Delta = 3.05$ at 78 K. These data indicate a low spin ($\delta = 0.21$) and a high spin ($\delta = 0.91$) species. The ratio of the area under the high spin spectrum to that under the low spin spectrum is about one at 78 K but it falls to about 0.2 at room temperature.

Together these data show that the solid FeterpyCl$_2$ contains tetrahedral FeCl$_4$ and octahedral Fe(terpy)$_2$ ions. Four coordinate iron(II) always displays a rather low Mössbauer fraction, accounting for the temperature dependence of the area ratio. In the bromo and iodo compounds the iron is five coordinate.

The complex Fe(2.2'bipyridyl)Cl$_2$ can be obtained in an orange and a red form. The former has spectral parameters of $\delta = 1.01$ with $\Delta = 3.68$, at 78 K. The large quadrupole splitting suggests five coordinate iron. It is composed of long chains of FeCl_bipy units linked by two of the chlorines (Fig.5.15).

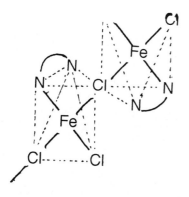

**Fig.5.15**

The more stable red form with parameters $\delta = 1.17$ with $\Delta = 1.73$ at 78 K contains six coordinate iron the octahedra being edge linked by pairs of chlorine atoms.

An unusual kind of isomerism is found for $Fe(SO_3CH_3)_2$. Two forms can be prepared both containing quasi-octahedral $Fe[O_6]$. At 293 K they give quadrupole splittings of 2.93 and 0.93 and at 80 K 3.37 and 1.42.

It is believed the difference lies in the distortion of the octahedra around the iron. The splitting of the $t_{2g}$ orbitals by tetragonal or trigonal distortion of the octahedra around $Fe^{2+}$ has already been discussed.( Section 3.6.2.). The large $\Delta$ form, $\alpha$, is trigonally compressed along the threefold axis and perhaps also rhombically distorted, the large negative value of $V_{zz}$ shows an orbital singlet ground state arising from trigonal compression. The low $\Delta$, $\beta$ form is trigonally elongated and the ground state is an orbital doublet. This is confirmed by the positive sign of Vzz.

The $\alpha$ form remains paramagnetic down to 4.2 K, but the $\beta$ form becomes antiferromagnetic below 23 K. The infra red spectra of the two forms also differ, the $\alpha$ form retains the $C_{3V}$ local symmetry of the $H_3CSO_3$ group, the $\beta$ form does not.

The compound $Fe[oC_6H_4\{P(C_6H_5)_2\}_2]I_2$, or $Fe[opdp]I_2$, gives both five and six coordinate species, The red dichloromethane solvate $[FeI(opdp)_2]I_2.CH_2Cl_2$ is diamagnetic and gives conducting solutions. Its Mössbauer spectral parameters are $\delta = 0.49$ and $\Delta = 2.43$, suggesting five coordinate iron. The crystal structure reveals two iron environments for this compound but only a single spectrum is seen. A yellow unsolvated non-conducting form can also be obtained. It is a high spin compound with a trans octahedral environment for the iron. Its spectrum has the parameters $\delta = 0.96$ with $\Delta = 2.29$.

### 5.6.3 Phase changes

Discontinuities in plots of either the isomer shift or the quadrupole splitting against the temperature indicate a phase change at the discontinuity.

The pyridine complex $FePy_2Cl_2$ has a polymeric structure with chlorine bridges and $D_{4h}$ symmetry at room temperature as shown in Fig. 5.16 a. The simple quadrupole split spectrum has parameters $\delta = 1.08$ with $\Delta = 0.57$ at 297 K. At 250 K there is a first order phase transition. Like many such transitions hysteresis occurs and at 195 K two quadrupole split pairs are seen with parameters $\delta = 1.13$ with $\Delta = 0.56$ and $\delta = 1.15$ with

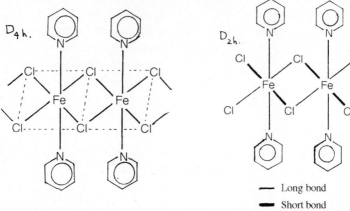

Fig.5.16  a                                    Fig.5.16  b

$\Delta = 1.14$. At 78 K only one doublet is seen, $\delta = 1.21$ with  $\Delta = 1.25$. The sign of $\Delta$
does not change. It appears that the chlorine bridging becomes asymmetric changing the
symmetry to $D_{2h}$  (Fig. 5.17 b). The corresponding iodo compound is monomeric and
tetrahedral, with parameters $\delta = 0.76$ with $\Delta = 0.94$ at 297 K.

Mössbauer spectroscopy can provide information about substances which, because
they are too nearly amorphous, fail to yield well defined X-ray diffraction data. This is
often the case with the products from thermal decompositions. For example if iron(II)
salicylate dihydrate is dehydrated at the lowest possible temperature and at very low
pressure, the product does not show the pair of quadrupole split doublets described in a
previous section: instead a sharp quadrupole split spectrum with parameters $\delta = 1.14$ and
$\Delta = 2.13$ at 293 K is found.

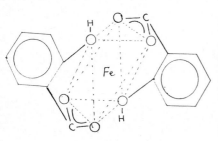

Apparently the salicylate is tridentate in this
form, the two ligands coordinating at the apices
of opposite faces of the octahedra round the iron
(See Fig.5.17). This leads to a low density
structure and  the product can be converted to the
more stable form by the application of pressure.

The crystal→glass transition can be observed
in the spectrum of iron(II) phosphate in phosphoric acid. At 209 K a quadrupole split doublet
is seen, $\delta = 1.3$ with $\Delta = 3.3$, but at 220 K

Fig.5.17

a very broad  single line is observed.  Diffusional broadening takes place in the glassy
phase formed at the higher temperature.

Transitions associated with the onset of rotation are often detected in Mössbauer
spectra. The iron(II) hexammins salts provide good examples. At room temperature
$Fe(NH_3)_6X_2$ salts have a fluorite like structure, space group Fm3m. For X = Cl a single,
line is found with isomer shift 1.12. (Fig.5.18.a)   This implies strict octahedral
symmetry at the iron atom which is only possible if the ammonia groups are rotating.

Below a transition temperature, $T_t = 100$ K, the structure becomes monoclinic and the spectrun changes to a quadrupole split pair, $\delta = 1.12$ with $\Delta = 1.50$. A static distortion of the octahedron takes place as free rotation of the ammonia moities ceases. (Fig. 5.18 b, c). The Mössbauer fraction also changes sharply at the transition temperature.

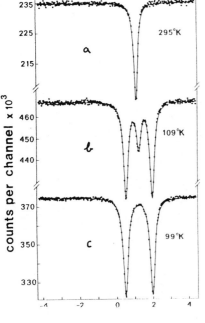

**Fig.5.18  a,b & c**

Similar transitions are found for the bromide, iodide, perchlorate and fluoborate which give $T_t$ values of 45, 47, 150 and 140 K, respectively. In $Fe(NH_3)_6(NO_3)_2$ there are two transition temperatures, at 228 and 99 K, associated with the onset of rotation of the ammonia and of the nitrate groups. Around 110 K both doublets may appear in the spectrum.

The spectra obtained from various hydrated iron(II) salts are especially interesting. The perchlorate $Fe(H_2O)_6(ClO_4)_2$ at 300 K a quadrupole split pair with $\Delta = 1.40$ is obtained but at 77 K the splitting is 3.40.

The octahedron around the iron is trigonally distorted. (See Section 3.6.2.). On applying a magnetic field it is found that the sign of $V_{ZZ}$ changes between the two temperatures, (Fig. 5.19). The transition is first order and shows marked hysteresis. (Fig.5.20) $V_{ZZ}$ is positive at room temperature and negative at 77 K. This implies a change from elongation to compression along the threefold axis. The ground state of the low temperature form is an orbital singlet and the high temperature form an orbital doublet (See 3.6.2). It can be expected that $|V_{ZZ}|$ for the latter form will be very approximately half that for the low temperature form.

(a)  At 300 K          (b) No field          (c) At 4.2 K
Field 2 T ‖ to         at 4.2 K.             Field 2 .5 T ‖
photon beam                                 to Photon beam.

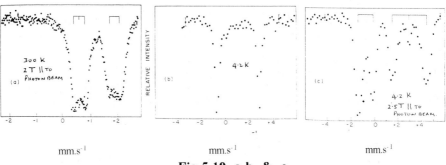

**Fig.5.19  a,b & c**

A similar effect has been found for $Fe(H_2O)_6(BF_4)_2$. In this compound there are very small changes in δ and Δ but appreciable changes in the transition temperature on using the deuterated species. The transition probably alters the hydrogen bonding in the crystals. This transition also shows noticeable hysteresis.

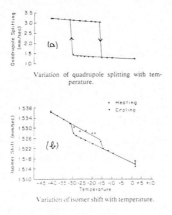

Variation of quadrupole splitting with temperature.

Variation of isomer shift with temperature.

## Fig.5.20

Similar transitions are seen in the spectra of iron(II) (NN'dicyclohexylthiourea)$_6$ $(ClO_4)_2$. In none of these transitions does the isomer shift alter appreciably, but the Mössbauer fraction is higher for the low temperature forms.

## 5.7  IRON IN HIGH OXIDATION STATES

In some compounds iron reaches formal oxidation states of 4, 5, 6 and possibly 8. Many of the compounds in this category are difficult to obtain in a pure state. They are often rather unstable and the presence of excess oxidant may be necessary for their survival. Further some are liable to disproportionation reactions, while others are markedly non-stoichiometric. Hence classical analysis is not very suitable for their investigation. Fortunately the isomer shifts of these compounds are characteristic of the oxidation state of the iron they contain and Mossbauer spectroscopy provides much information permitting the characterisation of such compounds.

It is convenient to consider the oxide/oxyanion compounds and the other complexes separately. Iron(IV) forms numerous compounds of both types, iron(V) very rarely gives complexes and iron(VI) only gives the oxyanion compounds.

### 5.7.1. Oxide and oxyanion compounds

In practically all these compounds the iron is in a high spin electronic configuration. $SrFeO_3$ , with a perovskite structure, contains iron(IV), $d^4$.   It gives a single line spectrum with δ = -0.17, suggesting a closely octahedral environment for the iron.

However the line width is rather large and there may be some unresolved quadrupole splitting. The compound becomes antiferromagnetic at 80 K with $B_{hf} = 33.1$ T at 4.2 K. This is notably lower than would be expected for four unpaired electrons and reflects the covalent contribution to the Fe-O bonding, which is also seen in the rather short bond length.

A range of compounds of formulae $SrFeO_{3-x}$ with $x < 0.28$ have been explored. In these the perovskite structure is tetragonally distorted. Using a sample whose analysis gave $x = 0.192$ the Mössbauer spectrum shows three components at room temperature, $\delta = 0.36$ with $\Delta = 1.32$, 30%; $\delta = 0.30$ with $\Delta = 0.50$, 20% and $\delta = -0.08$ with $\Delta = 0.27$, 50%.

Above 550 K delocalisation of the 3d electrons on the iron apparently takes place, and a single spectrum with isomer shift the weighted mean of the above three values was observed. The quadrupole splitting of 1.32 did not alter appreciably with change in x, although the proportion of this component of the spectrum increased with x, indicating a constant environment for the iron. This spectrum was attributed to Fe(III) in sheets of distorted tetrahedral sites arising from the oxygen deficiency, but with no oxygen vacancies close to the iron. The parameters of the other spectra were more dependent on x.

It was suggested that there were no sharply defined Fe(IV) sites and that perhaps the Fe(IV) in this system disproportionates, to give two species in which the oxidation state of the iron is rather above and rather below IV. At 4.2 K three magnetically split spectra could be seen.

The iron(VI) compound $K_2FeO_4$, $d^2$, $^3A_2$, also gives a single line spectrum with $\delta = -0.87$; a very low value. The compound dissolves in very alkaline solutions and the frozen solution gives an isomer shift of -0.7. The temperature dependence of the isomer shift indicates an effective recoil mass of $\approx 100$ suggesting $FeO_4$ units (See 3.5.1). At very low temperatures, $< 3.6$ K, the compound becomes antiferromagnetic with the low hyperfine field $B_{hf} = 14$ T, again reflecting covalence in the bonding.

Iron(V) is rather less stable and readily undergoes disproportionation. $K_3FeO_4$, $d^3$, $^4A_2$, gives a single line with isomer shift $\delta = -0.42$, rather less negative than expected due to the covalence in the Fe-O bonding.

The isomer shifts for these three compounds are characteristic of high spin compounds of the three oxidation states concerned. Other compounds of iron in the same oxidation state seldom differ in isomer shift by more than $\cong 0.15$, so there is little overlap of values for the three states.

Anodic oxidation of iron in 6 M NaOH solution produces an iron compound which gives a single line spectrum with an isomer shift of -1.62. This may be $FeO_4$, an iron(VIII) compound.

In the non-stiochiometric compounds of this type a quadrupole splitting often appears, due to the less symmetric environments of the iron atoms.

### 5.7.2. Complexes of iron(IV)

Only one or two complexes of iron(V) have been made and no Mössbauer data reported, but extensive data for complexes of iron(IV) are available. They generally involve rather nephelauxetic ligands, such as o.bis(dimethylarsino or phosphino)benzene or the di-alkyl dithiocarbamates. In most of these complexes the iron is in a distorted octahedral environment. The isomer shifts lie in the range -0.1 to o.27 at room temperature; the quadrupole splittings span a larger range from about 1.5 to 3.25.

The diarsine complex [Fe(diars)$_2$Cl$_2$](BF$_4$)$_2$, a D$_{2h}$ trans species, is a low spin S = 1, d$^4$ compound with parameters $\delta$ = 0.20 and $\Delta$ = +3.22 at 298 K, at 4.2 K the values are $\delta$ = 0.27 and $\Delta$ = +3.25. The positive quadrupole splitting suggests a $^3$B$_{2g}$ ground state, and the approximate temperature independence of $\Delta$ indicates the first unoccupied level must lie at least 2000 cm$^{-1}$ higher.

Most of the iron(III) dialkyl dithiocarbamates, Fe(R$_2$NCS$_2$)$_2$ can be oxidised to give cationic iron(IV) compounds. A typical salt is Fe(R$_2$NCS$_2$)$_3$(BF$_4$), where R is an alkyl or aryl group. The iron is present in a trigonally distorted environment with elongation along the C$_3$ axis. Typical Mössbauer paramaters for these compounds are $\delta \approx 0.1$ with $\Delta$ ranging from 2.06 to 2.52 at 295 K. The isomer shifts change very little with different R.

Upon application of a 6 T magnetic field at 4.2 K the di-iso Pr compound behaves as a rapidly relaxing paramagnet and the triplet group of lines appears at higher velocities showing that the sign of $\Delta$ is negative For R = cyclohexyl $\Delta$ changes from -2.00 at 300 K to -2.40 at 4.2 K. indicating a separation of the 3e level from the 3a level of about 24 cm$^{-1}$.

For comparison typical Fe(III)(R$_2$NCS$_2$)$_3$ compounds have isomer shifts close to 0.4 and quadrupole splittings in the range 0.3 to 0.6 at room temperature.

The spectra of the Ph$_4$As$^+$ salts of Fe(MNT)$_3^{2-}$, Fe(MNT)$_2$(i.MNT)$_3^{2-}$ and Fe(MNT)$_2$S$_2$C$_2$(OOCC$_2$H$_5$)$_2^{2-}$ give the parameters shown in Table 5.9. The ligands are shown in Fig.5.21.

Table 5.9

| Compound | Temperature | $\delta$ | $\Delta$ |
|---|---|---|---|
| Fe(MNT)$_3$ | 297 K | 0.012 | +1.53 |
| " | 77 K | 0.105 | +1.59 |
| " | 4.2 K | 0.114 | +1.61 |
| Fe(MNT)$_2$(i MNT) | 297 K | 0.012 | -1.58 |
| " | 77 K | 0.123 | -1.63 |
| " | 4.2 K | 0.133 | -1.64 |
| Fe(MNT)$_2$S$_2$C$_2$(OOCC$_2$H$_5$)$_2$ | 297 K | 0.023 | -1.75 |
| " | 77 K | 0.110 | -1.82 |
| " | 4.2 K | 0.125 | -1.83 |

For these trigonally distorted, S = 1, d$^4$ complexes one would expect an orbital singlet ground state, |d$_{xy}$> lying lowest, and therefore $\Delta$ should be negative. Magnetically perturbed spectra confirm that this is the case for the second and third complexes shown above, but for the first a positive sign is found. It seems unlikely that the sequence of

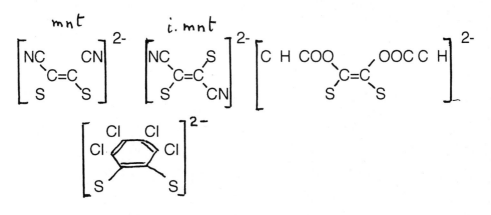

**Fig.5.21**

energy levels of the orbitals is different in this compound. A more reasonable explanation is that substantial participation of the iron 3d orbitals in $\pi$ bonding in the $(MNT)_3$ compound changes the balance of occupation of these orbitals and hence the EFG. This explanation is supported by UV/Vis spectroscopic evidence and by the Fe - S bond lengths which suggest that the $\pi$ bonding by the MNT is stronger than with the other two ligands.

It is interesting to note that reduction of the formally iron(IV) compound $Fe(S_2C_6Cl_4)_2$ gives $(Bu_4N)_2Fe(S_2C_6Cl_4)_2$ and both these five coordinate iron complexes have practically the same Mössbauer parameters, $\delta = 0.32$ and $\Delta = 3.15$ for the first, and $\delta = 0.32$ and $\Delta = 3.02$ for the second complex, so that the electronic environment of the iron must be almost the same in the two compounds. The reduction must mainly involve electrons on the ligands.

Perhaps the most significant result from the Mössbauer spectra of iron(IV) complexes concerns the tetraphenyl porphyrin and phthalocyanin derivatives shown in Fig 5.22 a & b.

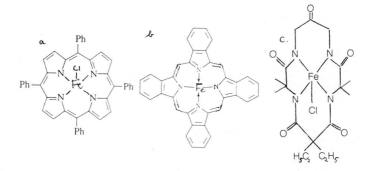

Porphyrin                    Phthalocyanin                    5 Coordinate complex

**Fig.5.22  a,b  &  c**

Several of these compounds can be made by the oxidation of the iron(III) compounds. Mössbauer, data are shown in Table 5.10.

The Mössbauer data show that in compounds 1 to 5 the iron is effectively iron(III). The oxidation leading to the formation of compounds 2,3 and 5 must involve the ligand, probably forming a delocalised cation by removal of a $\pi$ electron.

**Table 5.10**

| | Compound | $\delta$ | $\Delta$ | Temperature. |
|---|---|---|---|---|
| 1. | Fe(III)TPPCl | 0.42 | 0.46 | 4.2 K |
| 2. | FeTPPCl.ClO$_4$ | 0.45 | 1.27 | 77 K |
| 3. | FeTPPCl.SbCl$_6$ | 0.40 | 0.55 | 77 K |
| 4. | (FeTPP)$_2$O | 0.40 | 0.62 | 131 K |
| 5. | (FeTPP)$_2$O.ClO$_4$ | 0.39 | 0.57 | 77 K |
| 6. | (FeTPP)$_2$N.ClO$_4$ | 0.03 | 2.00 | 77 K |
| 7. | (FeTPP)$_2$C | 0.10 | 1.88 | 77 K |
| 8. | (FeTPP)$_2$N | 0.18 | 1.08 | 131 K |
| 9. | (FePhth)$_2$O | 0.36 | 0.44 | 77 K |
| 10. | (FePhth)$_2$N.PF$_6$ | -0.10 | 2.06 | 131 K |
| 11. | (FePhth)$_2$N | 0.06 | 1.76 | 131 K |

The other TPP complexes do contain iron(IV). Compounds 8 and 11 nominally contain Fe(III) and Fe(IV) atoms, but only one spectrum is seen and there must be valence delocalisation between the two iron atoms. (See Section 7.4 )

Five coordinate iron(IV) is also found in the complex with the ligand shown in Fig.5.22 c. The Mössbauer parameters reported for this complex are $\delta$ = -0.03 and $\Delta$ = 0.87 at 150 K.

## 5.8. IRON IN UNUSUAL SPIN STATES

In low symmetry environments when most, or all, of the degeneracy of the 3d orbitals is removed, the unusual spin states of S = 1 for iron(II) and S = 3/2 for iron(III) may be stabilised. It is necessary that the highest occupied 3d orbital lies well below the first excited level. Such compounds have been identified by their bulk magnetic properties. Care is needed to distinguish abnormal spin states from cooperative antiferromagnetic magnetic interactions between pairs or larger numbers of iron atoms, coupling of spins with other paramagnetic atoms in the molecule and spin cross over systems. Mössbauer spectra can be helpful in identifying these compounds when used in conjunction with other means of investigation.

### 5.8.1 Compounds of Iron(II) with S = 1

Many of these compounds fall into one or other of two groups. One group comprises compounds of the type (L-L)$_2$FeX$_2$. These have only C$_2$ symmetry.

Some data for this kind of complex are shown in Table 5.11.

As would be expected the isomer shifts for these compounds are much less than for related high spin iron(II) compounds: compare the data for the high spin compound 10. The shifts are close to, but generally a little above, values for S = 0, low spin, iron(II) compounds.   Compounds 4 - 6 are related examples of the latter type.

The quadrupole splittings are rather small and rather insensitive to temperature. This is implied by the conditions for the existence of such spin states outlined above. An isomer shift of about 0.3 and a small temperature independent quadrupole splitting seem characteristic of this group.

The electronic configuration of the iron in these compounds is not certain. Probably the energy sequence of the one electron orbitals is: $d_{z^2-y^2} >> d_{z^2} > d_{xz}$ and $d_{yz} > d_{xy}$.

This would suggest $\Delta$ might be negative, but the ligand contribution may dominate the EFG and lead to a positive value.

## Table 5.11

| No.  Compound | Temperature of measurement/ K | $\delta$ | $\Delta$ |
|---|---|---|---|
| 1. $FePhen_2F_2$ | 298 | 0.31 | 0.23 |
| " | 77 | 0.28 | 0.24 |
| " | 4.2 | 0.28 | 0.24 |
| 2. $FePhen_2(NCBH_3)_2$ | 301 | 0.40 | 0.18 |
| 3. $FePhen_2CarbothioamideCl_2$ | 300 | 0.26 | 0.53 |
| "    " | 78 | 0.22 | 0.53 |
| Data for related low spin (S = 0) complexes. | | | |
| 4. $FePhen_3(ClO_4)_2$ | 301 | 0.31 | 0.15 |
| 5. $FePhen_2(CN)_2$ | 298 | 0.18 | 0.58 |
| 6. $FePhen_2(CNBH_3)_2$ | 301 | 0.15 | 0.61 |
| Data for a related high spin (S = 2) complex. | | | |
| 7. $FePhen_2(CNS)_2$ | 298 | 1.03 | 2.52 |

Phen = 1.10. Phenanthroline.

EFG and lead to a positive value.

The isomeric compounds 2 and 6 are especially interesting. The low spin species 6 has rather a high magnetic susceptibility for an S = 0 species, but still much below the value for compound 2, which is close to that expected for S = 1.

For several years a number of compounds of the type $Fe(L-L)_2X_2$ with $X_2$ the anion of a dicarboxylic acid, for example the oxalato group, were believed to be members of this group.  But it has been shown that the red compounds examined had suffered aerial oxidation to yield compounds better formulated as $[FePhen_3]_2[FeOx_3]$ 1/2Ox or perhaps $[FePhen_3]_2[FeOx_3]OH$ and containing one high spin iron(III) atom and two low spin iron(II) atoms.

On cooling these compounds to 4.2 K, a magnetically split spectrum arising from the iron(III) appeared, in addition to the quadrupole split doublet from the iron(II). Working in the absence of air genuine violet $FePhen_2malonate.7H_2O$ has been obtained which exhibits pentuplet - singlet spin cross over, like the compound with two NCS groups (See Section 7.5).

It should be noted that Fe(Phen2carbothioamide)$_2$Cl$_2$ is a low spin (S = 0) complex and the compound 2 in Table 5.11 might contain some iron(III) from aerial oxidation.

Another group of compounds with S = 1 are based on a planar arrangement Fe[N$_4$]. Several compounds of the family shown in Fig.5.23 have been examined. With R' and R″ = CH$_3$ or H and X and Y = -CH$_2$CH$_2$-, giving 14 membered rings, the isomer shifts range from 0.33 to 0.46 and the quadrupole splittings from 1.99 to 3.06. With X and Y = -CH$_2$CH$_2$CH$_2$- and 16 membered rings, the isomer shifts were rather larger. Magnetic measurements confirmed S = 1.

Several other compounds based on Fe[N$_4$] coordination have been examined, but data are only available for measurements at room temperature in most cases.

The most extensively studied compound in this group is iron(II) phthalocyanine, for which Mössbauer parameters are, at 293 K, $\delta$ = 0.40 and $\Delta$ = 2.62; at 77 K, $\delta$ = 0.51 and $\Delta$ = 2.69 and at 4.2 K, $\delta$ = 0.49 and $\Delta$ = 2.70.

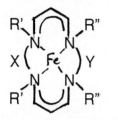

R' and R″ = H or CH$_3$

X and Y = -(CH$_2$)$_n$- or
-{C(CH$_3$)$_n$- n = 2 or 3

**Fig.5.23**

S = 1 state was identified by magnetic susceptibility measurements. As in the previous group the quadrupole splitting is relatively insensitive to temperature. It is not established whether the ground state is a $^3E_g$ state corresponding to a configuration

$$d_{xy}^2 \; ; (d_{xz},d_{yz})^3 \; ; d_{z^2}^1 \quad \text{or a } ^3B_{2g} \text{ state, with}$$

$$(d_{xz},d_{yz})^4 \; ; d_{xy}^1 \; ; d_{z^2}^1. \text{ Experimental electron}$$

density distributions agree with either configuration. The quadrupole splitting is of positive sign, but for both the above configurations the free ion would make a negative contribution to $\Delta$ so that the ligand contribution must dominate.

It may be noted that in none of these studies was the crystal form of the phthalocyanine used identified; most transition metal phthalocyanines, including iron, exist in two or more crystal modifications.

Iron(II) phthalocyanine can add two pyridine molecules to the iron to give a low spin complex, but on the application of a high pressure, 50 kbar, two spectra are found with parameters $\delta$ = 0.20 with $\Delta$ = 1.93 and $\delta$ = 0.34 with $\Delta$ = 2.78. The latter spectrum suggests the formation of an S = 1 form.

The macrocycle TMC = 1,4,8,11 tetra methyl 1,4,8,11 tetra azacyclotetradecane, Fig.5.24, gives a number of high spin iron(II) complexes, for example, [FeTMCBr]Br, containing five coordinate iron. These have isomer shifts from 0.87 to 0.91 and very large quadrupole splittings, 3.03 to 3.92.

In moist air [FeTMCN$_3$](BF$_4$) gives a compound with Mössbauer parameters $\delta$ = 0.35 and $\Delta$ = 0.80, [FeTMCNCS]BF$_4$ behaves in a similar manner. The products may be six coordinate compounds arising from the addition of water and may have S = 1.

The iron porphyrins, which may also have S = 1, will be considered in a later section with the biologically significant compounds (See Section 8.5).

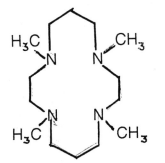

Fig.5.24

There are several S = 1 compounds that contain a Fe[N₄S₄] core. The ligand H₂tsalen shown in Fig.5.25, ethylene bis(thiosalicylideneimine), gives iron complexes with S = 1. It is a tetradentate ligand. The Mössbauer parameters for the iron complex are $\delta = 0.44$ with $\Delta = +2.20$. An electronic configuration like iron phthalocyanine would make a negative contribution to $V_{zz}$, so that the ligand contribution must be

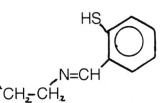

H₂TSalen = ethylene bis(thiosalicylideneimine

Fig.5.25

important. At very low temperatures an antiferromagnetic coupling of the spins of the iron atoms in pairs of molecules takes place by a super-exchange mechanism.

The high temperature forms in some spin cross over systems are S = 1 species. For example [Fe(P₄)X]BPh₄CH₂Cl₂, where P₄ is 1,4,7,10 tetra phosphadecane, gives $\delta = 0.20$ and $\Delta = 1.81$ at 298 K. Magnetic susceptability data show that the compound is only partially converted to the S = 1 state at this temperature, which accounts for the low isomer shift.

### 5.8.2 Compounds of Iron(III) with S = 3/2

Numerous iron(III) bis dialkyl dithiocarbamates, shown in Fig.5.26, have been reported to be S = 3/2 species. They contain five coordinate iron(III) and are of $C_{2v}$ symmetry. The low symmetry leads to unusually large quadrupole splittings for iron(III) complexes.

Some of these compounds exhibit fast spin cross-over 3/2 ⇔ 1/2.(See section 7.5).

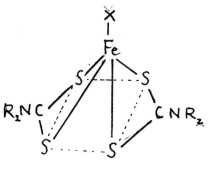

Fig.5.26

The isomer shifts for R = Ethyl and X = Cl, Br, I, or SCN are close to 0.50 while the quadrupole splittings range from 2.65 to 2.90. At low temperatures short range coupling of iron spins, effectively dimerisation, leads to antiferromagnetic properties when X = I. At low temperatures the chloro compound gives a magnetically split spectrum with $B_{hf} = 22$ T, a low value, indicating covalent character of bonding.

Some iron(III) dithiolate complexes have a similar pyramidal structure and are 3/2 spin species. Some data on such compounds are given in Table 5.12.

## Table 5.12

| Compound | FePyL$_2$ | | FePyL'$_2$ | | FePyL"$_2$ | | FePicL$_2$ | | FeisoquinL$_2$ | |
|---|---|---|---|---|---|---|---|---|---|---|
| | $\delta$ | $\Delta$ | $\delta$ | $\Delta$ | $\delta$ | $\Delta$ | $\delta$ | $\Delta$ | $\delta$ | $\Delta$ |
| 295 K. | 0.28 | 2.54 | 0.27 | 2.51 | 0.23 | 3.12 | 0.26 | 2.61 | 0.29 | 2.44 |
| 77 K. | 0.33 | 2.61 | 0.33 | 2.41 | 0.33 | 3.02 | 0.36 | 2.59 | 0.29 | 2.46 |

$$L = S_2C_2(CF_3)_2; \quad L' = S_2C_2(CN)_2; \quad L" = S_2C_6Cl_4;$$

Pic = 4 MePyridine; Py = pyridine; isoquin = isoquinoline.

Iron(III) porphyrin complexes of similar five coordinate configuration are also S = 3/2 compounds and give similar isomer shifts and large quadrupole splittings (See also Sections 7.5 and 8.5).

The Iron(III) phthalocyanines, FePhth.X, are five coordinate and generally have S = 3/2. Typical Mössbauer data are given in Table 5.13.

## Table 5.13.

| X = | Cl | Br | I | CF$_3$CO$_2$ | CCl$_3$CO$_2$ |
|---|---|---|---|---|---|
| $\delta$ = | 0.29 | 0.28 | 0.28 | 0.28 | 0.29 |
| $\Delta$ = | 2.94 | 3.12 | 3.23 | 3.08 | 3.07 |

Some display spin cross-over below room temperature.

Like most of the five coordinate iron complexes they give very large quadrupole splittings. The related six coordinate complexes are low spin species.

The NO adducts of the iron Schiffs base complexes have S = 3/2 at room temperature and shown 3/2 $\Leftrightarrow$ 1/2 spin cross-over at lower temperatures (See Section 7.5).

Some iron(III) N aryl biguanide complexes, FeL$_2$ClHCl,

L = NH$_2$(C=NH)NR(C=NH)NH$_2$, with R = phenyl, p.tolyl or o.ClC$_6$H$_4$, give isomer shifts about 0.43, but much lower quadrupole splittings around 0.75. These may also be intermmediate spin compounds.

For neither iron(II) nor iron(III) complexes can the isomer shift or quadrupole splitting give definitive evidence of abnormal spin states. But for iron(III) complexes a magnetic spectrum at low temperatures with B$_{hf}$ between 20 and 25 T provides strong evidence for a S = 3/2 state.

## 5.9 FIVE COORDINATE IRON COMPLEXES

Five coordinate complexes of both iron(II) and iron(III) show very large quadrupole splittings. The compounds $(R_2NCS_2)_2FeX$ give the values shown in Table 5.14 a & b.

### Table 5.14 a.

| X = | NCS | Cl | Br | I |
|---|---|---|---|---|
| R = Et | 2.55 | 2.63 | 2.78 | 2.87 |
| R = Me | 2.56 | 2.67 | 2.86 | 2.96 |

These are very large splittings for iron(III) compounds.

Data for a number of other five coordinate iron(III) compounds are given in Table 5.14 b.

**Table 5.14 b.**

1. $(Et_4N)_2[Fe(S_2C_2X_2)_2]_2$                    & 2. $(Et_4N)_2[Fe(S_2C_6Cl_4)_2]_2$.
   X = CN, $CF_3$ or Ph.                          All give $\Delta$'s from 2.37 to 3.02.
3. $(Ph_4P)Fe(S_2C_2X_2)_2L$                        & 4. $(Ph_4P)Fe(S_2C_6Cl_4)_2L$
   X = CN, $CF_3$ or Ph;     L = pyridine, have $\Delta$'s between 2.5 and 3.2.

Measurements all at 77 K.

The ligands are shown in Fig.5.21. The isomer shifts of all these compounds are close to 0.33.

Five coordinate iron(II) complexes also give some exceptionally large splittings. Examples are given in Table 5.15.

**Table 5.15.**

Sodium dibenzo 18 crown 6 salt of
FeTPP.OPh (TPP = tetraphenyl porphyrin)                    $\Delta$ = 3.95 at 128 K.
$[Fe(R_2NCS_2]_2$                                          $\Delta$ = 4.16 at 100 K.
$[FeL_2NCS]$ NCS ( L = cyclohexanone thiosemicarbazone)    $\Delta$ = 4.28 at  77 K.
Fe(2.2'bipyridyl)$Cl_2$ (5 coordinate form)                $\Delta$ = 3.68 at  78 K.
Fe($NH_2CSNHNH_2$)$_2$ $SO_4$                              $\Delta$ = 4.36 at  77 K.
$(Et_4N)_2Fe(C_5H_4CS_2)_2$                                $\Delta$ = 4.52 at  78 K.

All these are known to contain five coordinate iron. Several iron(II) macrocycle complexes also contain five coordinate iron and give large quadrupole splittings.

The structure of the dithiocarbamate dimer complex is shown in Fig.5.27. Compounds 1 and 2 in table 5.14 b are similar.(See also Section 5.8.1.) An orbital singlet ground state is needed for these high values.

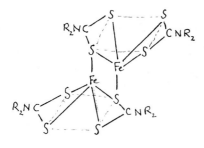

Separation of Fe

from 4 S atom planes

exaggerayed

**Fig.5.27**

The compound [Fe 1.7.CT Cl]$ClO_4$, where 1.7 CT is 5,5,7,12,12,14 Hexamethyl 1,4,8,11 Tetraazacyclotetradeca 1(14)7 diene, which has no more than $C_2$ symmetry around the iron, has $\Delta$ = +4.0 at room temperature. The free ion contribution to $V_{zz}$ is negative for a square based pyramid or a trigonal bipyramid so that the ligand contribution would need to outweigh greatly the free ion term to yield a positive $\Delta$.

The macrocycle complex Fe[16 ane $S_4$]$X_2$ is known to have a highly tetragonally distorted octahedral structure. But its quadrupole splitting for X = Cl is greater than 4. Unfortunately the sign of $V_{zz}$ is not available for this compound.

The complex tetrakis 1.8.naphthylpyridine iron(II) perchlorate, $FeL_4(ClO_4)_2$ L = $C_8H_6N_2$, an eight coordinate complex of $D_{2D}$, dodecahedral, symmetry about the iron has a quadrupole splitting of 4.54 at 78 K.

Unfortunately it is apparent from these data that a very large quadrupole splitting is not a sufficient condition for five coordination.

**Acknowledgements**

Fig.5.4 Reproduced with permission from Champion, A.R., Vaughan, R.W. and Drickhamer, H.G. (1967) *J.Chem.Phys.*, **47**, 2538

Fig.5.5 Reproduced with permission from Fiser, D.C. and Drickhamer, H.G.,(1971), *J.Chem.Phys.*, **54**, 4825

Fig.5.7 Reproduced with permission from Bancroft,G.M., Garrod,R.E.B., Maddock, A.G., Mays,M.J . and Prater, B.E.,(1970) *Chem.Comm.*, **1970**, 200.

Fig.5.11 Reproduced with permission from Mattievich, E.,and Danon,J.(1976) *Journ.Physique*, **C6**, **37**, 483.

Fig.5.12 Reproduced with permission from Ladrière, J. and Maddock, A.G.(1974) *Journ.Physique*, **C6**, **35**, 647

Fig.5.13 Reproduced with permission from Kostina,A.,*Acta Crystallographica* **B32**, 2427

Fig.5.14 Reproduced with permission from Nicolini,C. and Reiff, W,M. (1990) *Journ.Physique*, **C1**, **41**, 287.

Fig.5.18 Reproduced with permission from Asch,L.,Shenoy,G.K.,Freidt,J.M.. Adloff,J.P.and Kleinberger,R.R. (1975) *J.Chem.Phys.*,**62**, 2335.

Fig.5.19 Reproduced with permission from Ouseph.P.H.,Thomas,P.M.and Deszi,I. (1973) *Chem.Phys.Lett.*, **22**, 124.

Fig.5.20 Reproduced with permission from Ouseph,P.H., Thomas,P.M. and Deszi,I (1974) *J.Phys.Chem.Solids*, **35**, 604

# 6

# Further consideration of magnetic effects

Magnetic effects play a very important role in the Mössbauer spectra recorded from iron and its compounds. In most of the compounds there are unpaired 3d electrons on the iron atoms, and the compounds are paramagnetic at a sufficiently high temperature. In such compounds a magnetic field will exist at the iron nucleus and this gives rise to a variety of effects which may determine the details of the spectra, particularly in the presence of an external magnetic field.

The splitting of the nuclear energy levels by the magnetic field has already been treated in Section 1.4.5 but the magnitude and origin of the internal magnetic fields has not been considered. In addition the relaxation effects have only been mentioned briefly.

## 6.1 HYPERFINE MAGNETIC FIELDS.

It is convenient to explore this matter in terms of the magnetic field at the nucleus rather than the nuclear energy levels that ensue. The flux density of the magnetic field at the iron nucleus $B_{int.}$ is comprised of (i) any external magnetic field to which the absorber is subjected $B_{ext.}$, and (ii) the field produced by the atom of which the nucleus forms part. This field is affected by the more distant atoms in the solid, which influence the distribution and occupation of the electronic levels of the iron atom. This contribution is called the **hyperfine field** $B_{hf}$ and is defined by the relation $g_n \mu_n B_{hf} = -A \cdot <S>$, where $\mu_n$ is the nuclear magneton, $e\hbar / 2m_p c$, $g_n$ is the nuclear g factor; $<S>$ is the expectation value of the electron spin; $m_p$ the mass of the proton; and $A$ is a tensor determining the magnetic coupling in the system.

Thus $B_{int} = B_{ext.} + B_{hf}$, so that if there is no external field $B_{int.} = B_{hf}$ and

$B_{hf} = -A \cdot <S>/g_n \mu_n$ or $+ B \overset{o}{\,hf} <S>/S$. $B_{hf}$ is the saturation hyperfine field

corresponding to $<S> = S$ the electronic spin. $B_{hf} \Rightarrow B \overset{o}{\,hf}$ as $T \Rightarrow 0$ K.

It is apparent that $<S>$ will depend on the distribution of electrons between the levels shown in Fig.3.7.b & c. Thus the ligand field and spin orbit coupling have a role in determining $B_{hf}$ and in this way the ligands are involved.

The hyperfine field arises from three main components: (i) the Fermi contact term, $B_F$, which is given by $-\mu_0 \mu_b <r^{-3}> \kappa <S>/2\pi$, where $\mu_b$ is the electronic Bohr magneton, $\mu_b = e\hbar / 2m$, m is the mass of the electron, $\mu_0$ is the permeability of a vacuum, and $<r^{-3}>$ is the expectation value of $r^3$ for the 3d electrons of iron. This field is due to the perturbation of the density of spin up and spin down s electrons at the iron nucleus because of polarisation by the unpaired 3d electrons on the iron. $\kappa$ is a constant determining the interaction. $B_F$ is isotropic.

(ii) A second contribution comes from the orbital motion of the 3d electrons on the iron, $B_0$. This is given by $\mu_0\mu_b<r^{-3}><L>/2\pi$. For a full or half filled d shell $<L>$, the expectation value of the orbital angular momentum, is zero and in most complexes of iron the ligand field reduces $<L>$ to a small value. The residual $<L>$ is largely due to spin orbit coupling. This leads to a value for $<L>$ of $(g - 2)<S>$, where g is the electronic g factor for the complex, so that $B_0 = \mu_0\mu_b<r^{-3}>(g - 2)<S>/2\pi$.

(iii) Finally a field arises from the dipolar interaction of the 3d electron spin with the nucleus. Expressing this as a magnetic flux density, $B_D$, one finds:

$B_D = \mu_0\mu_b <r^{-3}>[3 \ \mathbf{r} \ (\mathbf{S} \bullet \mathbf{r})/r^2 - \mathbf{S}]/2\pi = \mu_0\mu_b<r^{-3}><3\cos^2\theta - 1><S>/2\pi$ Here **r** is the vector determining the position of the d electron and $\theta$ the angle this vector makes with the principal axis of the EFG. Now $V_{zz}/e$ is proportional to $<3\cos^2\theta - 1><r^{-3}>$ so one can put $B_D = 2\mu_0\mu_b\varepsilon_0 V_{zz}<S>$ where $\varepsilon_0$ is the permittivity of the vacuum. But $\mu_0\varepsilon_0 = 1/c^2$ so that $B_D = 2\mu_b V_{zz}<S>/c^2$. $V_{zz}$ is the principal component of the EFG. If $V_{zz}$ is 0 then $B_D$ must also be 0.

Thus $B_{hf}$ = the vector sum $B_F + B_0 + B_D$

$$= \mu_0\mu_b<r^{-3}>[ -\kappa + (g-2) + <3\cos^2\theta - 1>]<S>/2\pi.$$

It will be noted that $B_0$ and $B_D$ are of opposite sign to $B_F$. The contributions $B_0$ and $B_D$ are usually small for $Fe^{3+}$ and become zero in a cubic environment. $B_{int}$ will generally be insensitive to changes in the $Fe^{3+}$ environment. By contrast for $Fe^{2+}$ all three terms may be significant, thus $B_{int}$ for $FeF_2$ is -33 T while for $FeBr_2$ it is +3 T; the wide range of values makes it useful in exploring $Fe^{2+}$ sites in solids.

## 6.2 ORDERING IN MAGNETIC SOLIDS

The above contributions to the magnetic field at the iron nucleus are affected by the surroundings of the atom, since these determine the energies and occupation of the iron orbitals which in turn determine $<S>$.

But another kind of interaction with the surroundings can also occur. If the solid contains other magnetically active atoms, including other iron atoms, there will be a magnetic interaction which can be expressed as $B_{EX}$. the exchange field. The energy of this interaction, in the simplest case, is given by $-2JS_1 \bullet S_2$. J is called the **Coulomb exchange integral**, and for direct interaction depends on the overlap of orbitals on the two atoms. Another kind of interaction, leading to similar effects, can take place through the intermediary of a small atom such as oxygen or fluorine. One form of this **superexchange interaction** is shown in Fig.6.1.

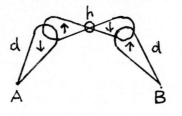

**Fig.6.1**

Some spin up electron density on O is transferred to an empty 3d orbital of suitable symmetry on A. Because of the overlap of orbitals on O and B spin down density on

B rises  and a transferred field is produced on B.  This mechanism is very dependent on the geometry of the AOB ensemble and the covalency of the A-O and B-O bonds. Neither process is effective if the iron and other magnetically active atoms are well separated in the lattice giving a **magnetically  dilute  solid**.

The superexchange mechanism also permits a field to be produced at the nucleus of a diamagnetic atom. In this way magnetic splitting of a tin spectrum can arise from iron atom neighbours. Such transferred fields can be as great as 20 T.

These exchange interactions can give rise to **magnetic   ordering**. If the exchange energy $-2ZJ\ \mathbf{S}_1 \cdot \mathbf{S}_2$ becomes greater than kT it will be energetically favourable for the nuclear magnetic fields and the associated nuclear magnetic moments to align in one of the arrangements shown in Fig. 6.2.  Each iron atom is interacting with Z near neighbour atoms.

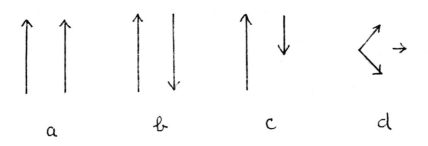

a                    b                    c                    d

**Fig.6.2**

(a) **Ferromagnetic  coupling**. Magnetisation even in absence of an applied field. Exchange integral for the coupling of two sub-lattices of iron atoms is negative.

(b) **Antiferromagnetic  coupling**. Exchange integrsal is positive. Magnetic moments of all thr iron atoms the same,but orientation antiparallel.

(c) **Ferrimagnetism**. Exchange integral positive but magnetic moments of atoms on the two sub-lattices different.

(d) **Canted  antiferromagnetic** situation. Moments of all iron atoms the same but the moments on the two sub-lattices are not exactly antiparallel. With canted spins the material displays antiferromagnetic characteristics in one direction and weakly ferromagnetic behaviour normal to this direction.

The condition for magnetic ordering given above does not imply that all paramagnetic solids will order magnetically at a sufficiently low temperature. The exchange energy must also exceed the vibrational zero point energy of the solid and the crystal field energy at 0 K. It is also possible that the compound may have a singlet diamagnetic ground state occupied at low temperatures.

Above the ordering temperature the resultant spin is changing its orientation rapidly, <S> is zero for other than very short periods. Below this temperature S tends to align in relation to the magnetic field and although still changing its orientation it does so less frequently. $<S>$ assumes a finite value which grows rapidly as the temperature falls and $B_{hf}$ grows. Ultimately when T approaches 0 K $<S>$ approaches the value S.

In the case of a compound with $Fe^{3+}$ in a cubic environment and in the absence of an external magnetic field, $B_0$ and $B_D$ are zero and $B_{int} = B_F = B_{hf} = B_{hf}^{\circ} <S>/S$.

The variation of $B_{hf}$ with temperature under these conditions reflects the change in $<S>$ which is proportional to the magnetisation. The function $<S>/S$ in terms of T is the **Brillouin function**, established by bulk magnetic measurements. Fig. 6.3 Shows a plot of $Bhf/B_{hf}^{\circ}$ against $T/T_n$ for $\alpha$ $Fe_2O_3$. $T_n$ is the temperature at which magnetic ordering occurs and the solid curve is the Brillouin function for S = 5/2.

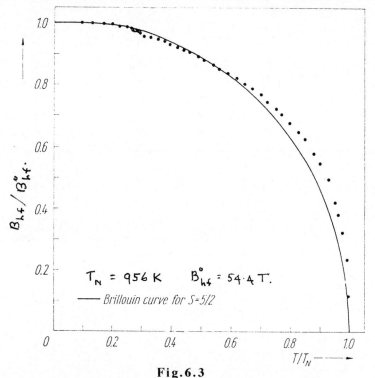

**Fig.6.3**

In this way the value of $B_{hf}$ can be established for this category of iron(III) compounds without making measurements at extremely low temperatures.

The $B_{hf}$ could be as great as 70 T in these compounds, but it is reduced by the covalency of the bonding. The effect is larger for the more covalent tetrahedral $FeX_4$ compounds.

Table 6.1 shows some data for iron(III) compounds.

Because of the $B_D$ and, particularly, the $B_0$ contributions to $B_{hf}$ no such simple relation exists for the iron(II) compounds, indeed $B_{hf}$ for such compounds is sometimes quite small.

Table 6.1.

| Bhf for octahedral FeX6 | | Bhf for tetrahedral FeX4 |
|---|---|---|
| $X = F^-$ | 62 T | |
| $X = H_2O$ | 58 T | |
| $X = O^{2-}$ | 55 T | 50 T |
| $X = Cl^-$ | 48 T | 47 T |
| $X = Br^-$ | | 42 T |
| $X = NCO^-$ | | 39 T |

Because of the $B_D$ and particularly the $B_O$ contributions to $B_{hf}$ no such simple relation exists for the iron(II) compounds, indeed $B_{hf}$ for such compounds is sometimes quite small.


## 6.3  RELAXATION  EFFECTS

The appearance of a hyperfine interaction in a Mössbauer spectrum requires that the time associated with the interaction be shorter than the mean life time of the $^{57m}$Fe excited state, $\tau_n$. For the case of a magnetic interaction this demands that $\tau_L$ or $2\pi/\omega_L$ be less than $\tau_n$. On a classical picture the nuclear magnetic moment must precess in the magnetic field several times during the mean life of the excited nucleus. The Larmor frequency, $\omega_L = g_n\mu_nI_zB_{hf}/\hbar$ and the nuclear magneton $\mu_n = e\hbar/2m_pc$, where $m_p$ is the mass of the proton. Hence $\tau_L = 4\pi m_pc / g_neI_zB_{hf}$.

As shown above the magnetic field at the nucleus of the Mössbauer atom arises from the resultant spin of the orbital electrons and this is usually a time dependent quantity. Suppose the life time of the state with spin S is $\tau_r$, if this relaxation life time is less than $\tau_L$ the nuclear magnetic moment will only experience a field corresponding to the average value of S. For paramagnetic solids this is zero and no magnetic splitting will be seen. If $\tau_r$ is substantially longer than $\tau_L$, spectra corresponding to all values of S will appear.

The orders of magnitude of these quantities are $\tau_n \approx 10^{-7}$ s, $\tau \approx 10^{-8}$ to $10^{-9}$ s while $\tau_r$ can range from less than $10^{-10}$ to more than $10^{-6}$ s.

Fig.6.4.a & b gives calculated and observed spectra showing the effect of $\tau_r$ as it changes from a very short to a long value.

The relaxation time $\tau_r$ is determined by two processes, spin-spin relaxation and spin-lattice relaxation. The first of these processes involves spin exchange with other paramagnetic species in the lattice, especially other atoms of the same kind. In a paramagnetic iron compound the rate of spin-spin relaxation will depend on the concentration of iron in the compound, and it will be slow in large complexes where the iron atoms are well separated from each other.

The spin-lattice process involves spin orbit coupling and is very slow in the absence of resultant orbital angular momentum on the atom. Unlike the spin-spin process it is temperature dependent proceeding more slowly at lower temperatures.

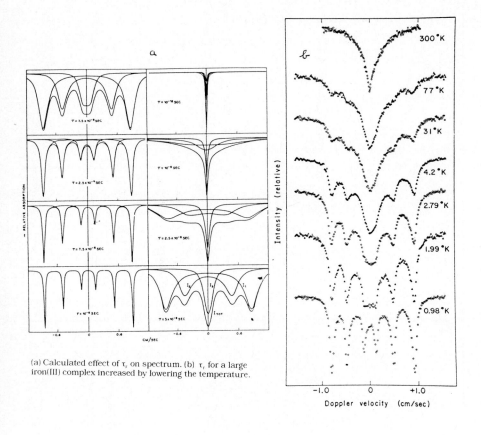

(a) Calculated effect of $\tau_r$ on spectrum. (b) $\tau_r$ for a large iron(III) complex increased by lowering the temperature.

The right hand figure shows experimental data for a large iron(III) complex

**Fig.6.4**

## 6.4. MAGNETIC SPECTRA FROM PARAMAGNETIC SOLIDS

In the absence of magnetic ordering, a paramagnetic iron compound will give a magnetically split Mössbauer spectrum if $\tau_r$ substantially exceeds $\tau_L$. Even in large $Fe^{2+}$ complexes the spin-lattice relaxation is generally fast enough at very low temperatures so that the above condition cannot be fulfilled. But some complexes undergo magnetic ordering at low temperatures, these systems will be considered later.(Section 6.5).

In high spin iron(III) compounds because there is no substantial orbital angular momentum the spin-spin relaxation is the dominant process. If this is reduced by using large complexes and the residual spin-lattice relaxation reduced by lowering the temperature, some compounds give magnetically split spoectra, so that $\tau_r$ exceeds $\tau_L$.

In an axial ligand field the $Fe^{3+}$ atom has three low lying energy states.(See Fig.5.6). These are three Kramers doublets that can be represented as $|\pm 5/2\rangle$, $|\pm 3/2\rangle$ and $|\pm 1/2\rangle$. They have $S_z = 5/2$, $3/2$ and $1/2$ respectively. Since there is no preferred direction for $S_z$ the nuclear moment is affected by S; unlike the ordered case, to be considered in Section 6.5, where a dependence on $\langle S_z\rangle$ is involved. The separation of these levels is small and except at very low temperatures they are all populated. In principle this should lead to three magnetic spectra. However the spin-spin relaxation times for the three states are different, the $|\pm 5/2\rangle$ and $|\pm 3/2\rangle$ states relaxing much more slowly than the $|\pm 1/2\rangle$ state, so that only spectra due to these two states may be seen.

An example is found in the spectrum of 0.14% $Fe_2O_3$ in $Al_2O_3$ shown in Fig.6.5. Pure $Fe_2O_3$ gives a magnetically split spectrum at room temperature (See Fig.1.11.). This is due to ferromagnetic ordering and a rather high Curie temperature. The diamagnetic dilution in the alumina reduces spin-spin relaxation and avoids magnetic ordering.

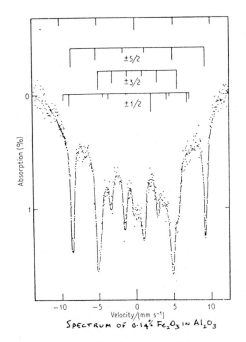

SPECTRUM OF 0.14% $Fe_2O_3$ IN $Al_2O_3$

Stick diagrams show positions and line intensities
predicted for the three low lying states.

## Fig.6.5

Usually $\tau_r$ increases as the temperature falls because relaxation takes place more slowly. But exceptions can be found. When a compound is at such a temperature that nearly all the $Fe^{3+}$ is in the $|\pm 1/2\rangle$ state, a rise in temperature reduces the population of this state which leads to an increase in $\tau_r$. This is because the spin-spin relaxation is slower since it takes place most efficiently with other iron atoms in the same state and the relaxation of both the $|\pm 3/2\rangle$ and $|\pm 5/2\rangle$ states are slower. The effect is revealed by broadening of the lines in the Mössbauer spectrum.

Consider the effect of gradually increasing $\tau_r$, by cooling or isomorphous diamagnetic dilution:, the compound will eventually enter a region where $\tau_r$ approaches $\tau_L$ and the lines of an initially sharp quadrupole split spectrum will broaden. This region may extend over a substantial range of temperatures. Now the Larmor time $\tau$ is shorter for the $\pm 3/2$ state of the excited $^{57m}Fe$ than for the $\pm 1/2$ state. Hence the $\pm 3/2 \Leftrightarrow \pm 1/2$ line will be affected first. This commonly leads to rather asymmetric quadrupole split spectra for $Fe^{3+}$ compounds (Fig. 6.6). It can serve to identify the $\pm 3/2 \Leftrightarrow \pm 1/2$ line and thus determine the sign of $\Delta$ . The lines in such spectra can be sharpened by subjecting the absorber to a magnetic field.

## 6.5  MAGNETICALLY ORDERED SYSTEMS

Solids are magnetically anisotropic. The axes suggested by the modes of ordering shown in Fig. 6.2 are not necessarily simply related to the crystal axes. Crystals will display one or more **easy axes** for magnetisation and the quantisation of magnetic energy states will refer to such an easy axis. The difference in energy of the magnetically ordered solid with the spins parallel and perpendicular to the easy axis is known as the **magnetic anisotropy**. It can be associated with a field $B_A$.

The transition from the paramagnetic to the magnetically ordered state usually takes place over a narrow range of temperatures. For a ferromagnetic compound the transition temperature is known as the **Curie temperature** $T_C$, and for an antiferromagnetic compound it is called the **Néel temperature** $T_N$.  Hysteresis effects are quite common.

The change to the ordered state and changes in the kind of ordering are often associated with small changes in the isomer shift and the quadrupole splitting.

Stress in the absorber, such as may be introduced by grinding the compound before preparation of the absorber, can alter the transition temperature and modify the ordering.

The ordering temperature, $T_N$ or $T_C$, for the absorber can often be determined most easily by measuring the change in counting rate with temperature at a source velocity corresponding to a maximum absorption in the magnetic spectrum.

### 6.5.1 Characteristics of the different ordered states.

Ferrimagnetism leads to two superimposed magnetically split spectra in the absence of an externally applied field. These arise because ferrimagnetism requires two lattices of iron atoms with different opposing $B_{hf}$. Fig.6.7 shows a spectrum from a ferrimagnetic substance.

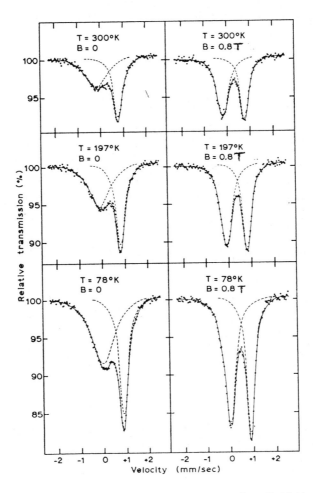

Spectra of polycrystalline $FeCl_3$ $6H_2O$ at different temperatures in applied fields of 0 and 0.8 T.

**Fig.6.6**

Although there are several ferromagnetic alloys ferromagnetic compounds are uncommon. They give six line spectra in the absence of any applied field. The $B_{hf}$ are randomly orientated and the line intensities for a powder absorber are in the ratio 3 : 2 : 1 :: 1 : 2 : 3.    Two other small terms enter into the expression for $B_{int}$ in these systems, a demagnetization term and a Lorentz field term. They are rather small and fortunately cancel each other if the absorber is composed of spherical particles.

In a ferromagnetic material the iron atoms are present in strongly aligned domains. On application of a growing external magnetic field, applied perpendicular to the γ beam axis, the domains at first align along the easy axis (for simplicity a single easy axis will be assumed) but as the field grows they align with the magnetic field. The line intensity pattern becomes 3 : 4 : 1 :: 1 : 4 : 3. and $B_{int} = B_{hf} + B_{ext}$. In this way the sign of $B_{hf}$ can be determined.

In an antiferromagnetic material the spins are aligned in relation to the easy axis. The application of an external magnetic field along the easy axis and parallel to the $\gamma$

Spectrum of
$Fe_2F_5.2H_2O$
at 20 K.

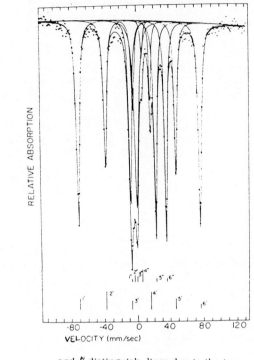

and ″ distinguish lines due to the two sublattices.

**Fig.6.7**

beam gives $B_{int} = B_{ext} \pm B_{hf}$. Lines 2 and 5 of the magnetic spectrum will not appear. (See Fig.6.8)

If another orientation of the crystal is used such that the angle between the easy antiferromagnetic axis and the direction of the applied field is $\theta$   $B_{int}$ will become $B_{ext} \cos\theta \pm B_{hf}$. Hence the direction of the easy axis can be determined. A canted antiferromagnet behaves in a similar manner.

A powder Mössbauer spectrum does not determine the easy axis directly. But it does enable one to determine the angle between the easy antiferromagnetic axis and the principle axis of the EFG.

Alternatively if one measures the line intensities as a function of the orientation of a single crystal absorber, in the absence of an external field, the ratios are given by

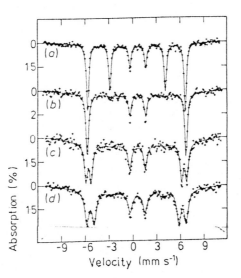

Spectra of $K_2FeF_5$ at 4,2 K

(a) Powder; (b) Single crystal

in absence of magnetic field.

(c) As in (b) but with 2 T field

parallel to the photon beam.

(d) As in (c) but with 3 T field.

**Fig.6.8**

$3 : 4\sin^2\theta\ /(1 + \cos^{2\theta}) : 1 :: 1 : 4\sin^2\theta\ /(1 + \cos^2\theta) : 3$. Where $\theta$ is the angle
between the antiferromagnetic axis and the $\gamma$ beam.

### 6.5.2  **Magnetic  field  induced  ordering**

The relaxation of the electronic spin is so fast in many $Fe^{2+}$ compounds that they
remain paramagnetic down to extremely low temperatures, even below 1 K . But the
application of a magnetic field will often lead to magnetic ordering at a much higher
temperature. Figs.6.9 a & b compare spectra for a single crystal of $RbFeCl_3$ with and
without an external magnetic field. The field was applied parallel to the chains of
$FeCl_6$ octahedra, which share opposite faces in the crystal (along the c axis of the
crystal) at 4.2 K.

The spectrum seen does not have the familiar six line pattern because the electric
quadrupolar interaction is so large that it cannot be treated as a small perturbation of
the magnetic splitting. It is closer to the reverse situation shown in Fig.1.10 c where
the magnetic interaction is considered as the perturbation.

The magnetic field splits the $|\pm 1>$ level shown in Fig.3.7 lowering the $|+1>$
component so that mixing of states takes place and ordering can occur.

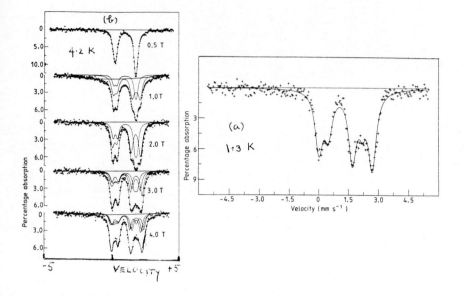

Spectra of a single crystal of $CsFeCl_3$     (a)   At 1.3 K with photon beam perpedicular to principalpal axis of EFG.    (b) In magnetic fields of various strengths, field applied perpendicular to principal axis of EFG'

### Fig.6.9 a & b

The ordering temperature can also be increased by another mechanism. It will increase with the magnitude of the product of $<S_1>$ and $<S_2>$. Now if the applied field reduces the fluctuations in spin, tending to alignment, these mean values will increase and $T_N$ will fall. A field perpendicular to the easy axis will have this effect. Fig. 6.10 shows how the ratio $T_N(B)/T_0(0)$ varies with the field for $K_2FeF_5$..

## 6.6   MAGNETIC DIMENSIONALITY

The magnetic properties of solids are generally anisotropic. The exchange integral J is direction dependent and one can envisage exchange coupling and ordering in 1,2 or 3 dimensions. The different situations are shown in Table 6.2.

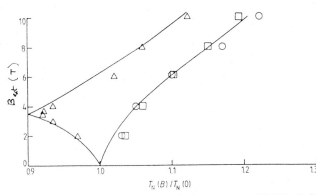

The observed variation of the Néel temperature of $K_2FeF_5$, $T_N(B)$, as a function of the applied magnetic field, $B$. The symbols $\bigcirc$, $\triangle$ and $\square$ represent the values obtained with the field parallel to the $a$, $b$ and $c$ axes respectively.

**Fig.6.10**

**Table 6.2**

| Magnetic dimensionality | Exchange J | Spin components | Type |
|---|---|---|---|
| 3 | $J_x \approx J_y \approx J_z$ | $S_x \approx S_y \approx S_z$ | Heisenberg |
| 2 | $J_x \cong J_y > J_z$ | $S_x \cong S_y > S_z$ | x - y |
| 1 | $J_x \cong J_y < J_z$ | $S_x \cong S_y < S_z$ | Ising |

This description oversimplifies the situation because it has been shown that long range ordering in one dimension cannot take place unless $J_x$ and $J_y$ are greater than zero, that is to say some interaction in the other dimensions takes place.

For a compound with axial symmetry at the iron atom one can express the anisotropy in the form $DS_z$, where $S_z$ is the component of spin along the symmetry axis. The Heisenberg type corresponds to a very small value of D, and the spins are weakly confined. The x - y type correspond to D > 0 with weak confinement of spins in the x - y plane and the Ising type corresponds to D < 0 with strong confinement of spins along the symmetry axis.

Magnetic dimensionality is clearly affected by the crystal structure and the nature and number of paths between iron atoms permitting the superexchange process. Not surprisingly the value of $B_{hf}^{\circ}$ falls as the dimensionality decreases for the same kind of ligand atoms.

Iron(III) fluoride and its complexes provide examples of these three kinds of magnetic ordering. $FeF_3$ is composed of a continuum of $FeF_6$ octahedra, each fluorine being shared between two iron atoms. This gives rise to a three dimensional antiferromagnet. $B_{hf}^{\circ}$ for the compound is 62.2 T.

$KFeF_4$ is composed of planes of $FeF_6$ octahedra sharing opposite edges and

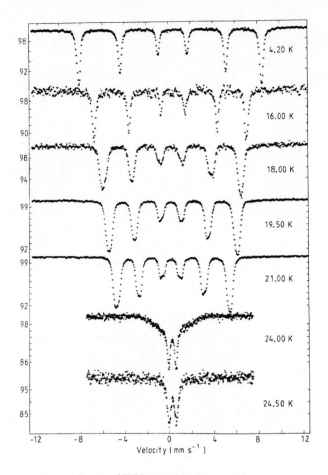

Spectra of FePO$_4$ at various temperatures.

**Fig.6.11  a**

behaves as an effectively two dimensional antiferromagnet. Its B$_{hf}$ value is 53.4 T.

In K$_2$FeF$_5$ there are chains of FeF$_6$ octahedra sharing opposite apices. The iron atoms in the chains are strongly antiferromagnetically coupled and there is a similar, but very much weaker interaction between chains -- effectively an uni-dimensional antiferromagnet. The B$_{hf}^{\circ}$ value for this compound is 41.0 T.

At temperatures just a little below T$_N$ a theoretical treatment suggests that

$$B_{hf}/B_{hf}^{\circ} = (1 - T/T_N)^{\beta}$$ with β close to 1/3 for three dimensional ordering and 1/8 for two dimensional ordered systems. This provides a usful means of establishing dimensionality.

## 6.7  MAGNETIC PHASE CHANGES

The magnetic ordering that takes place at and below T$_N$ or T$_C$ may change at still lower temperatures. Magnetic phase changes are quite common. Sometimes they are

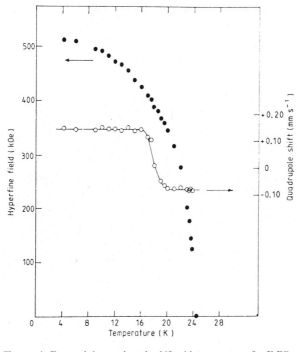

Changes in $B_{hf}$ and the quadrupole shift with temperature for $FePO_4$ .

**Fig.6.11   b**

accompanied by changes in the quadrupole interaction and, or, isomer shift.   Such changes may take place as sharp first order transitions, often displaying hysteresis, or by a continuous second order transition.

Some data for anhydrous iron(III) phosphate are shown in Fig. 6.11.a and b. $T_N$ for this compound lies just below 24 K and the compound remains antiferromagnetic down to 4.2 K. Around 16 K the lines in the spectrum become much sharper and the difference of the separation between lines 1 and 2 and lines 5 and 6 which measures $\varepsilon = e^{\sim}qQ(3\cos^2\theta - 1)/8$ change from $-\varepsilon/2$ above, to $\varepsilon$ below, 16 K. This clearly shows a change in quadrupole interaction but since $\theta$ is not available it cannot be expressed in terms of $\Delta$. (See Fig. 6.11.b)

Neutron diffraction data show that the spins are ordered ferromagnetically within the [001] plane but adjacent planes interact strongly antiferromagnetically. As the temperature is lowered spin canting leads to broader lines and at about 16 K the spins reorientate around the c axis. $B_{hf}$ changes smoothly throughout the temperature range 24 - 4.2 K. A related change, the Morin transition, will be described in section 7.1.

Changes in magnetic ordering can also be effected by the application of an external magnetic field, indeed at a sufficiently high field the spins of the sublattices both align with the field and their antiferromagnetic order breaks down. The material reverts to paramagnetic behaviour. As a magnetic field, applied parallel to the easy antiferromagnetic axis, is increased at first the spins of the two sublattices are aligned in relation to this axis, but eventually a reorientation of spins takes place --- **Spin-flop.**  The effect is associated with the magnetic anisotropy of the material.

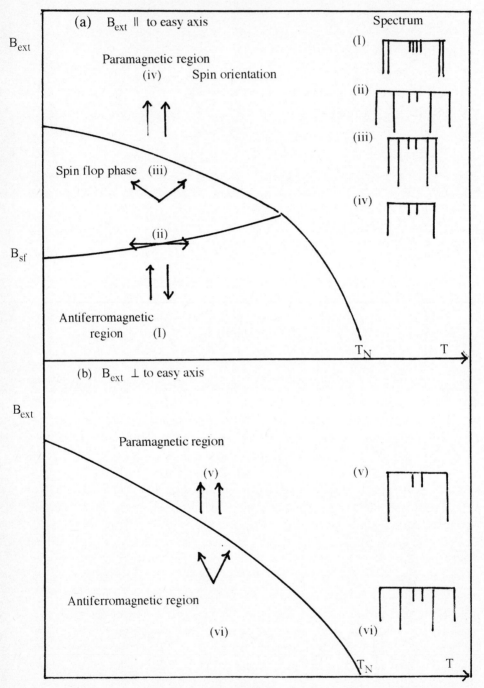

Spin flop behaviour of an uniaxial antiferromagnet.  Magnetic field parallel to photon beam

**Fig.6.12**

In the antiferromagnetic region, as shown in Fig.6.12, the field enhances <S> for one sublattice and reduces it for the other, antiparallel, sublattice. The magnetic energy is given by $-1/2 \; \chi_{||} \; B^2$ where $\chi_{||}$ is the magnetic susceptibility of the material parallel to the easy axis. Spin flop leads to a reorientation of the spins perpendicular to the easy axis. The magnetic energy now becomes $K - 1/2 \; \chi_{\perp} \; B^2$, $\chi_{\perp}$ being the susceptibility perpendicular to the easy axis. Now below $T_N$ $\chi_{\perp}$ becomes independent of temperature, but $\chi_{||}$, which has the same value as $\chi_{\perp}$ at $T_N$, decreases with temperature to reach 0 at 0 K and for temperatures below $T_N$ will be less than $\chi_{\perp}$. Hence there will be a transition field, $B_{sf}$, at which it is energetically favourable for the spin change to take place.

On further increasing the field spin canting towards the direction of the field grows steadily, and eventually the spins of the two sublattices become aligned with the field; the transition to the paramagnetic state has taken place.

The spin flop field, $B_{sf}$, is related to the exchange and anisotropy fields by the expression:- $B_{sf} = [2(B_E + B_A)B_A]^{1/2}$ or approximately $[2B_A B_E]^{1/2}$.

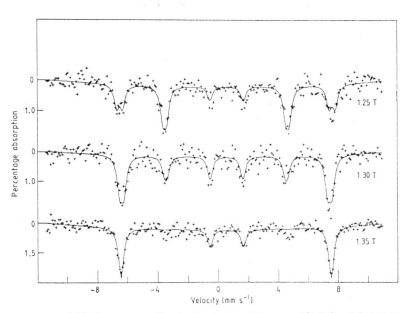

Mössbauer spectra of an *ab*-plane single-crystal sample of $Cs_2FeCl_5 . H_2O$ at 4.2 K with magnetic fields applied parallel to the *a* axis at values near the spin-flop field. In this experimental arrangement the $Fe^{3+}$ spins flop into the γ-ray beam direction and the phase change is most clearly observed through the intensity of the $\Delta m = 0$ lines of the spectra.

## Fig.6.13

The spin flop is clearly revealed in the Mössbauer spectrum, as can be seen in Fig. 6.13, which shows results for a single crystal of $Cs_2FeCl_5 \; H_2O$. With the applied field parallel to the easy axis of the single crystal absorber and normal to the γ beam, below $T_N$ the spectrum has the 3 : 4 : 1 intensity pattern. As the field increases

Bext will add to the $B_{hf}$ arising in one sublattice and subtract from the other so that the lines become doublets.

But at the spin flop field, about 1.3 T, the spins of the sublattices both become normal to the applied field and parallel to the $\gamma$ beam, so that a 3 : 0 : 1 intensity pattern develops; the $\Delta m = 0$ lines (2 and 5) disappear. Further increase in the field leads to spin canting towards the direction of the field and gradual re-emergence of these lines ending eventually in transition to a paramagnetic phase.

The analogous spin flop behaviour in $K_2FeF_5$ can be seen in spectra (e) to (g) in Fig.6.8.

## Acknowledgements

Fig.6.3   Reproduced with permission from: van der Woude, F. (1966)
*Phys.Stat.Solidi*, **17**, 417.

Fig.6.4   Reproduced with permission from: Wickman, H.H. (1966) in Mössbauer
Effect Methodology, **2**, 39. Ed. Gruverman, I.J. Pub. Plenum Press.

Fig.6.5   Reproduced with permission from: Johnson, C.E., Cranshaw,T.E. and
Ridout, M.S., (1964) *Proc.Intern.Conf.Magnetism*, Nottingham.

Fig,6.6   Reproduced with permission from: Thrane, M.F. and Trumpy, G. (1970)
*Phys.Rev.*, **B, 1**,  163.

Fig.6.7   Reproduced with permission from: Brown, D.B., Wong, H. and
Reiff, W. M.(1977) *Inorg.Chem.*, **16**, 2425.

Fig.6.8   Reproduced with permission from: Gupta, G.P., Dickinson, P.E.D. and
Thomas, M.F. (1978) *J.Phys.C*, **11**, 215.

Fig.6.9   Reproduced with permission from: Baines, J.A., Johnson,C.E. and
Thomas, M.F. (1983) *J.Phys. C.*, **16**, 3579.

Fig.6.10 Reproduced with permission from: Boerrsma, F., Cooper,D.M.,
de Junge,W.J.M., Dickinson, P.E.D., Johnson, C.E. and Tinus, A.M.C.,
(1982) *J.Phys, C*, **15**, 4141.

Fig.6.11 Reproduced with permission from: Battle, P.D., Chetman, A.K.,
Gleitzer, C., Harrison, W.T.A., Long, G.J. and Longworth, G. (1984)
*J.Phys. C*, **15**, L919.

Fig.6.13 Reproduced with permission from: Johnson, C.E. and Thomas, M.F.
(1987) *J.Phys. C*, **30**, 91.

# 7

# Further Features of Iron Spectra

## 7.1 OXIDES AND HYDROXIDES

Mössbauer spectroscopy is a very powerful technique in the investigation of the properties, especially the magnetic characteristics, of the iron oxides, hydroxides and mixed oxides. It also provides a means of identification of these compounds in soils and superficial deposits. The most satisfactory results are obtained when it is used in conjunction with other techniques, such as X-ray diffraction, magnetic susceptibility measurements etc. It is generally desirable to record spectra over a wide range of temperatures and to subject the data obtained to a detailed computer analysis and comparison with calculated spectra.

All these compounds are liable to substitutional impurities and may be present as small crystallites, both factors affect their magnetic properties and their Mössbauer spectra.

### 7.1.1 $\alpha$ Fe$_2$O$_3$, Haematite

Haematite, or $\alpha$ Fe$_2$O$_3$, approximates to an hexagonal close packed array of oxygen atoms with two thirds of the octahedral holes occupied by iron atoms. But there is some trigonal distortion of the octahedra. Magnetic ordering begins at a very high temperature, $T_N = 965$ K. Above 260 K the compound is weakly ferromagnetic, below this temperature it becomes antiferromagnetic.

Considering first the low temperature state, an absorber composed of single crystals mounted so that their trigonal 111 axes are parallel to the photon beam gives a spectrum in which the $\Delta m_I = 0$ lines are missing Fig.7.1.a. The 3:0:1::1:0:3 type spectrum shows that the A and B sublattices spins are collinear and opposed to each other giving the antiferromagnetic behaviour. Their moments are parallel to the 111 axis. If an external magnetic field is applied along the 111 axis, the outer lines from the A and B sublattices become separated by 2B$_{ext}$. and the $\Delta m = 0$ lines begin to reappear. In this temperature region $\varepsilon = -0.22$.

If the absorber is allowed to warm to above 260 K the moments rotate through 90⁻, becoming parallel to the [111] plane, so that an approximately 3:4:1:1:4::3 spectrum develops, Fig.7.1.b. There is also some spin canting above 260 K, so that the Fe$_2$O$_3$ develops weakly ferromagnetic properties. A similar change can be induced by applying a stronger magnetic field, Fig.7.1.c. The transition is one of spin-flop (See Section 6.7) The separate A and B sextet lines produced at weaker fields coalesce and a single sextet with narrower lines appears. The thermally induced spin-flop transition at $T_M$ is called the **Morin transition**. At $T_M$ $\varepsilon$ changes more or less abruptly to +o.12. The isomer shift at 298 K is 0.38, any discontinuity in δ at 260 K is small. There is a small kink in the plot of $B_{hf}./B\overset{o}{hf}$ .against $T/T_N$ as can be seen in Fig.6.3. The change can also be seen

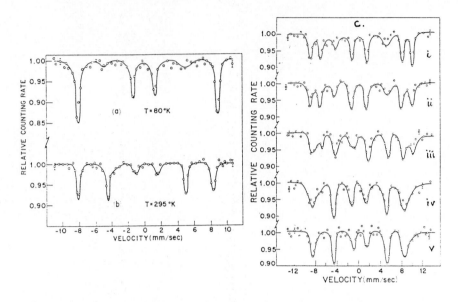

Spectrum of crystal of $Fe_2O_3$ with [III] crystal axis parallel to photon beam; (a) below (80 K) and (b) above (295 K) the Morin transition temperature. (c) effect of increasing (i → v ) external field parallel to photon beam. Shows spin-flop above (iii).

## Fig.7.1

clearly in the powder spectra, taken above and below 260 K, shown in Fig.7.2. Note the difference in the sign of the difference between the separations of lines 1 and 2 and lines 5 and 6 in the two spectra.

The Morin transition temperature is sensitive to impurities and imperfections in the crystals of the absorber, so the transition often extends over a range of temperature. This is commonly the case with mineral samples because haematite is isomorphous with corundum and some replacement of iron by aluminium commonly occurs.

### 7.1.2 γ $Fe_2O_3$, Maghemite

γ $Fe_2O_3$ can be regarded as an inverse spinel based on a cubic close packed array of oxygen atoms. As with haematite the octahedral B sites are trigonally distorted. One ninth of the cation sites are vacant. As shown in Table 6.1, the iron in the tetrahedral A sites and the octahedral B sites have different $B_{hf}$. The A and B sublattices align in opposite directions and this leads to a ferrimagnetic material. The Curie temperature is high but cannot be established accurately because maghemite transforms to haematite above about 600 K.

The moments of the iron atoms in the A and B sublattices are not very different and the room temperature Mössbauer spectrum is a six line spectrum, but the lines are somewhat broad. If a small magnetic field is applied to the absorber the lines broaden

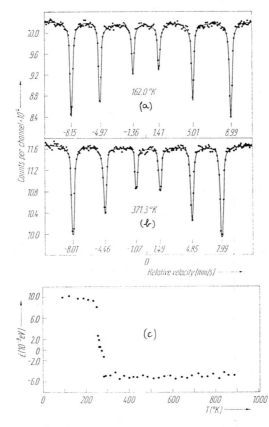

Spectrum of $Fe_2O_3$ below (162 K) and above (371.3 K) the Morin transition
($T_M \approx 260$ K) and Quadrupole interaction, $\varepsilon$, as a function of temperature.

### Fig.7.2 a & b

still more, since the field adds to one lattice and subtracts from the other. A computer analysis of the spectrum then yields $\overset{\circ}{B}_{hf}$ (A sites) = 48.8 T and $\overset{\circ}{B}_{hf}$ (B sites) = 49.9 T at room temperature. The isomer shifts are 0.27 for the tetrahedral A sites and 0.41 for the octahedral B sites. The intensities of the two six line spectra suggest that the cation vacancies are largely in the B sublattice.

### 7.1.3 Wustite

The non-stoichiometric oxide $Fe_{1-x}O$, where x is less 0.15, has a face centred cubic structure with enough $Fe^{2+}$ ions to effect charge balance. It is paramagnetic at room temperature, but the Néel temperature is about 200 K, dependent on x. The room temperature spectrum can be analysed in terms of two quadrupole split doublets, but with rather broad lines; the parameters obtained are $\delta = 0.95$ with $\Delta = 0.44$ and $\delta = 0.90$ with $\Delta = 0.79$ at 295 K. Some valence delocalisation may occur. At 500 K the nonstoichiometric phase becomes unstable and a broad single line spectrum is found, $\delta = 1.08$, suggesting $Fe^{2+}$ in a cubic environment. The spectrum becomes more complex the larger the value of x. At 4.2 K wustite gives two six line spectra which yield $B_{hf}^{\circ}. = 38.6$ T for the $Fe^{2+}$ and 51 T for the $Fe^{3+}$.

Mössbauer spectroscopy does not have a very high concentration sensitivity and small amounts of lattice defects seldom produce easily identifiable features in the spectrum.

### 7.1.4 Superparamagnetic behaviour

This is a convenient point to introduce superpara- magnetic behaviour since it has an important effect on the Mössbauer spectra of the iron oxides and hydroxides, especially in naturally occurring materials. Since the magnetic ordering of a lattice or a sublattice is a cooperative effect, it is to be expected that as the crystallites in the absorber become smaller a change in magnetic properties will take place.

For crystallites of volume V the magnetic anisotropy energy will amount to KV. The value of K may be such that KV is about kT for 300 K for roughly spherical particles of diameter 20 nm. A departure of the spin direction by an angle θ from the easy axis of magnetisation requires an energy of $KV\sin^2\theta$. When kT/KV is small thermal fluctuations of the spin direction in relation to the easy axis will take place and the average $B_{hf}.$ over the precession time $\tau_L$ will be $B_{hf}^{max.}$ $<\cos\theta> \approx B_{hf}^{max.}$ .(1- kT/2KV). It is these fluctuations that account for the temperature dependence of $B_{int}$ following the Brillouin curve. The mean life before θ changes is short compared to $\tau_L$. A magnetic spectrum will be found.

On the other hand if kT/KV is large, typically ≥ 0.2, θ may reach 180⁻, that is, complete spin reversal. The average value of $B_{hf}.$ over $\tau_L$ becomes zero. The frequency of spin re-orientation is still fast compared to $\tau_L$. The substance now behaves as a normal paramagnet. Hence with an absorber composed of very small crystallites of uniform size, at low temperatures a magnetic spectrum will be obtained. But on measuring the spectrum at increasing temperatures, above some temperature $T_B$, lower than the bulk $T_C$ or $T_N$, a paramagnetic spectrum, like that found for the material above the ordering temperature, will be seen. This is known as **superparamagnetism** and $T_B$ is called the **blocking temperature**, Fig. 7.3.

In practice absorbers are seldom of uniform particle size and the blocking temperature is not sharply defined. However if the particle size distribution is not too broad $T_B$ gives an estimate of the mean particle size.     In Fig.7.3 the full line shows the behaviour of a powder of uniform particle size, the dotted line a powder with a narrow range of particle sizes. It is very important to take superparamagnetic effects into account when exploring dispersed iron oxides and hydroxides in soils and sediments.

The iron atoms in the surface of crystals will experience different magnetic coupling from those in the interior. Hence if the particles in the absorber have a very high specific surface its magnetic behaviour will be modified and its Mössbauer spectrum changed. The spin alignment of the surface and interior iron atoms are different. This is called a **speromagnetic** effect. Compression of such materials may also lead to changes in the spectrum because of magnetic particle - particle interactions.

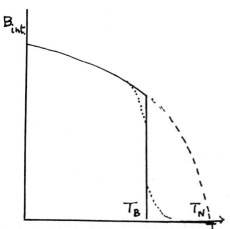

Hyperfine field as a function of temperature. Full line --- powder of uniform particle size; Dotted line --- small range of particle sizes; Dashed line large particles.

**Fig.7.3**

### 7.1.5  Hydroxides

The Mossbauer spectra of the iron hydroxides are complicated both by the effects of impurities in naturally occurring materials and by the effects of superparamagnetism in dispersed or finely powdered samples. Nevertheless the spectra have proved of considerable use in establishing the magnetic properties of these substances and detecting their presence in soils and sediments.

7.1.5.1 *Goethite*. $\alpha FeOOH$ is rhombohedral space group Pbnm. It is composed of double rows of distorted octahedra formed by four oxygen atoms and two hydroxide groups around each iron atom. Sharing at apices and edges leads to every oxygen atom interacting with three iron atoms. Successive layers connect by sharing apices and by hydrogen bonding. Both Fe - Fe and Fe - O - Fe magnetic coupling are possible, but the antiferromagnetic properties show the latter is the more important.

Below $T_N$ the spectrum is a single sextet, but unless the sample is very pure the lines may be asymmetric and wide. The easy axis of magnetisation lies parallel to the c axis of the crystals. The difference in the signs of $\Delta$ above and $\varepsilon$ below $T_N$ and the ratio of their magnitudes suggests that the principal axis of the EFG lies in the ab plane. $B_{hf}$ as a function of temperature follows the Brillouin curve for $S = 5/2$.

The conversion to $\alpha Fe_2O_3$ by heating can be followed by Mössbauer spectroscopy.

The anisotropy constant for Goethite $K = 10^3 J\ m^{-3}$.

7.1.5.2 *Lepidocrocite*.. In $\gamma$ FeOOH the FeO/OH octahedra form layers that are held together by hydrogen bonding. The chains of octahedra are parallel to [001]. It is the only member of this group of compounds that is paramagnetic at room temperature in the pure state. However the other hydroxides often give superparamagnetic spectra at room temperature because of small particle size. Magnetic interaction between the layers is weak and Bhf. is low compared to the related compounds. Below 73 K the compound is antiferromagnetic with the spins aligned parallel to the [001] plane.

7.5.1.3  *Akaganeite*. $\beta$ FeOOH has a very open structure with channels that can accommodate ions such as $F^-$ or $Cl^-$, or water molecules. Indeed the structure may not be stable in the absence of occupants of the channels. Natural occurring samples of akaganeite usually contain $F^-$ or $Cl^=$, and synthetic products usually have excess water

At room temperature the Mössbauer spectrum is usually dominated by two quadrupole split pairs, with broad lines, superimposed on a broad absorption with incipient magnetic splitting. At lower temperatures the magnetic component begins to dominate and its lines sharpen. A best fit at 4.2 K is obtained for three sextets. (See Table 7.1)

A summary of reported iron hydroxide Mössbauer parameters for the pure bulk materials (except for $\beta$ FeOOH) is given in Table 7.1

### Table 7.1.

| Compound | T/K | $T_{N \text{ or } C}$ | $\delta$ | $\Delta$ or $\varepsilon$ * | $B_{hf}$ | Magnetism |
|---|---|---|---|---|---|---|
| 1. $\alpha$ FeOOH,Goethite | 420 | 403 | 0.35 | 0.6 | --- | P |
| "    " | 295 | 403 | 0.37 | -0.3 | 38.2 | A |
| "    " | 4.2 | 403 | 0.48 | -0.25 | 50.6 | A |
| 2. $\beta$ FeOOH,Lepidocrocite | 295 | 73 | 0.37 | 0.55 | --- | P |
| "    " | 4.2 | 73 | 0.49 | 0.02 | 45.7 | A |
| "    " | 4.2 | 73 | 0.51 | 0.06 | 44.0 | A |
| 3. $\gamma$ FeOOH,Akaganeite | 295 | 300# | 0.38 | 0.55 | --- | P |
| "    " | 295 | 300# | 0.37 | 0.95 | --- | P |
| "    " | 4.2 | 300# | 0.50 | -0.02 | 48.9 | A |
| "    " | 4.2 | 300# | 0.48 | -0.24 | 47.8 | A |
| "    " | 4.2 | 300# | 0.49 | -0.81 | 47.3 | A |
| 4. $\delta$ FeOOH Ferroxyhite | 4.2 | 370# | 0.45 | 0.12 | 53.0 | F |
| "    " | 4.2 | 370# | 0.48 | 0.06 | 50.8 | F |
| 5. $\delta$' FeOOH | 295 | 570 | 0.37 | -0.13 | 47.2 | A |
| " | 10 | 570 | 0.49 | -0.12 | 52.3 | A |

\# = TN or Tc poorly defined. * $\Delta$ for paramagnets, $\varepsilon$ for magnetically ordered materials. A - Antiferromagnetic P - Paramagnetic F - Ferrimagnetic.

7.1.5.4 *Ferroxyhite*. $\delta$ and $\delta$' FeOOH probably only differ in the disorder of the occupation of the octahedra by $Fe^{3+}$. Ferroxyhite is the only ferrimagnetic hydroxide. It does not occur naturally and macrocrystalline material is hard to prepare, most samples being superparamagnetic. At 4.2 the spectrum can be interpreted in terms of two sextets arising from the two sublattices. The isomer shifts suggest that the iron has the same coordination number in both sublattices. A one tesla magnetic field applied perpendicular to the $\gamma$ ray beam leads to a 3:4:1:1:4:3 type spectrum, due to the ferrimagnetic character of ferroxyhite.

$\varepsilon$ FeOOH is only formed at high pressures.

Representative spectra of these compounds are shown in Fig.7.4.

### 7.1.6 **Effect of substitutional aluminium.**

Aluminium can replace the iron in both the oxides and the hydroxides over a wide range of atom fractions. The introduction of the diamagnetic aluminium changes the lattice dimensions and distortion of the octahedral iron environment. In addition it reduces the extent of magnetic coupling between iron atoms. Both these effects lead to changes in the Mössbauer spectra. Aluminium containing haematite and goethite have been studied extensively.

To identify the effect of the aluminium care must be taken to use materials of approximately the same size distribution to avoid differing superparamagnetic effects. This requirement is less important if only spectra measured at very low temperatures are to be compared.

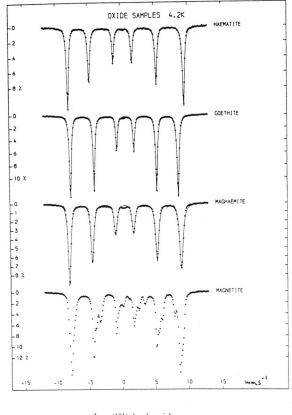

Iron(III) hydroxides

**Fig.7.4**

Both $T_N$ and $T_M$ decrease as the atom percentage of aluminium, $x = [100 \, Al \, /(Al + Fe)]$, increases. Magnetic relaxation becomes more pronounced and line widths increase as $x$ increases. The environment of the iron atoms varies in the aluminium substituted compounds. Fig.7.5 shows the profound effect of aluminium on goethite spectra at room temperature.

At 4.2 K normal sextets are obtained but the line widths increase with $x$.

The electron density at the nuclei of the iron atoms is not appreciably affected by the aluminium so that the isomer shift remains nearly constant. The value of $\varepsilon$ however changes because of the change in lattice dimensions and the iron environment. For haematite spectra at 77 K $\varepsilon$ increases with $x$ up to about 3%, but further increase in $x$ leads to a decrease.

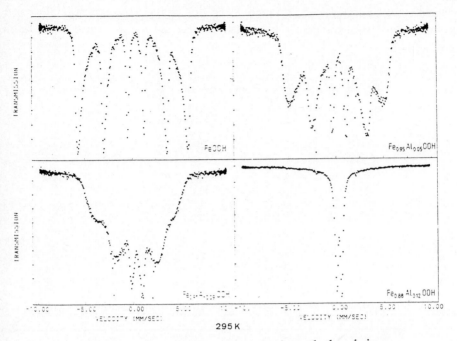

295 K

Spectra of Goethites with substituted aluminium.

**Fig.7.5**

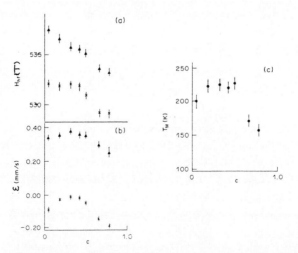

(a) Saturation hyperfine field; (b) Quadrupole coupling and (c) Morin
temperature as functions of the atomic fraction of substituted aluminium
in Haematite.

**Fig.7.6**

Substitution of iron by aluminium increases the Mössbauer fraction, f. The lattice shrinks and the phonon spectrum and the Debye temperature change. For low values the change is roughly linear with x. These changes for aluminium substituted haematite are shown in Fig. 7.6.

As expected the field at the nucleus, $B_{int}$, decreases as x increases. For measurements on $(Al/Fe)_2O_3$ at 77 or 4.2 K the field is linearly related to x for x ≤ 20. Since $B_{int}$ can be measured rather accurately this provides a useful method of determining x in the range 1 to 10. At 4.2 K $B_{int}$. = $53.60 - 0.42x$. for samples of constant particle size of the mixed oxide.

## 7.2 SPINELS AND MIXED OXIDES

There is a very extensive literature on the Mössbauer spectroscopy of these compounds. The technique supplements crystallographic studies and provides information on the often complicated magnetic properties of the mixed oxides. In addition the industrial importance of the ferrites stimulates interest.

### 7.2.1 Spinels

The spinels are a numerous group of mixed oxides of general formula $AB_2O_4$. They are based on a face centred, cubic close packed, oxide lattice with the B occupying octahedral holes and the A tetrahedral sites. The unit cell contains 32 oxide ions. The cations occupy one eighth of the tetrahedral A sites and one half of the octahedral B sites. Thus a quarter of the interstitial holes are occupied.

In normal spinels the divalent species occupy the A sites and the trivalent species the B sites. A metal atom on an A site will be denoted by (M) and on a B site by [M]. Thus a normal spinel can be represented by $(M)[M_2]O_4$. In the inverse spinels one has $(M')[MM']O_4$. These represent limiting cases, and in many spinels the cation distribution gives $(M_{1-\lambda}M'_\lambda)[M_\lambda M''_{2-\lambda}]O_4$. The $\lambda$ measures the degree of inversion of the spinel. A further complication arises because the spinels are not always stoichiometric.

Actual spinels may be rather different from this idealised structure: many are tetragonal or orthorhombic at low temperatures, but suffer a phase change to a cubic form above some transition temperature. The tetrahedral A sites then provide cubic symmetry. The octahedral B sites are always trigonally distorted.
An important feature is that spinels may be magnetically ordered at room temperature since their Neel or Curie temperatures are often rather high.

The value of Mössbauer spectroscopy in the study of the spinels is best illustrated by enumerating some of the problems presented by normal and inverse spinels and exploring how far Mössbauer spectroscopy can solve them.

#### 7.2.1.1 *Cation distribution in normal spinels.*

In the simplest case the spectrum will identify a normal spinel. Figure 5.1 shows the isomer shifts to be expected for $Fe^{2+}$ and $Fe^{3+}$ in the A and B sites. A single spectrum will be obtained for either $(Fe)[M_2]O_4$ or $(M)[Fe_2]O_4$. The $Fe^{2+}$ in a tetrahedral site should suffer a Jahn Teller distortion, and display a substantial quadrupole splitting. This is sometimes observed, but in other cases a single line is found. This may be due to the lattice rigidity imposing cubic symmetry at the A sites or it may be due a rapid dynamic Jahn Teller effect leading to an average of cubic symmetry. The spectrum readily identifies the oxidation state and environment of the iron atoms in these cases.

Three kinds of normal iron spinels are found. In $(Ge)[Fe^{2+}]O_4$ the iron in the trigonally distorted B site shows a large quadrupole splitting, $\Delta = -2.9$. The $d_{z^2}$ orbital lies lowest giving a $^5A_{2g}$ state. The quadrupole splitting falls at higher temperatures as the $^5E_g$ state, which lies about $1140$ cm$^{-1}$ higher, begins to be populated. The isomer shift is $1.2$.

Iron(III) occupies the B sites in $(Zn)[Fe_2]O_4$ and the analogous cadmium compound. The quadrupole splittings at 295 K are $-0.34$ and $0.28$ for the zinc and cadmium compounds respectively. Both spinels have isomer shifts of $0.36$, and give sharp line spectra.

Iron(II) occupies the A sites in $(Fe)[Cr_2]O_4$ and in the analogous compound with vanadium instead of chromium. The chromium spinel changes from cubic to tetragonal below 135 K and the vanadium compound below 140 K. At room temperature both compounds give a single line spectrum with an isomer shift of $0.9$. In the tetragonal phase a quadrupole split spectrum is obtained but the splitting begins substantially above the transition temperature. This may indicate that a dynamic Jahn-Teller effect is responsible for the single line spectrum, and that this slows down above the transition temperature.

Except in the above mentioned temperature region these normal spinels give sharp line spectra with normal line widths.

### 7.2.1.2 *Magnetic spectra in normal spinels.*

Normal spinels generally show magnetic ordering at rather low temperatures. The exchange interactions between the iron atoms in the A or B sites, A-A or B-B, prove to be much weaker than the A-B interactions. By contrast the partly inverse spinel $MgFe_2O_4$, where A-B interactions can occur, is ferrimagnetic at room temperature. $(Zn)[Fe_2]O_4$ and $(Ge)[Fe_2]O_4$ remain paramagnetic down to below 20 K. If both cationic components are paramagnetic A-B exchange interactions arise, thus $(Fe)[V_2]O_4$ becomes ferrimagnetic at 109 K.

### 7.2.1.3 *Inverse spinels*

Inverse spinels give more complex spectra and new features must be cosidered. The inverse spinel $(Fe)[NiFe]O_4$ is ferrimagnetic at room temperature and the fields at the nuclei of the iron atoms on the A and B sites sufficiently different, so that the spectrum at 4.2 K is composed of two sextets. (Fig.7.7.a)

The outer sextet is shown by the isomer shift to arise from the B sites. Ferrimagnetic properties could arise in various ways, the simplest being for the moments on the two sublattices to be collinear and in opposition, the Néel model. This interpretation is

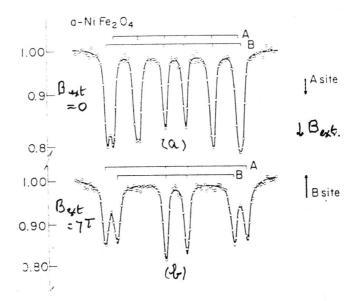

**Fig.7.7 a & b**

confirmed on applying an external magnetic field of 7 T parallel to the photon beam. The spectrum shown in Fig.7.7.b is obtained. The $\Delta m_I = 0$ lines disappear completely showing the the ordering is collinear without canting. The outer lines of the sextets are now well separated, the external field adding to the A field and subtracting from the B.

The $B_{hf}^{\circ}$. for the A and B sites are -50.4 and -54.7 T respectively. The areas under the outermost lines of the two sextets in Fig.7.7.b. are practically equal, there being equal numbers of iron(III) ions on the A and B sites.

The line widths in these spectra are bigger than usual. This may be due to disordered occupation of the B sites which will lead to the A site cations having varying numbers of next nearest neighbour nickel and iron ions.

The separation of some of the lines of the two sextets by an external parallel magnetic field also permits the proportions of iron on the A and B sites to be estimated, from the areas under the lines, in partially inverse spinels. A value for $\lambda$ can be obtained.

The three cation spinels enable one to explore several further interactions.

### 7.2.1.4. *Next nearest neighbour effects*

(Fe)[$Cr_2$]$O_4$ gives a sharp single line above about 150 K. Substantial substitution of chromium by aluminium can be made preserving the normal spinel structure. Such Fe[$Cr_{1-x}Al_3$]$O_4$ show quadrupole splitting with substantial line widths; the lines can be computer fitted by three quadrupole split pairs, Fig.7.8.a and b.

The splitting arises from oxide ions with differing environments of next nearest neighbour cations; 3Cr; 2Cr,1Al; 1Cr,2Al and 3Al. One can regard these different oxide ions as different ligands and associate with each a partial quadrupole splitting. The proportions of each of the rather numerous $FeO_p'O_q''O_r'''O_s''''$   (p+q+r+s = 4) can be

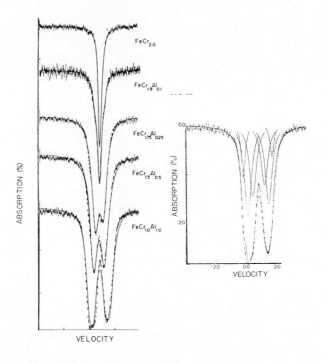

ABSORPTION (%)

VELOCITY

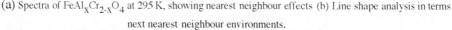

(a) Spectra of $FeAl_xCr_{2-x}O_4$ at 295 K, showing nearest neighbour effects (b) Line shape analysis in terms next nearest neighbour environments.

### Fig.7.8 a & b

calculated for a given value of x assuming random distribution of Al and Cr on the B sites. The data then enable one to estimate the p.q.s. values. Allowing for the limited resolution, only three quadrupole splittings are likely to be identified. These, together with their relative proportions, agree well with the experimental data, as can be seen in Fig.7.8.b.

The ferrimagnetic spinel $(Co_xFe_{1-x})[Co_{1-x}Fe_{1+x}]O_4$ is often substantially inverse. Line 1 of the B sextet is very broad. The B site iron atoms will have various proportions of next nearest neighbour cobalt and iron A site atoms. Since x is small the commonest environments will be 6Fe; 5Fe,1Co; 4Fe,2Co and 3Fe,3Co, denoted by $B_1$, $B_2$, $B_3$ and $B_4$ sites. Knowing x and assuming random occupation of the A sites, the proportion of each type of B site can be calculated. It is convenient to measure the spectrum in an external magnetic field applied parallel to the photon beam to separate the A and B sextets as much as possible. The first line in the B sextet is then analysed in terms of four Lorentzians with their areas constrained to the values calculated for the x found for the sample under study.

Fig.7.9 shows the line, with a 1.7 T applied field, on a sample for which x = 0.24. The $B_{hf}^{\circ}$. obtained are $B_1$ 51.5; $B_2$ 49.9; $B_3$ 47.5 and $B_4$ 44.5 T.

The replacement of an iron atom by cobalt as a next nearest neighbours reduces $B_{hf}^{\circ}$.

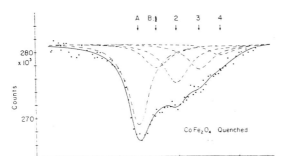

Analysis of line 1 of spectrum of quenched $CoFe_2O_4$ in terms of next nearest neighbour effects. 1.7 T field applied to enhance separation of A and B sub-lattice lines. Iron on tetrahedral A sites not much affected

**Fig.7.9**

by about 0.8 T. This difference must reflect the difference in the superexchange interactions Co(A) - O - Fe(B) and Fe(A) - O - Fe(B). These interactions can also be investigated by introducing diamagnetic cations.

### 7.2.1.5 *Thermal equilibria in spinels*.

The energy difference between the inverse and normal structures of a spinel are not very large so that, on heating to a high temperature, a partly inverse configuration will develop. If the material is cooled very quickly the high temperature distribution of cations may be, at least partly frozen in. This will be revealed in the Mössbauer spectrum especially in an external magnetic field separating the A and B sextets. Such an effect is readily seen on heating $CoFe_2O_4$ to 1300 K for some time and then cooling rapidly to room temperature by quenching in water. For this reason the spectra of spinels are dependent on the thermal history of the sample. Within some temperature range it is possible to explore the kinetics of the slow return to the the equilibrium structure.

### 7.2.1.6 *Spin canting*.

A ferrimagnetic spinel, with collinear opposed moments in the two sublattices, in an external magnetic field parallel to the photon beam, will not show any $\Delta m_I = 0$ lines (lines 2 and 5 in the sextets). Zinc substituted in a spinel invariably enters the A sites and a series of spinels of the general formula $Co_{1-x}Zn_xFeO_4$ can be made. The spectrum for a sample with x = 0.4 is shown in Fig.7.10(b) in a field of 5.9 T.

The degree of inversion is readily obtained from the areas of the 1 or 6 lines of the A and B sextets. The spectrum should be compared with that of the cobalt spinel shown in Fig.7.10(a) when it will be seen that the proportion of iron on the A sites, giving the outer line has decreased.

The spectrum of a sample with x = 0.6 is shown in Fig.7.10(c). The $\Delta m_I = 0$ can now be seen, and for x > 0.6 they become prominent. These spectra indicate that spin canting is occurring when x > 0.5, the individual iron atom moments on the B sites, are

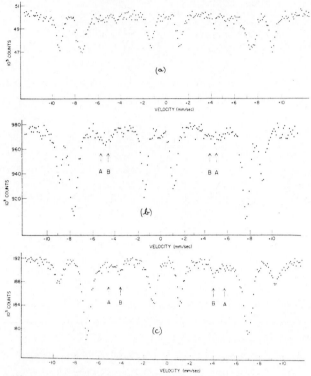

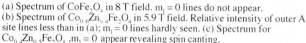

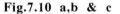

(a) Spectrum of $CoFe_2O_4$ in 8 T field. $m_1 = 0$ lines do not appear.
(b) Spectrum of $Co_{0.6}Zn_{0.4}Fe_2O_4$ in 5.9 T field. Relative intensity of outer A
site lines less than in (a); $m_1 = 0$ lines hardly seen. (c) Spectrum for
$Co_{0.4}Zn_{0.5}Fe_2O_4$ , $m_1 = 0$ appear revealing spin canting.

## Fig.7.10 a,b & c

at an angle to the applied magnetic field and the direction of the sublattice magnetisation.
This effect arises as the iron atoms on the B sites acquire increasing numbers of zinc next
nearest neighbours on A sites. If there were a single value to this angle one could write
for line 1 of the B spectrum:- $B = [ B_{ext.} + B_{hf.} + 2 B_{ext.}B_{hf.}\cos\theta ]^{1/2}$ or in terms of
the areas under the lines: $\theta = \sin^{-1}[3A_{2,5}/2A_{1,6}(1 + 3A_{2,5}/4A_{1,6})]$, the subscripts
referring to lines in the sextets.

However examination of line 2 in the B sextet suggests a number of B values
(Fig.7.11)

A more elaborate calculation can be made and the lines analysed in terms of different
fields and values of $\theta$ for different next nearest neighbour environments.

These studies of spinels are often complicated by superparamagnetic behaviour when
powders are used.

Although the magnetically split spectra yield isomer shifts they only give
$\varepsilon = \Delta_0(3\cos^2\theta - 1)/4$ and in general $\theta$ is not known, except when single crystal
absorbers are used. Very often $\varepsilon \approx 0$.

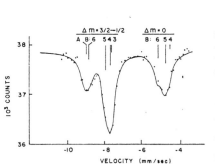

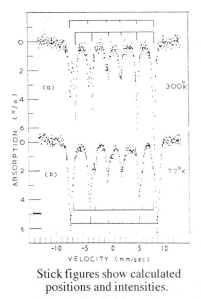

Analysis of lines 1 & 2 in spectrum of $Co_{0.2}Zn_{0.8}Fe_2O_4$ in 5 T field, in terms of the fields due to 3,4,5 & 6 zinc neighbours on A sites.

**Fig.7.11**

Spinels usually permit measurements of their spectra at high temperature, because of their large Mössbauer fractions, and above $T_N$ or $T_C$ a direct determination of $\Delta$ becomes possible.

Another possibility is to subject the absorber to a strong radiofrequency field, above 100 Mhz. This suppresses the magnetic splitting because the field the nucleus experiences changes more rapidly than the Larmor precession frequency.

7.2.1.7 *Magnetite and electron hopping*. More work has been reported on Magnetite and substituted magnetites than on any other group of spinels. Nonetheless there are still several features of the Mössbauer spectra of magnetite tha tare not completely understood.

Magnetite is an inverse spinel. A complicating factor in its study is that it is prone to nonstoichiometry. It undergoes a phase transition at 119 K changing from orthorhombic at lower to cubic at higher temperatures. Several other changes accompany this transition, the Verwey transition, TV. Mössbauer spectra taken above and below this transition temperature are shown in Fig.7.12.

Stick figures show calculated positions and intensities.

**Fig.7.12**

Magnetite is ferrimagnetic, $T_C = 839$ K, the combined moments of the atoms on the B sites opposing those on the A sites. The easy direction of magnetisation is along [111]. Below the Verwey transition the resistivity of magnetite increases by two orders of magnitude. Above the transition the spectrum can be analysed satisfactorily in terms of two sextets with lines with areal ratios 2 : 1. The hyperfine fields are $B_{hf}$ (A)=48.8 and $B_{hf}$ (B)=46.1 T. The line widths for the B spectrum are substantially greater than for the A spectrum.

The quadrupole interaction is zero for the A sites and 0.05 for the B sites at 310 K. The isomer shifts provide a clue to the nature of the Verwey transition: above $T_V$ the A spectrum gives $\delta = 0.27$ and the the B spectrum $\delta = 0.67$. Now the B sites accomodate both $Fe^{2+}$ and $Fe^{3+}$ ions and this isomer shift is appropriate for a valence delocalisation, either by rapid electron hopping between iron ions or by delocalisation into a 3d band stucture.

Below the Verwey temperature ordering of the iron on the B sites takes place, the $Fe^{2+}$ occupying alternate [100] planes. The electron delocalisation ceases and the spectra demand several sextets to give a good fit. Measurements of the Mössbauer fraction in the region of $T_V$ show complex changes and this and other data suggest $T_V$ is not sharp and a number of different changes may be involved.

In non-stoichiometric or impure samples $T_V$ becomes poorly defined or absent and in superparamagnetic small crystallites it cannot be seen.

Extensive work has been carried out on substituted spinels derived from magnetite, diamagnetic $Ga^{3+}$ replacing $Fe^{3+}$ or $Zn^{2+}$, or other diamagnetic $M^{2+}$ ions replacing $Fe^{2+}$, to discover more about the valence delocalisation and the magnetic interactions between the A and B lattices.

### 7.2.2 **Perovskites and superconducting mixed oxides**.

Similar studies to those outlined for the spinels have been made on the rather simpler perovskites, mixed oxides of composition $MM'O_3$ and cubic, tetragonal or orthorhombic structure.

Several of the high temperature superconductors are mixed oxides with orthorhombic perovskite-like structures and Mössbauer spectroscopy seems likely to throw some light on their interesting behaviour. The 1.2.3. type such as $YBa_2Cu_3O_{7-x}$ do not contain a suitable element but the Y can be partly substituted by [157]Eu which gives good spectra.(See Section 9.3.1) In addition small amounts of [57]Fe or [119]Sn can be introduced as probes. Still smaller amounts of [57]Co can also be used to provide emission spectra, but the interpretation of such spectra is somewhat less reliable.(See Section 10.4).

The spectra from the iron or tin doped materials agree with crystallographic and other evidence that both dopants occupy $Cu(1)$ sites when x is very small, but enter both sites as x increases.

The Mössbauer spectra change when the material is cooled into the superconducting temperature region. Within this region discontinuous changes in Mössbauer parameters of the spectra indicate phase changes. Perhaps more important, the temperature dependence of the isomer shift becomes anomalous. The Mössauer fraction, which is related to the change in the area under the absorption lines with temperature, also undergoes changes. (See Sections 1.2 and 4.7). These reflect changes in the phonon spectrum of the solid and consequently the Debye temperature, which are of importance in determining superconducting properties.

### 7.3  **LOW OXIDATION STATE COMPOUNDS**

The majority of iron compounds in low oxidation states are diamagnetic, predominantly covalently bonded, compounds in which the iron acquires eighteen electrons in the valence shells. Their isomer shifts only span a narrow range of low values, generally lying between 0.2 and -0.2. Iron in formal oxidation states of -1, 0 or 1 are not distinguishable by the isomer shift. Quadrupole splitting is largely due to ligand asymmetry since there is no ordinary valence contribution. Splittings range from 0 up to about 3.2. They are helpful in distinguishing 4 or 6 coordinate iron sites, which give values < 1.5, from 5 coordinate iron, which generally gives splittings in excess 2.0.

The Mössbauer fraction is small for most of these compounds and spectra are best measured at 80 K or below.

### 7.3.1 Iron Carbonyls and Derivatives

There is a large body of data on the Mössbauer parameters for these compounds. However because of the small range of isomer shifts, the technique is not well suited to distinguishing different iron sites in polynuclear carbonyl derivatives. For example the cis and trans forms of $Cp(CO)Fe[PPh_3]_2Fe(CO)Cp$ give very similar spectra. But iron atoms of different coordination numbers are usually distinguishable.

A selection of these extensive data is given in the Tables 7.2.a,b,c & d.

Mössbauer spectroscopy contributed to the identification of the structure of $Fe_3(CO)_{12}$ in the solid state. A detailed analysis of its apparently three line spectrum shows that it is comprised of two quadrupole split doublets. The inner pair is not resolved.

At low temperatures the areal ratio of the two pairs is close to two, but it decreases as the temperature of measurement of the spectrum is increased. However if correction is made for the differing Mössbauer fractions at the two iron sites the ratio remains close to two. This is in agreement with the structure shown in Fig.7.13. The outer doublet arises from the two iron atoms involved in the CO bridging. The lines in this doublet are not of equal area. Rotating the absorber shows that this is not due to any texture effect in the absorber (See Section 3.4.3.).

### Table 7.2. a,b & c

**a.**  Carbonyls

| | $\delta$ | $\Delta$ |
|---|---|---|
| 1. $Fe(CO)_5$ | -0.09 | +2.57 |
| 2. $Fe_2(CO)_9$ | 0.1 | +0.42 |
| 3. $Fe_3(CO)_{12}$ | A 0.08 | -0.08 |
| | B 0.13 | +1.12 |

Salts of carbonyl anions

| | $\delta$ | $\Delta$ |
|---|---|---|
| 4. $Na_2Fe(CO)_4$ | -0.18 | 0 |
| 5. $[Et_4N]Fe(CO)_4H$ | -0.17 | 1.35 |
| 6. $[Et_4N]_2Fe_2(CO)_8$ | -0.08 | 2.24 |
| 7. $[Et_4N]Fe_2(CO)_8H$ | 0.07 | 0.50 |
| 8. $[PyH][Fe_3(CO)_{11}H]$ A | 0.02 | 0.16 |
| B | 0.04 | 1.41 |

Ratio A : B areas 1 : 2.

**b.**  $LFe(CO)_3$

| L | $\delta$ | $\Delta$ |
|---|---|---|
| 9. $PPh_3$ | -0.10 | +2.63 |
| 10. $AsPh_3$ | -0.06 | 3.20 |
| 11. $SbPh_3$ | -0.07 | 3.16 |
| 12. $P(OMe)_3$ | -0.14 | 2.28 |
| 13. $P(OPh)_3$ | -- | -- |

$L_2Fe(CO)_3$

| L | $\delta$ | $\Delta$ |
|---|---|---|
| 14. $PPh_3$ | -0.14 | +2.71 |
| 15. $AsPh_3$ | -0.06 | 3.19 |
| 16. $SbPh_3$ | -0.02 | 3.22 |
| 17. $P(OMe)_3$ | -0.22 | 2.28 |
| 18. $P(OPh)_3$ | -0.20 | 2.60 |

**Table 7.2 c**

| c.<br>L-L | L-LFe(CO)$_3$<br>δ | A | | L-LFe(CO)$_4$<br>δ | A | | L-LFe$_2$(CO)$_8$<br>δ | A |
|---|---|---|---|---|---|---|---|---|
| 19. f$_4$fos | -0.06 | 2.34 | 23. | -0.07 | 2.68 | 27. A | 0.03 | 0.66 |
| | | | | | | B | 0.06 | 1.30 |
| 20. f$_4$Ars | -0.05 | 2.83 | 24. | -0.05 | 2.82 | 28. A | 0.02 | 0.64 |
| | | | | | | B | 0.02 | 1.44 |
| 21. f$_4$AsP | -0.05 | 2.47 | 25. | -0.08 | 2.09 | | | |
| 22. f$_8$AsP | -0.05 | 2.61 | 26. | -0.08 | 2.58. | | | |

f$_4$fos ---- 2 CF$_3$ groups with X = PPh$_3$;  f$_4$Ars ----- 2 CF$_3$ groups and X = AsMe$_3$
f$_8$AsP ---- 4 CF$_3$, with Ph$_3$P and Me$_3$As groups.

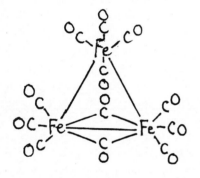

**Fig.7.13**

The difference in line areas must be due to different f factors for different directions of recoil, the Goldanski-Karyagin effect (See Section 3.4.1.).

A decrease in the isomer shift in a series of similar compounds can either arise from increased donation into the iron 4s orbital or from reduced 3d occupation because of back donation Fe→L, or from both causes. Data on a large number of LFe(CO)$_4$ compounds suggests the increase in 4s occupation is the more important factor.

The single doublet found for compound 7 (Table 7.2.a) suggests the two iron atoms are equivalent so that there must be a hydrogen bridge between them. It can be seen that the higher coordination number of the iron in this compound leads to a much lower quadrupole splitting as compared to compound 6. Fig.7.14 a and b show the structure of compounds 7 and 8.

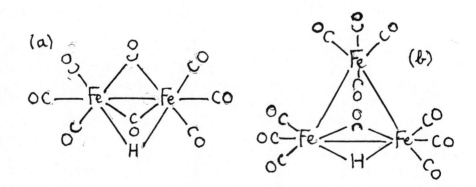

**Fig.7.14 a & b**

The similarity of the spectra of Fe$_3$(CO)$_{12}$ and Fe$_3$(CO)$_{11}$H shows that a bridging carbonyl has been replaced by hydrogen.

Compounds 9 to 26 apparently all contain five coordinate iron, so that the ligand must be monodentate in the compounds 23 to 26. A number of compounds of the type $L_2MFe(CO)_4$ where L is a nitrogen base and M is Zn or Cd all give quadrupole splittings of less than one. This indicates a polymeric structure with M - Fe chains. In compounds 27 and 28 and others similar, the two iron atoms are easily distinguishable, but the quadrupole splitting for the B atoms is low for five coordination. A possible structure is shown in Fig. 7.15.a.

A number of these compounds have been measured in frozen solutions in non-coordinating solvents. The spectra are only slightly different from those of the solids.

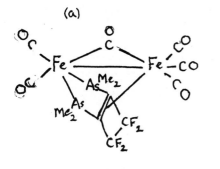

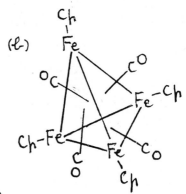

**Fig.7.15 a & b**

The compound shown in Fig.7.15 b is interesting since it can be oxidised to a cationic species of the same structure. The isomer shifts for the neutral and cationic complexes are practically the same, 0.66 and 0.67 indicating valence delocalisation. The electron must be removed from the molecular orbitals associated with the $Fe_4$ tetrahedra. Such behaviour is treated in Section 7.4.

Low quadrupole splittings show that the iron atoms in complexes of the type $Fe_2(CO)_5(C_2Ph)PPh_2$ interact with the acetylenic system.

### 7.3.2 Other Low Oxidation State Compound

The appearance of $Fe^+$ in $^{57}Co$ labelled sodium chloride is mentioned in Section 10.3.

$(R_2NCS_2)_2FeNO$ compounds are formally iron(I) compounds and have a single unpaired electron. They are low spin $d^7$ compounds. The Mössbauer parameters for the $R = C_2H_5$ compound are $\delta = 0.28$ and $\Delta = 1.89$ at 300 K. The iron has no low lying electronically excited states and the quadrupole splitting is unchanged at 4.2 K.

### 7.4 MIXED VALENCE COMPOUNDS: DELOCALISATION

If a compound contains iron atoms in more than one oxidation state and the environment of the different iron atoms are not too different, electron delocalisation may take place. The iron atoms can be regarded as oscillating between the two, or more, localised valences; effectively they assume a weighted mean valence, determined by the proportions of iron atoms in the different valence states. One of the earliest examples of such behaviour has already been mentioned in Section 7.2.1.7; the Verwey transition.

The characteristics of the Mössbauer spectra that will be displayed by such compounds will depend on the mean life time of the iron atom in the localised valence state. Suppose that the compound contains equal numbers of iron atoms in two oxidation states. In the absence of any delocalisation it will be supposed that the spectrum consists of two absorption lines centred at $\omega_1$ and $\omega_2$, expressing their positions as angular frequencies ( $\omega = v(\delta)E_\gamma /c\hbar$ ). The intensity distribution as a function of $\omega$ for the two line spectrum is given by: $I(\omega) = \dfrac{k\left[\Gamma + i(\omega - \varpi) + \omega_r\right]}{\left[\Gamma + i(\omega - \varpi)^2 + (\Delta\omega^2)\right]}$ where $\varpi = 1/2(\omega_1 + \omega_2)$ and $\varpi$

$= 1/2(\omega_1 - \omega_2)$. $\Gamma$ is the theoretical line width expressed as a frequency. $\omega_r$ is the frequency of change of oxidation state of the iron atoms, so that the mean life time in one valence state is given by $\tau = 2\pi/\omega_r$. When $\omega_r \Rightarrow \square 0$ this expression becomes:

$$I(\omega) = \frac{k}{2}\left\{\left[\frac{1}{\Gamma + i(\omega - \varpi) - i\Delta\omega}\right] + \left[\frac{1}{\Gamma + i(\omega - \varpi) + i\Delta\omega}\right]\right\}$$

which corresponds to two lines of theoretical line width located at $\omega + \Delta\omega$ and $\omega - \Delta\omega$, that is at $\omega_1$ and $\omega_2$. The valences are now completely localised giving distinguishable iron atoms.

If $\omega_r$ becomes much greater than $\Delta\omega$ then: $I(\omega) = \dfrac{k}{\left[\Gamma + i(\omega - \varpi) + (\Delta\omega)^2 / \omega_r\right]}$ this

corresponds to a single line located at $\varpi$, but with a line width given by $\Gamma + (\Delta\omega)/\omega_r$. As $\omega_r$ increases this tends towarde the value $\Gamma$ .

Thus as $\omega_r$ grows from a very small value, giving localised valences on the Mössbauer time scale, the lines in the spectrum broaden and collapse and a new line centred on $\varpi$ appears. The new line is broad while $\omega_r < (\Delta\omega)^2$, but as $\omega_r$ becomes much larger than $(\Delta\omega)^2$ it tends towards the theoretical line width.

Spectra due to the individual valence states will always ensue if $2\pi/\omega_r > \tau_N$. The appearance of an averaged quadrupole split spectrum introduces another limit to $\omega_r$ since it will require $\tau_N > \tau_r > \tau_q$, the last term being the time needed for quadrupole splitting to occur.

In the $\omega_r$ region where line broadening occurs, $\omega_r \approx (\Delta\omega)$, the line width can be used to estimate $\omega_r$.

## 7.4.1  **Thermally Activated Delocalisation**

In systems where the barrier to transfer of an electron from an iron atom in one valence state to an iron atom in another is small, transition from localised to delocalised valence states becomes temperature dependent

A schematic representation of electron exchange in mixed valence iron compounds is shown in Fig, 7.16.

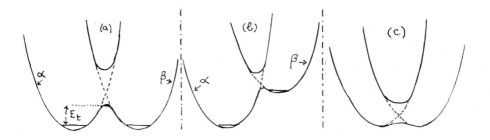

**Fig.7.16**

The figure shows the potential energy of a system of two iron ions as a function of a single coordinate corresponding to the metal- ligand distance in Fe(A) - X - Fe(B). The two iron ligand environments are assumed to be the same. The component curves, full lines, represent the Fe(A)II with Fe(B)III configuration, curve $\alpha$, and the reverse situation curve $\beta$. From orthogonality considerations, if transitions from one state to the other can occur, the two curves cannot cross and the system follows the full lines.

In the situation shown in Fig.7.16a an optical stimulation of electron transfer between A and B is possible, but thermal delocalisation is energetically unlikely. In an asymmetric coordination situation curve $\alpha$ will be displaced vertically in relation to curve $\beta$.

Fig.7.16.b represents a situation where there is appreciable electron vibrational coupling. If $E_t$ becomes comparable with kT at some temperature thermally activated delocalisation will occur. At low temperatures tunnelling may contribute to electron transfer but will not affect the rate of exchange on the Mössbauer time scale appreciably.

Fig.7.16.c shows a case of strong coupling. This leads to delocalisation at very low temperatures.

### 7.4.2 Mixed Valence in Three Dimensional Lattices

**7.4.2.1 *Fluorides*.** A number of mixed valence fluoride complexes of iron have been described. The red $Fe_2F_5\,2H_2O$ orders ferrimagnetically at 48 K. Below this temperature the change in the spontaneous magnetization with temperature fits the Brillouin function for S = 1/2. In an applied magnetic field of 6 T, parallel to the gamma beam, two spectral sextets are found at 20 K, eleven of the lines being resolved.

At room temperature a three line spectrum is seen, but on cooling to 55 K two doublets are resolved. An analysis of the room temperature spectrum, in terms of two doublets, yields parameters $\delta$ = 1.34 with $\Delta$ = 2.44 and $\delta$ = 0.44 with $\Delta$ = 0.65, indicating localised $Fe^{2+}$ and $Fe^{3+}$ atoms.

The yellow $Fe_2F_5\ 7H_2O$ gives two well resolved quadrupole split doublets with $\delta = 1.25$ and $\Delta = 3.33$ and $\delta = 0.44$ with $\Delta = 0.59$ which are very close to the parameters found for $Fe(H_2O)_6^{2+}$ and $FeF_5H_2O^{2-}$ in other compounds.

It seems likely that if the iron atoms are six coordinate and bidentate ligands are not involved, the valences are usually localised.

7.4.2.2 **Phosphates.** In two mixed valence phosphates of iron delocalisation does occur. The compound $Fe_2(PO_4)O$ exists in two modifications: the $\beta$ form has a Néel temperature of 410 K. At 495 K the spectrum is a rather asymmetric quadrupole split doublet with parameters $\delta = 0.61$ and $\Delta = 2.94$. Sustantial delocalisation has taken place.

At room temperature the spectrum comprises two sextets, one due to $Fe^{2+}$, the other to $Fe^{3+}$, and a doublet arising from the valence delocalised iron. The transition from localised to delocalised valences often extends over some range of temperatures.

At 4.2 K three sextets are seen. Their parameters are $\delta = 0.455$ and 55 T, $\delta = 0.53$ and 53 T, arising from two $Fe^{3+}$ sites and $\delta = 1.52$ with 30 T clearly reflecting the $Fe^{2+}$.

Another phosphate, $Fe_9(PO_4)O_8$, has been reported to give a delocalised spectrum at above 373 K.

7.4.2.3 **Prussian and Turnbull's blue.** An early success of Mössbauer spectroscopy was to demonstrate that these two compounds are essentially the same. Both materials are appreciably non-stoichiometric, and reproducibility of spectra could be expected to be poor. The spectrum obtained for absorbers of natural isotopic composition consists of two rather broad lines. Using samples prepared from one of the following: ferrous sulphate, ferric sulphate, ferrocyanide or ferricyanide enriched in $^{57}Fe$, it was shown that the spectrum was composed of a single line and a quadrupole split doublet. The parameters obtained are shown in Table 7.3.

**Table 7.3.**

| | | | |
|---|---|---|---|
| Turnbull's blue | $Fe^{3+}$ | $\delta = 0.51$ | $\Delta = 0.49$ |
| " | FeII | $\delta = -0.07$ | $\Delta = 0$ |
| Prussian blue | $Fe^{3+}$ | $\delta = 0.49$ | $\Delta = 0.57$ |
| " | FeII | $\delta = -0.08$ | $\Delta = 0$ |

All measured at 77 K

Clearly both compounds contain high spin $Fe^{3+}$ cations and ferrocyanide anions. Soluble Prussian blue was also shown to contain $Fe^{3+}$ cations.

7.4.3 **Mixed valence binuclear complexes**

Delocalisation has been observed in a number of binuclear iron complexes.

A few examples will be given.

7.4.3.1 **$L_2Fe_2(\mu OH)_3(ClO_4)_2 2CH_3OH$**, L = NN'N"tri Me triazacyclononane is a paramagnet down to 4.2 K. The iron environment is quasi-octahedral, the octahedra sharing a face, with OH bridges. Spin coupling gives an S = 9/2 ground state. Even at this low temperature delocalisation seems complete. A single spectrum, $\delta = 0.74$ and $\Delta = 2.14$, is seen. The rather broad lines may indicate that $\omega_r$ is close to $\Delta\omega$.

7.4.3.2 $Et_4N.Fe_2(Salmp)_2$. The ligand and complex are shown in Figs.7.17 a & b.

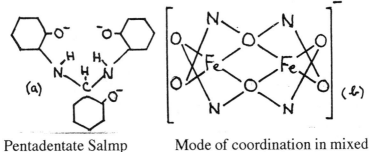

<table>
<tr><td>(a)</td><td>(b)</td></tr>
<tr><td>Pentadentate Salmp</td><td>Mode of coordination in mixed valence complex of Salmp.</td></tr>
</table>

**Fig.7.17 a & b**

Ferromagnetic coupling of spins on the iron atoms gives an S = 9/2 ground state. At temperatures below 100 K the valences are localised and at 1.5 K the spectrum consists of two sextets of equal intensities. Over a range of temperatures above 100 K localised and delocalised spectra coexist. At 247 K the parameters obtained are $Fe^{2+}$ $\delta = 0.99$ and $\Delta = 2.24$, 20%; $Fe^{3+}$ $\delta = 0.44$ and $\Delta = 0.88$, 20%; Fe (delocalised) $\delta = 0.71$ and $\Delta = 1.08$, 60%.

### 7.4.4 Mixed Valence Polynuclear Iron Carboxylates

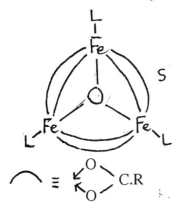

S = Uncoordinated solvent.
L = N donor ligand.

**Fig.7.18**

One of the earliest compounds in which valence delocalisation was studied as a function of temperature was the compound shown in Fig. 7.18 with $R = CH_3$, L = pyridine and S = pyridine. The trinuclear planar $Fe_3O$ units are stacked on a three-fold axis with the solvent S molecules inserted between the $Fe_3O$ planes.

These compounds provide an excellent opportunity to explore the factors influencing the electron-vibronic coupling that determines the temperature region in which delocalisation takes place. The solvate, S, the ligand, L, and the R in the carboxylate group can all be changed while preserving essentially similar complexes.

Spectra at different temperatures for the compound with S = L = 3 Me Pyridine and $R = CH_3$ are shown in Fig,7.19.   At sufficiently high temperatures, > 190 K, only a doublet spectrum with $\delta = 0.743$ and $\Delta = 0.520$ is found. At low temperatures two quadrupole split doublets, with intensities in the ratio 2 : 1, are found. The parameters at 117 K, $\delta = 1.221$ with $\Delta = 1.726$ and $\delta = 0.542$ with $\Delta = 0.989$, clearly indicate localised $Fe^{2+}$ and $Fe^{3+}$ respectively. The isomer shift for the delocalised valence state at 117 K, $\delta = 0.870$, is close to the weighted average from the above values.

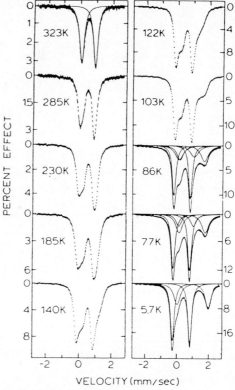

Fig.7.19

Variable-temperature $^{57}$Fe Mössbauer spectra for [Fe$_3$O-(O$_2$CCH$_3$)$_6$(3-Me-py)$_3$](3-Me-py)

It is interesting to note the effect of the time constant of the method of investigation on the occurrence of delocalisation. At both 200 K and 80 K the life time of a particular oxidation state is long compared with the time constant for infra red absorption measurements and delocalisation will not be observed. Similarly the slow process of an X-ray diffraction measurement will record equivalent iron atoms over the whole temperature range. Characteristic times for different techniwues are shown in Table 7.4.

### Table 7.4.

Measurement times for different techniques.

1. Diffraction methods, X-ray, electron and neutron.      $10^{-18}$ s.

   But these average over the vibrational motion.
2. Electron and X-ray spectrosdcopy,      $10^{-18}$ s.
3. Vibrational spectroscopy, I.R. and Raman.      $10^{-13}$ s.
4. Microwave rotational spectroscopy.      $10^{-10}$ s.
5. Electron spin resonance spectroscopy.      * $10^{-4}$ to $10^{-9}$ s.
6. Mössbauer spectroscopy.      * $10^{-6}$ to $10^{-9}$ s.
7. Nuclear magnetic resonance spectroscopy.      * $10^{-2}$ to $10^{-8}$ s.
8. Nuclear quadrupole resonance spectroscopy      * $10^{-2}$ to $10^{-8}$ s.

   * These depend very much on the particular system studied.

Thus in the case of Mössbauer spectroscopy the mean life of the excited state is the important factor.

Like spin cross-over the delocalisation process is very much affected by the identity of S and L in the complex shown in Fig.7.18. In the absence of S no delocalisation is observed up to room temperature for the compound with L = pyridine and R = $CH_3$.

Heat capacity measurements show phase changes in the compound concerned in Fig.7.19. Changes occur at 111, 112, 186 and 191 K. The last two are closely connected with the delocalisation and seem to be related to the onset of disorder of the solvent molecules. In the case of the benzene adduct at temperatures giving localised valences the plane of the benzene ring appears to be perpendicular to the threefold axis of the $Fe_3O$ units. During the phase changes disorder and rotation of the benzene takes place. The delocalisation mechanism involves electron vibronic coupling and is therefore affected by the phonon and vibration spectra of the sample. These are sensitive to the identity of S and L as well as R.

The delocalisation transition is pressure dependent. At room temperature the chloroform adduct of $Fe_3O(OAc)_6Py_3$ gives a spectrum composed of one quadrupole split doublet at a pressure of 20 k.bar but at 80 k.bar localisation of valences occurs and two spectra, due to localised valence states, are added to the first, which decreases in intensity.

Spectra due to localised and delocalised valence iron atoms commonly coexist over some range of temperatures. However there is almost no data on single crystals and the substantial temperature range for the transition may reflect the range of sizes of crystallites in the absorber or of the defects in these crystals.

### 7.4.5 Biferrocenes and Related Compounds

The mixed valence compounds of the types shown in Fig.7.20 have been studied extensively to learn more about the coupling mechanism.

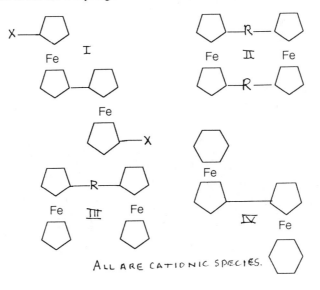

**Fig.7.20**

ALL ARE CATIONIC SPECIES.

The isomer shifts for localised low spin Fe(II) and Fe(III) are too little different to be useful in distinguishing localised and delocalised valence states so that attention is focussed on the quadrupole splittings.

In most cases thermally initiated delocalisation does not take place sharply but extends over some range of temperatures. At a sufficiently low temperature the localised valence form gives two quadrupole split doublets, the splittings being in the ranges characteristic of the two oxidation states of the iron. If the crystal structure indicates more than two iron sites, additional doublets will be found. In the temperature range where delocalisation is taking place the two splittings converge until, at a temperature high enough for complete delocalisation, a single quadrupole split pair is seen, as shown ihn Fig.7.21.

A set of spectra at different temperatures for the type I complex with X = -CH$_2$CH$_3$ is shown in Fig.7.22. The iron atom valences are entirely localised at 115 K and substantially delocalised at 287 K.

Like the spin cross-over systems, thermal delocalisation can be influenced by several factors. Biferrocene trichloroacetate, type I with X = H, shows very little delocalisation at room temperature and the I$_3^-$ salt of the chloro derivative, type I with X = Cl, does not delocalise below 310 K. The iodo and bromo compounds however are delocalised down to 4.2 K. For the I$_3^-$ salts of the compounds with X = H; -CH=CH$_2$ ; -CH$_2$CH$_3$; -CH$_2$CH$_2$CH$_3$; and -CHPh delocalisation is essentially complete at 365; 320; 275; 245; and 170 K respectively. It has been suggested that the more the rings tilt from a parallel

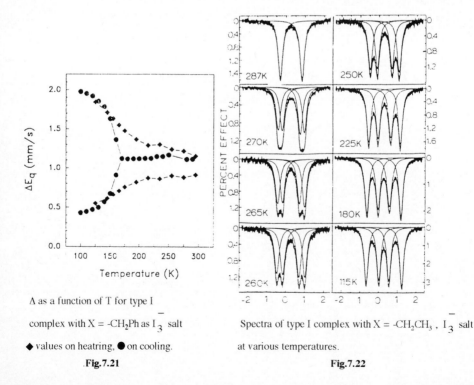

Λ as a function of T for type I

complex with X = -CH$_2$Ph as I$_3^-$ salt

◆ values on heatring, ● on cooling.

**Fig.7.21**

Spectra of type I complex with X = -CH$_2$CH$_3$ , I$_3^-$ salt

at various temperatures.

**Fig.7.22**

parallel orientation, the lower the temperature necessary for delocalisation. With the compound shown in Fig.7.23 a single quadrupole splitting is found between 77 and 300 K, $\varepsilon = 1.614$ at 77 K.

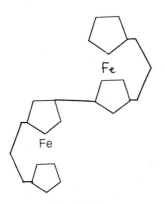

**Fig.7.23**

The type I compounds with X = Cl, Br and I all have trapped valences when examined by electron spin resonance, which in this case proves to be a faster measurement technique. Less substantial changes also lead to differences in delocalisation behaviour: the $I_3^-$ salts of type I componds with X = n butyl and X = benzyl exist in more than one crystal form. The two forms behave quite differently. The benzyl complex in the triclinic form, space group P1, is substantially delocalised from room temperature down to about 25 K. The other form, platelets of space group P21/n, has trapped valences at 300 K.

The delocalisation can also be seen in ethanol glasses containing the biferrocene complexes.

With the exception of the afore mentioned trichloroacetate the other observations refer to the $I_3^-$ salts. Changing the co-anion generally changes the characteristics of the thermal delocalisation. Even the change from $I_3^-$ to $I_5^-$ (possibly $I_3^-$ with lattice iodine) leads to differences. Reproducibility is a problem in these studies because delocalisation is affected by the defects in the solids and hence by the precise details of the preparation and treatment of the compounds.

The type I complex with X = -CH$_2$Ph and as the $I_3^-$ salt gives one quadrupole split doublet at 270 K when present as small crystals, but if the crystals are ground to a fine powder substantial valence trapping takes place and three quadrupole doublets are seen, corresponding to the Fe(II) and Fe(III) atoms and the delocalised valence atoms,

Heat capacity measurements recording $C_p$ as a function of the temperature have been made for some type I compounds. A number of phase change peaks usually appear together with a pronounced peak at the onset of delocalisation.   For the X = H compound, as the $I_3^-$ salt, the entropy change on delocalisation at about 328 K, obtained from these data, is 1.77 J $K^{-1}mol^{-1}$, notably less than that associated with the electronic change, $Rln2 = 5.76$ J $K^{-1}mol^{-1}$.

There is some evidence that delocalisation is connected with the onset of a further degree of freedom of the anion in the $I_3^-$ salts.

For type III compounds with R vacant (direct bond) and as the $PF_6^-$ salt, $\Delta = 1.719$ at 300 K and delocalisation persists down to very low temperatures. The Fe - Fe distance in the compound is 398 pm. When R = -C≡C- the iron-iron separation is 650 pm and $\Delta = 1.52$ at 298 and 1.61 at 78 K. Both compounds seem to be delocalised even on the very fast infrared time scale. With a type II compound and R = -C=C- however the valences are trapped at 78 K, $\Delta = 0.49$ and $\Delta = 2.18$.

Delocalisation of valences can also be found involving iron in oxidation states I and II in compounds of the type: $[Fe_2(\ \mu_2.\eta^{10}C_{10}H_8)(Arene)_2]^+$. With the arene hexamethyl benzene delocalisation persists down to very low temperatures. [$C_{10}H_8$ = 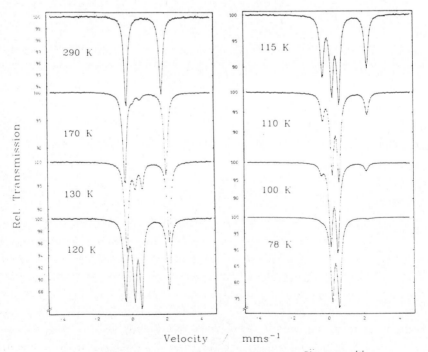  ]

## 7.5  SPIN CROSS-OVER

Magnetic measurements made in the 1930's showed that some iron compounds changed from a high spin electronic configuration at around room temperature to a low spin congiguration at low temperatures. Such magnetic measurements had the disadvantage that they did not directly determine the amounts of the iron species present at the different temperatures. Mössbauer spectroscopy can do so and it has been used very extensively to elucidate the behaviour of spin cross-over systems. Provided the mean life time of the two spin states involved in the HS⇔LS equilibrium exceeds the mean life time of $^{57m}$Fe, spectra arising from both spin states will be seen.

Fig. 7.24 shows how the Mössbauer spectrum of the spin cross-over compound Fe(2 pic)$_3$Cl$_2$.EtOH changes with temperature. If the spectra are measured with a thin absorber and corrections applied for the difference in the f factors for the high and low spin forms, the areas under the absorption lines are proportional to the amounts of the species responsible for these lines present in the absorber.

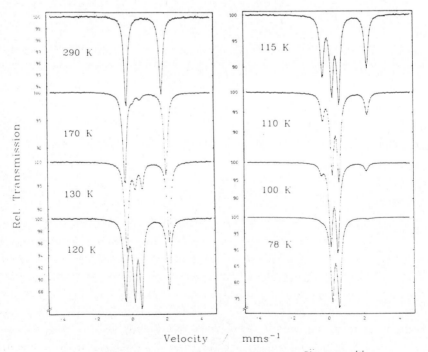

Spectra of Fe(2.pic)$_3$Cl$_2$.CH$_3$OH at various temperatures. Shows transition from entirely low spin at 78 K to entirely high spin at 290 K.

**Fig.7.24**

The compound giving the data in Fig.7.24 changes from being completely high spin at 298 K to completely low spin at 80 K, so that one can easily determine the areas corresponding to 100% high or low spin forms.

## 7.5.1 Spin Cross-over in Iron(II) Compounds

With iron(II) compounds the high spin spectrum can easily be identified by the large isomer shift and, usually, by a  large quadrupole splitting. The low  spin  form generally shows a low quadrupole splitting and in the case of some compounds the splitting is not resolved.

For Fe(phen)$_2$(NCS)$_2$ the Mössbauer parameters at 160 K for the high spin form are $\delta = 0.96$ and $\Delta = 3.03$, and  for  the  low spin form  $\delta = 0.43$  and  $\Delta = 0.35$.

Although most of the compounds concerned are of lower symmetry it is convenient to explain spin cross-over in terms of an octahedral environment for the iron. Thus with iron(II) compounds the high spin form will have an $e^{*2}.t_{2g}^4$ configuration, $^5T_2$, and the low spin $t_{2g}^6$ a $^1A_1$ state. Hence the high spin state can be regarded as an excited state of the low spin form.  But the cross-over process cannot be regarded as a simple thermal excitation since, as Fig.7.24 shows, the low spin form can be completely converted to the high spin form: the Boltzmann distribution does not permit such a population inversion.

A very large number of iron(II) compounds and rather fewer iron(III) compounds showing spin cross-over have been investigated, of which a selection of the more prominent iron(II) compounds are shown in Table 7.5.

### Table  7.5.

| No. | Compound | $T_{50}$ | No. | Compound | $T_{50}$ |
|---|---|---|---|---|---|
| 1. | Fe(Phen)$_2$(NCS)$_2$ | 174s | 8. | Fe(4.7Me$_2$Phen$_2$(NCS)$_2$ | 121s |
| 2. | Fe(Phen)$_2$(NCSe)$_2$ | 232s | 9. | Fe(2MePhen)$_3$(PF$_6$)$_2$ | 130 |
| 3. | Fe(bipy)$_2$(NCS)$_2$ | 212s | 10. | Fe(pythiaz)$_2$(ClO$_4$)$_2$ | 160b |
| 4. | Fe(HBPz$_3$)$_2$ | 393b | 11. | Fe(5ClPhen)$_2$(NCS)$_2$H$_2$O | b |
| 5. | Fe(HB[3.5MePz]$_3$)$_2$ | 220 | 12. | Fe(paptH)$_2$(NO$_3$)$_2$.H$_2$O | 254 |
| 6. | Fe(2.pic)$_3$Cl$_2$ EtOH | 122 | 13. | Fe(pyim)$_3$(ClO$_4$)$_2$.2H$_2$O | 193b |
| 7. | Fe(2.pic)$_3$Cl$_2$ MeOH | 153 | 14. | Fe(pyben)$_3$Br$_2$ | b |

b = broad         s = sharp.

phen = 1.10 Phenanthroline.      bipy = 2.2' bipyridyl          pyl =

2.pic = 2 picolylamine =           CH$_2$NH$_2$           pz = pyrazolyl =

pyim = 2-(2.pyridyl)imidazole = HN          pythiaz = 2.4 bis (2 pyridyl)thiazole =

pyben = 2-(2 pyridyl)benzimidazole =   pyl

paptH = 2-(2 pyridylamino).4.(2 pyridyl)thiazole =     pyl-N-                pyl

7.5.1.1 *Spin cross-over as a function of temperature.*
   A typical curve showing the proportion of the high spin spectrum as a function of the temperature of the measurement is shown in Fig. 7.25. The change in the proportion of high spin species present as a function of temperature follows one of the three patterns shown in Figs.7.26 a,b & c. For some compounds the transition is rather sharp with the characteristics of a first order phase transition as in (Fig.7.26 b). These are identified by an s in Table 7.5. With other compounds the transition extends over a range of temperatures (Fig.7.26 a).These are distinguished by a b. In both cases it is convenient to characterise the transition by $T_{50}$, the temperature at which equal amounts of each spin state are present. In both cases it is possible to obtain Mössbauer spectra in which both spin states can be identified. A third category exists where it appears to be impossible to convert the material completely to the low or, more rarely, the high spin state. (Fig.7.26c).

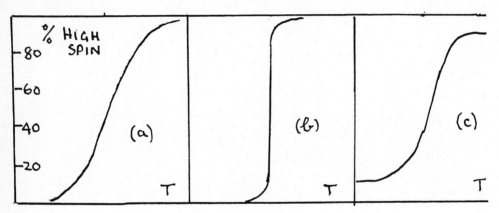

**Fig.7.26  a,b  &  c**

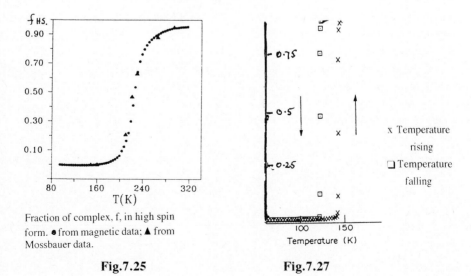

Fraction of complex, f, in high spin
form. ● from magnetic data; ▲ from
Mossbauer data.

**Fig.7.25**                          **Fig.7.27**

Although the reproducibility of the results is good for measurements on a single sample as absorber, different preparations of the sample and treatments such as grinding the material or introducing defects in other ways, leads to substantially different curves, a material giving an (b) type curve develops a type (a) curve In addition the transitions, especially the sharp transitions, show hysteresis effects, The curve measured with the sample at progressively increasing temperatures, being displaced towards higher temperatures than the curve obtained with decreasing temperatures.

The effects of hysteresis are shown in Fig.7.27 for $Fe(NCS)_2$ (4.4'bis 1.2.4 triazole)$_2$.$H_2O$.

Another difficulty arises with compound (3) in Table 7.5, this seems to be formed in more than one crystal modification and the different forms give quite different spin cross-over characteristics.

### 7.5.1.2 *Factors affecting spin cross-over.*

Since spin cross-over depends on the energy separation between the iron 3d orbitals in the complex, modification of the ligands by substitution may lead to a radical change in cross-over properties.

Substitution of H by $CH_3$ makes 2MePhen a stronger σ donor than Phen, but it also leads to steric interference with bonding to the metal and $Fe(2MePhen)_2(NCS)_2$ is high spin over the temperature range 298 to 77 K. The unsubstituted phenanthroline analogue shows cross-over within this range indicating a weaker ligand field in the 2Me substituted complex. Similarly the compounds $Fe(2MePhen)_3X_2$ with X = I, $PF_6$, $ClO_4$ show spin cross-over, while the corresponding $Fe(phen)_3$ salts are all low spin.

As might be expected different isomeric forms of a complex, such as fac and mer isomers of $FeA_3B_3$ species, usually show different spin cross-over characteristics. Both forms are found with some $Fe(pyim)_3^{2+}$ and $Fe(pyben)_3^{2+}$ salts. The fac form can be distinguished since its low spin form has a very small quadrupole splitting (See Table 3.6).

Not surprisingly many compounds displaying spin cross- over also show thermochromic effects.

### 7.5.1.3 *Anion and solvation effects.*

Even when the anions are not coordinated to the iron they affect spin cross-over behaviour.     Fig.7.28.a   shows the widely differing behaviour of some salts of the $Fe(pyben)_3$ cation.    Different solvates also display very different spin cross-over characteristics.   In the case of $Fe(2pic)_3Cl_2$ the dihydrate is low spin over a wide temperature range below room temperature.   However the monohydrate shows spin cross-over.  As can be seen in Fig.7.28.b the methyl and ethyl alcohol solvates also show different transition curves.   Anhydrous Fe 4.4'bis(1.2.4 triazole)$_2$(NCS)$_2$ is always high spin but the monohydrate shows spin cross-over with $T_{50}\downarrow \approx$ 123 K and $T_{50}\uparrow \approx$ 144 K, Fig.7.27.   The arrows indicate values of $T_{50}$ for heating, ↑ , and for cooling, ↓ ,the compound.

Some of the effect of hydration may be due to hydrogen bonding modifying the donor properties of the ligands. The different solvates of $Fe(2 pic)_3Cl_2$ have different crystal structures and the   behaviour of different crystal forms of the same compound is sometimes quite different.

Direct evidence that the transition is dependent on the lattice properties as well as the iron complex involved is provided by the data on $Fe(2\ pic)_3Cl_2\ CH_3OH$ and the deuterated analogue shown in Fig.7.28. Deuteration leads to different values of $T_{50}$ for the two forms. All these effects show that the importance of the rest of the lattice in determining spin cross-over.

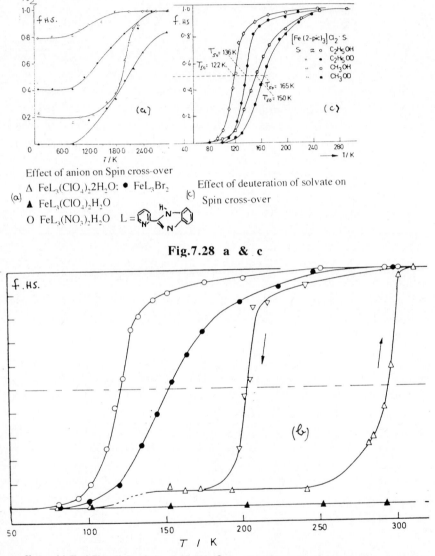

Effect of anion on Spin cross-over

(a)  Δ  $FeL_3(ClO_4)_2 2H_2O$:  ●  $FeL_3Br_2$
     ▲  $FeL_3(ClO_4)_2 H_2O$
     O  $FeL_3(NO_3)_2 H_2O$    L =

(c)  Effect of deuteration of solvate on Spin cross-over

**Fig.7.28  a  &  c**

Solvate effects with $Fe(2\ Pic)_2Cl_2 S$.  O $S = C_2H_5OH$;  ● $S = CH_3OH$;  ∇ $S = H_2O$ on heating, Δ on cooling
▲  $S = 2H_2O$.          **Fig.7.28  b**

Interesting situations arise with some polynuclear iron complexes. In the linear trinuclear complex $Fe_3(Rtr)_6(H_2O)_6(CF_3SO_3)_6$ where Rtr is $4.C_2H_5.1.2.4$ triazole, the terminal iron atoms are always high spin, but the central atom displays spin cross-over, $T_{50} \approx 200$ K.

In $FeL_2(NCS)_2$, where L = bromazepan, ( See inset ) there are two iron sites, one showing spin cross-over while the other is always low spin.

### 7.5.1.4 *Nature of the spin cross-over process.*

The cross-over process is reversible. Independent measurements of the rates of the spin change in iron(II) complexes showing spin cross-over, using fast kinetic relaxation techniques, have shown that this condition will generally be fulfilled. However iron(II) complexes displaying fast cross-over on the Mössbauer time scale have been found. An example is N.N.N'.N' Tetra bis(2- pyridylmethyl ethylenediamine iron(II) perchlorate.

The continuous, type (a) transitions can be treated fairly satisfactorily as a chemical equilibrium, LS⇔HS, taking into account that the solid is not an ideal solution.

For compounds where the transition is not associated with a change in crystal structure, typical enthalpy and entropy changes for the reaction are $\Delta H \approx 12$ kJ.mol$^{-1}$, $\Delta S \approx 50$ J.K$^{-1}$ mol$^{-1}$ The entropy term is much greater than the contribution from the electronic change and reflects the changes in vibrational and lattice frequencies taking place. But the sharp transitions cannot be treated satisfactorily in this way.

The effect of the surrounding lattice was investigated in experiments on crystals in which part of the iron was isomorphously replaced by another divalent transition metal cation. Results for $Fe_xMn_{1-x}(phen)_2(NCS)_2$ and the analogous cobalt solid solutions are shown in Figs.7.29.a & b.

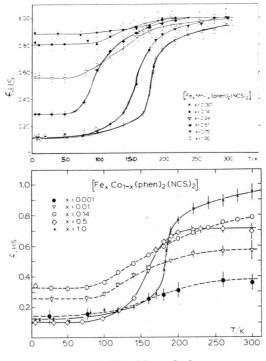

**Fig.29 a & b**

Introduction of $Mn^{2+}$, whose ionic radius is greater than that of high spin $Fe^{2+}$, favours the high spin form. $T_{50}$ decreases as x decreases, the residual high spin species at very low temperatures increases. The radius of $Co^{2+}$ lies between the radii of high and low spin iron(II) and substitution of $Co^{2+}$ for iron(II) tends to stabilise the low spin state. The spin cross-over behaviour of the iron complex depends on its crystal environment; the more spacious lattice of the manganese complex favours the larger high spin iron species, the more compact cobalt or nickel complexes favour the smaller low spin form.

The abrupt transitions have more the character of a first order phase change, involving cooperative phonon spin coupling leading to cross-over in small domains of molecules. The steeper the curve relating the fraction of high spin form to the temperature, the larger the number of molecules in the domains. The curve for $Fe(phen)_2(NCS)_2$ implies about 70 molecules in the domains while those for the $Fe(2\,pic)_3^{2+}$ salts indicate about 3 or 4 molecules.

Crystallographic studies have shown that the transition is sometimes, but not always, associated with a change in crystal symmetry. In all cases the Fe-L bond lengthens by up to 10% on going from the low to the high spin form. The unit cell volume increases by between 3 and 4 %. Repeated cycling between the high and low spin states may lead to fragmentation of crystals and loss of water from hydrates.

### 7.5.1.5 *Photo effects*

Mössbauer spectroscopy provides the best means of studying light induced excited state spin trapping, (**LIESST**). The complex $Fe(ptz)_6(BF_4)_2$ (Ptz = 1.propyl tetrazole) exhibits spin cross-over, the low temperature, low spin, form giving a single absorption line. The high spin room temperature form is easily distinguishred by a spectrum showing substantial quadrupole splitting. A noteworthy feature is its thermochromic character; the high spin form is almost colourless but the low spin form is strongly coloured. On irradiating the compound with light from a xenon arc lamp at 15 K the low spin form is completely converted to the high spin form, which is indefinitely stable at this temperature.

The irradiation excites the low spin $^1A_1$ form to an excited $^1T_1$ state. System crossing enables this state to relax to a $^3T_1$ and thence to the $^5T_2$ state. Conversion of this state back to the $^1A_1$ state requires some activation energy which is not available at 15 K. However if the metastable $^5T_2$ is allowed to warm to above about 40 K the transition takes place. Similar results have been obtained with other spin cross-over compounds including $Fe(phen)_2(NCS)_2$ and $Fe(2.pic)_3Cl_2$ EtOH.

A kinetic study of the relaxation of the excited $^5T_2$ state in the $[FeN_4]$ type complex: $[Fe(2\,methylPy)_2Py.tren](ClO_4)_2$ showed evidence of a slow transition below 4.2 K, suggesting a tunnelling mechanism.

### 7.5.1.6 *Pressure effects*

High to low spin cross-over can be induced by compression. $Fe[HB(3.5\,(CH_3)_2\,Pz)_3]_2$ seems very sensitive and a pressure of 4 kBar leads to equal

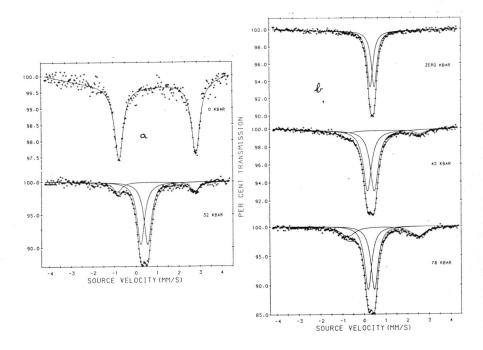

Pressure induced spin cross-over in (a) Fe[HB(3.4.5. methyl pyrazoyl)$_3$]$_2$
at room temperature. HS $\rightarrow$ LS: (b) at low temperature LS $\rightarrow$ HS.

### Fig.7.30 a & b

proportions of high and low spin forms. The transition is fully reversible.(Fig. 7.30)

Pressure induced spin cross-over can occur with compounds that do not show temperature dependent cross-over. Fe(dppe)$_2$X$_2$, where dppe = diphenylphosphinoethane and X is I or Br, are high spin between 298 and 4.2 K, although their solvates display cross-over. A pressure of 7 to 9 kBar applied to the chloro compound leads to the formation of the low spin state. The spectra of the two spin states are well separated. Complete conversion does not appear to be possible.

Other changes accompanying spin cross-over include an increase in the Mössbauer fraction, f, as the proportion of the low spin form increases, indicating different Debye temperatures for the two forms. The second order Doppler effects also differ. The sign of the quadrupole splitting changes, the high spin form usually having the positive sign.

Most of the compounds showing spin cross-over in the solid state also do so in solution, but such changes are best investigated by spectrophotometry or magnetic measurements.

Spin cross-over can be observed when the compounds are adsorbed on a substrate such as silica.

### 7.5.2   Other Spin Changes

A few cases of S =1 $\Leftrightarrow$ S = 0 have been studied. They all involve five coordinate iron(II). $FeLCl_2.H_2O$ where L is the tridentate ligand 1.10 Phen 2 carbothioamide.

The complex gives an S = 1 quadrupole split pair at room temperature, with $\delta = 0.22$ and $\Delta = 0.53$, and another doublet at low temperatures with $\delta = 0.23$ and $\Delta = +1.33$, the lines overlapping to some extent.

Another well substantiated example is $[FeP_4X]BPh_4$, where X = Br or I and $P_4$ is hexaphenyl 1,4,7,10 tetra phosphadecane. Unlike the previous example. this is an instance of rapid spin cross-over, a single spectrun is seen but its parameters change smoothly with temperature. At 4.2 K $\delta = 0.13$ and $\Delta = 2.25$ while at 298 K $\delta = 0.20$ and $\Delta = 1.81$. Magnetic measurements confirm that spin cross-over is responsible.

### 7.5.3 Spin Cross-over in Iron(III) Compounds

A rather smaller number of cases of cross-over in iron(III) complexes have been investigated. Most of them involve the S = 5/2 $\Leftrightarrow$ S = 1/2, or $^6A_1 \Leftrightarrow {}^2T_2$ change. The rate of this change is generally faster than the S = 2 $\Leftrightarrow$ S = 0 change and in many cases the life time of the individual states is too short for two spectra to be seen. A weighted average spectrum is found; weighted according to the proportions of the two spin states present. The results can be compared with magnetic susceptibility data. In a few compounds $\tau_r$, the life time of the spin state may be sufficiently close to $\tau_n$ the mean life of $^{57m}Fe$ that substantial line broadening will occur. The complex bis (N Me ethylenediamine salicylaldiminato iron(III) hexafluorophosphate seems to be an example.

A spectrum showing a quadrupole split pair, the splitting changing with the temperature and therefore the proportion of the LS and HS present, sets rather narrow limits for $\tau_r$ since in this case $\tau_n > \tau_r > \tau_q$, where $\tau_q$ is the quadrupole measurement time. Now $\tau_n \approx 1.4.\ 10^{-7}$ s and for a quadrupole splitting of 1 mms$^{-1}$ $\tau_q \approx 8.6\ 10^{-8}$ s.

Fast kinetic studies have confirmed that $\tau_r$ may be of this order of magnitude. Electron paramagnetic resonance, for which the precession time corresponding to $\tau_q$ is about $10^{-9}$ s, shows signals due to both HS and LS forms Bis (N methyl ethylene-diamine) salicylaldiminatoiron(III) is an example of such behaviour.

Many of the iron(III) dithiocarbamates, $Fe(S_2CNR_2)_3$, show fast cross-over. Typical spectra are shown in Fig.7.32.

Cases of slow cross-over in iron(III) complexes are much less common. Most are complexes of the type $Fe[N_4O_2]$. Examples are bis N 8 quinolyl salicylaldimine Fe(III) chloride and bis acetylacetonato trisethylenediamine iron(III) tetraphenyl borate. In both cases the identity of the anionic component is important; other salts show fast cross-

Fig.7.31

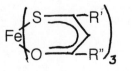

over. The complexes shown in Fig.7.31 with R' = $CH_3$ and R" = $C_6H_5$ or R' = $C_6H_5$ and R" = $CH_3$ also show slow cross-over.

Perhaps the clearest demonstration of the importance of the lattice in spin cross-over is found in the case of $FeLB(Ph)_4$ where L is the ligand produced when trisethylenetetraamine

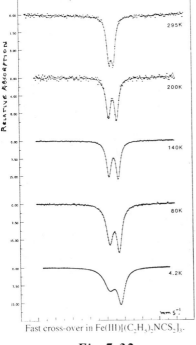

Fast cross-over in Fe(III)[(C₂H₅)₂NCS₂]₃.

**Fig.7.32**

is condensed with salicylaldehyde. The complex exists in two crystal modifications. One form is high spin over the temperature interval 320 - 78 K, whereas the monoclinic form changes from high to low spin within this range.

### 7.5.4 $3/2 \Leftrightarrow 1/2$ cross-over

The best established examples of $3/2 \Leftrightarrow 1/2$ cross-over are the NO adducts of the Schiff base complexes of iron. The cross-over is fast on the Mössbauer time scale and averaged spectra are seen. For FeSalphen NO the isomer shift changes smoothly from 0.29 at 300 K to 0.15 at 100 K while the quadrupole splitting runs from 0.198 to 1.738 over the same temperature interval. $T_{50}$ is about 180 K. The NO adduct $[FeTMC.NO.NCCH_3](BF_4)_2$ also appears to undergo fast $3/2 \Leftrightarrow 1/2$ cross-over.

## 7.6  IRON SPECIES IN SOLUTION, ON ADSORBENTS AND IN MATRICES

### 7.6.1  Adsorbed and Ion Exchanged Iron

Mössbauer spectroscopy is a valuable technique for determining the oxidation state, coordination and location of adsorbed iron species.

#### 7.6.1.1  *Iron in zeolites.*

Two kinds of investigation can be distinguished:  (i) determination of the state of iron present as a component in the zeolite structure, usually in natural zeolites, and (ii) the state of adsorbed or exchanged iron ions in iron free, usually synthetic, zeolites.

Both iron(II) and iron(III) can be adsorbed from solution by zeolites, and the products can be dehydrated and the reactions of the iron species in the zeolite with other adsorbates explored.  Unless otherwise specified the data given here refers to spectra measured at 298 K.  $Fe^{2+}$ exchanged into $NH_4$ Zeolite A, while fully hydrated, gives two spectra with Mössbauer parameters: $\delta = 1.26$ and $\Delta = 2.54$, 43% and $\delta = 1.25$ with $\Delta = 1.91$, 57%. The line widths of all the spectra in this section are rather large, typically 0.3 to 0.4 mms$^{-1}$ suggesting that the environment of the iron is somewhat variable.

Both these spectra show the $Fe^{2+}$ to be located in nearly octahedral coordination. Taking account of structural data on this zeolite, it has been concluded that the outer doublet arises from $Fe^{2+}$ situated at the six ring sites separating the sodalite cages from the large cavities. The iron is in a trigonally distorted octahedron, coordinated to three oxygens of the zeolite structure and three oxygens from water molecules. The proportions of the inner doublet are markedly dependent on the pH at which the iron was introduced. This spectrum is due to $Fe^{2+}$ at similar sites to those already mentioned but at which one cordinated water is replaced by an OH group.

Upon dehydrating at $100^-$C two spectra are seen with parameters: $\delta = 1.13$ and $\Delta = 2.30$, 27% and $\delta = 0.84$ with $\Delta = 0.48$, 73%. Both are undoubtedly due to iron(II) species,the latter arising from a less symmetric environment and lower coordination number for the iron.

In moist air or oxygen an iron(III) spectrum with parameters $\delta = 0.40$ and $\Delta = 0.70$ develops, undoubtedly due to six coordinate $Fe^{3+}$: indeed direct adsorption of $Fe^{3+}$ on this zeolite gives a spectrum with $\delta = 0.38$ and $\Delta = 0.87$. Reducion in situ then gives a spectrum in which $\delta = 1.05$ with $\Delta = 2.31$.

Treatment of the dehydrated $Fe^{2+}$ Zeolite A with ethanol or acetonitrile vapour changes the Mössbauer spectrum, showing that the $Fe^{2+}$ cannot be in the sodalite cages where they would be inaccessible to the new ligands. The spectra of the alcohol treated dehydratedzeolite were very similar to the those for the fully hydrated material. However re-hydration does not simply regenerate the fully hydrated spectrum. A fully hydrated sample whose spectrum measureed at 100 K gave the parameters $\delta = 1.37$ and $\Delta = 3.15$ as well as $\delta = 1.37$ with $\Delta = 2.31$ on rehydration after dehydration at $100^-$C gave $\delta = 1.35$ and $\Delta = 2.85$. More significantly the area under the absorbtion peaks in the spectrum decreases by about 20% after rehydration, indicating that some change in the zeolite structure has taken place.

Plots of dlog A/dT provide information about the mechanical coupling of the iron ions to the zeolite lattice.

Since the iron is always oxygen coordinated in these systems the isomer shift distinguishes tetrahedral and octahedral coordination of the iron. $Fe^{3+}$ generally occupies tetrahedral sites so that the isomer shift is $\leq 0.32$ . In octahedral coordination the shift is generally $\geq 0.4$ mms.

Different types of zeolites give quite different environments. Thus $Fe^{2+}$ exchanged into potassium L zeolite is found in two kinds of site, A and E. The spectra obtained comprise two quadrupole split doublets, the A site gives $\delta = 1.12$ and $\Delta = 2.04$, 50%; the E site gives $\delta = 0.95$ and $\Delta = 0.72$, 50%. These data show the A site is a not much distorted octahedral site, but that the E site is a lower symmetry site of lower coordination number for the iron. The structure of this zeolite suggests that this is a planar four coordinate site.

On adsorbing ethanol, the proportion of iron in the E site decreases as shown by the following spectral parameters for the three quadrupole split doublets that appear: $\delta = 1.08$ and $\Delta = 1.82$, 44%; $\delta = 1.22$ and $\Delta = 2.76$, 41% and $\delta = 0.96$ with $\Delta = 0.71$, 15%. The iron in the E site coordinates alcohol giving another six coordinate species and some of the A site iron also attaches alcohol. With still larger amounts of adsorbed alcohol the four coordinate species disappears completely. More bulky alcohols, such as iso butyl alcohol, this kind of reaction may only be able to take place to a limited extent.

Similar studies have been made of iron adsorbedon a variety of other substrates.

### 7.6.1.2 *Iron on ion exchange resins.*

Cationic and anionic iron species attached to ion exchangers are easily investigated by Mössbauer spectroscopy. Iron$^{2+}$ cations on, for instances Dowex 50, give quadrupole split spectra resembling the spectra of hydrated iron(II) salts, but with broader lines, like the spectra described in the next section. Iron(III) give more complex spectra showing evidence of hydrolysed and dimeric or polymeric species. Magnetically split spectra may also be obtained especially in dilute systems.

### 7.6.2 **Frozen Solutions of Iron Compounds**

Mössbauer spectra can be measured using frozen solutions of iron compounds, so that it may be possible to learn something about the species in the solutions. There are however both experimental and interpretational difficulties with the technique.

First it is necessary to ensure that freezing does not simply lead to the separation of small crystallites of the solute, which is especially important in the case of solutions of iron salts. In addition any effects arising from the crystal structure of the frozen solvent must be avoided. These two objectives are most often reached by very rapid freezing to produce a glassy solid.

One must be cautious of facilitating the production of a glassy solid by the addition of another solvent, since this may change the nature of the iron species in the solution, for instance by producing mixed solvates with cationic iron solutes.

Iron(III) compounds often yield slowing relaxing, or even clearly magnetically split, spectra at room temperature. The dilution of the iron in the frozen solid leads to slow spin-spin relaxation.

Finally there is the problem of verifying the relation of the spectrum observed to the iron species in the solution. Taking as an example frozen aqueous solutions of an iron(II) salt of a weakly complexing anion one finds that:-

    i/ Over a reasonable range of low concentrations
       of the salt the spectrum remains the same.
    ii/ In the pH range 0 to 3 the spectrum is
       unchanged.
    iii/ On the addition of a complexing anion a
       different spectrum is obtained.

These observations suggest that the spectrum in (i) and (ii) is in fact due to the hydrated $Fe^{2+}$ ion. Addition of other solvents may give another, mixed solvate, spectrum, but because water is strongly solvating a rather high concentration of the second solvent may be necessary.

The quadrupole split spectra obtained from such frozen solutions have about twice the line width found for solid hydrated iron(II) perchlorate but roughly the same splitting. This suggests the iron ions are in varying environments, but substantially similar to the that in the hydrated iron(II) salts.

The addition of another ligand can be investigated in this way. A solution of the five coordinate Fedtc$^\prime$X in a non-coordinating solvent will on addition of a donor solvent show a profound change in spectrum as it changes from a five to a six coordinate species.

The technique has been used to explore in greater detail the species produced by the hydrolysis of $Fe^{3+}$ solutions.

Similar studies can be made of frozen solvent extracts of iron compounds. These can be very helpful in establishing the identity of the extracted species.

### 7.6.3 Iron in Rare Gas Matrices

Iron gives a monatomic vapour. If a beam of krypton or xenon containing a low concentration of iron vapour is directed on to a liquid helium cooled beryllium target a matrix containing iron atoms is formed. The Mössbauer spectrum can be measured in transmission in the usual way.

At great dilution, iron to rare gas about 1 to 150, a single line spectrum is seen, $\delta = -0.75$. At a ratio of 1 to 50, and also if the matrix is allowed to warm slightly, a quadrupole split pair of sharp lines is found in addition to the single line. These arise from a dimer which gives $\delta = -0.13$ and $\Delta = 4.1$. On still longer annealing other $Fe_n$ species appear. Slightly different Mössbauer parameters are found for other rare gas matrices.

A similar procedure permits the measurement of the spectra of matrix isolated molecules such as $FeCl_2$.

Low temperature reactions in a matrix have been explored by co-depositing iron and tin in an argon matrix at 4.2 K. By varying the concentrations of the two elements and annealing at up to 20 K evidence for $FeSn$, $FeSn_2$, $FeSn_3$ and $Fe_2Sn_3$ was obtained. Their Mössbauer parameters are given in Table 7.6.

### Table 7.6.

| Compound | FeSn | $FeSn_2$ | $FeSn_3$ | $Fe_2Sn_3$ |
|---|---|---|---|---|
| $^{57}$Fe parameters $\delta / \Delta$ | 0/3.26 | 0.12/2.06 | 0.2/2.7 | 0.07/0.61 |
| $^{119}$Sn parameters $\delta / \Delta$ | 2.0/4.32. | --- | 2.9/1.8 | --- |

Interesting results have been obtained incorporating iron in CO, $NH_3$, $N_2$, $C_2H_4$ and $C_3H_5$ matrices. Compounds of the type FeX, X = $NH_3$, $C_2H_4$ and $C_3H_6$, have isomer shifts from 0.5 to 0.6 and large quadrupole splittings, > 2. Compounds of the type $FeX_3$ form with $N_2$, $C_2H_4$ and $C_3H_6$, The last two have similar isomer shifts to the analogous monomeric compounds but lower quadrupole splittings. In the nitrogen matrix however the parameters are -0.78/2.70. Now the isomer shift of iron atoms in a rare gas is about -0.75 so that the interaction of the nitrogen with the $Fe_2$ unit must lead to an $|\Psi(0)|^2$ close to that on an iron atom. The electronic configuration of the iron in these compounds is not yet clear.

The formation of a number of these compounds is dependent on annealing the matrix at above 4.2 K for some time.

By incorporating $Fe(CO)_5$ in a rare gas matrix and photolysing the carbonyl, spectra for species of the type $Fe(CO)_n$ can be measured. Reactions of the fragments with for example $C_2H_4$ can be investigated by laying down both iron carbonyl and ethylene in the matrix.

### 7.6.4 **Iron containing intercalated systems.**

A number of solid compounds can incorporate foreign molecules in their lattice. The precise mode of interaction of the foreign molecules with the substrate lattice is not thoroughly understood.  It is believed that the bonding is of the nature of a charge transfer interaction.  Mössbauer spectroscopy can contribute to the clarification of such interactions since substrates and intercalates containing suitable elements are well known.

The substrates can be divided into the charge donors, such as graphite and the charge acceptors such as $FeOCl$.  In some cases the incorporation proceeds towards some stoichiometric proportions, as is the case for the graphite/ iron(III) chloride system; in others there is no real evidence for this, as is the case for graphite/iron(III) bromide.

At room temperature the graphite/iron(III) chloride system shows a single line spectrum with an isomer shift somewhat above that found for anhydrous iron(III) chloride. At 80 K two additional spectra are seen, a quadrupole split iron(III) spectrum and another doublet due to iron(II).  The change on cooling is reversible.  These results may indicate a temperature dependent electron hopping between the graphite and the iron(III).  At higher temperatures an irreversible change takes place and two quadrupole split doublets due to iron(II) are seen.

Graphite intercalates with $SnCl_4$, $SbCl_5$ and $EuCl_3$ have also been studied by Mössbauer spectroscopy.  A reversible reduction is seen with $SbCl_5$, as with $FeCl_3$, but not with the other two chlorides.

FeOCl yields intercalates with most organic amines and phosphines. At low temperatures an iron(II) spectrum is seen, at room temperature a quadrupole split spectrum with a somewhat larger isomer shift than found for normal FeOCl.  These results are compatible with reversible electron transfer from the intercalate to the substrate and rapid electron hopping or delocalisation of valence between iron(II) and iron(III) in the substrate.  An interesting system is FeOCl/Ferrocene where both reduction of sustrate and oxidation of intercalate can be observed.

### Acknowledgements

Fig.7.1   Reproduced with permission from Blum,N., Freeman,A.J., Shaner,J.W. and
Grodzin,L. 1965)  *J.Applied Phys.*, **36** 1169.

Fig.7.2   Reproduced with permission from van der Woude,F., (1966)
*Phys.Stat.Solidi,* **17**, 417.

Fig.7.4   Reproduced with permission from Longworth,G. and Tite,M.S. (1977)
*Archeometry,* **19**, 3.

Fig.7.5   Reproduced with permission from Murad,E. and Schwertmann,U. (1983)
*Clays, Clay Minerals,* **18**, 301.

Fig.7.6   Reproduced with permission from de Grave, E., Bolen,L.H.,Vochten,R. and
Vandenberghe,B.E., (1988) *J.Mag.Magnet.Mater.,* **72**, 141.

Fig.7.7   Reproduced with permission from Chappert,J. and Frankel,R.B. (1967)
*Phys.Rev.Letts.,* **19**, 570.

Fig.7.8   Reproduced with permission from Osbourne,N.E., Fleet,M.E. and
Bancroft,G.M. (1983) *Solid State Comm.,* **47**, 623.

Fig.7.9 Reproduced with permission from Sawatsky,G.A., van der Woude, F. and
Morrish,A.H. (1969) *Phys.Rev.,* **187**, 747.

Fig.7.10 Reproduced with permission from Pettit,G.A. and Forester,D.W. (1971)
*Phys.Rev.,* **4B**, 3912.

Fig.7.11 Reproduced with permission from Pettit,G.A. and Forester,D.W. idem.

Fig.7.12 Reproduced with permission from Banerjee,S.K., O'Reilly,W. and
Johnson,C.E. (1967) *J.Applied Phys.,* **30**, 1289.

Fig.7.19 Reproduced with permission from Wohler,S.E., Witterbort,R.J.,
Seung,M.Oh., Takeshi Kambara, Hendrickson,D.N., Inniss,D. and
Strause,C.E. (1987) *J.Amer.Chem.Soc.,* **109**, 1073.

Fig.7.21 Reproduced with permission from Webb,R.J., Geib,S.J., Staley,D.L.,
Rheingold,A.L. and Hendrickson,D.N. (1990) *J.Amer.Chem.Soc.,*
**112**, 5031.

Fig.7.22 Reproduced with permission from Teng Yuan Dong, Hendrickson,D.N.
Kumiko Iwai, Cohn,M.J., Geib,S.J., Rheingold,A.L., Izumi Motoyama and
Saroru Nakashima. (1985) *J.Amer.Chem.Soc.,* **107**, 7996.

Fig.7.24 Reproduced with permission from Koppen,H., Moller,E.W., Koehler,C.P.,
Spiering,H., Meissner,E. and Gütlich,H. (1982) *Chem.Phys.Letts.,* **91**, 348.

Fig.7.25 Reproduced with permission from Lemercier,G., Rousseksou,A.,
Seigneuric,S., Varret,F. and Tuchagues,J.P. (1994)
*Chem.Phys.Letts.,* **226**, 259.

Fig.7.27 Reproduced with permission from Vreugdenhil,W.M., van Diemen,J.H.,
de Graff, R.A.G., Haasnoot,J.G., Reedijk, J., van der Kraan, Kahn,O. and
Zarembowitch,J. (1990) *Polyhedron,* **9**, 2971.

Fig,7,28a, b & c. (a) Reproduced with permission from Sams,J.R.and Fsin,T.B.
(1976) *J.Chem.Soc.(Dalton)* **1976**, 488. (b) Reproduced with permission from
Sorai,M., Ensling,J., Hasselbach, K.M. and Gülich,P. (1977)
*Chem.Phys.* **20**, 197. (c) Reproduced with permission from Gülich,P.,
Koppen, H. and Steinhauser, H.G. (1980) *Chem.Phys.Letts.,* **74**, 475.

Fig.7.29 Reproduced with permission from Ganguli, P., Gütlich,P. and Muller,E.W.
(1982) *Inorg.Chem.,* **21**, 3429.

Fig.7.30 Reproduced with permission from Long,G. and Hutchinson,B.B. (1989)
*Inorg.Chem.,* **26**, 608.

Fig.7.32 Reproduced with permission from Fiddy,J.M., Hall,I., Grandjean,F.,
Russo,U. and Lang,G.J. (1987) *Inorg.Chem.,* **28**, 4138.

**8**

# CEMS and Applications in Mineralogy and Biochemistry.

## 8.1 CONVERSION ELECTRON MöSSBAUER SPECTROSCOPY, (CEMS).

Besides measurements in a transmission mode Mössbauer spectra can be recorded by observing the back scattered 14.4 keV photons by placing the detector in front of the absorber but outside the direct photon beam from the source. With suitable collimation and screening of the detector, the spectrum will appear as peaks rising above a rather low background. However the efficiency of such an arrangement is rather low because of internal conversion.

Another possibility is to record the conversion electrons produced by the de-excitation of the $^{57m}Fe$ formed in the absorber. Only about 10% of the $^{57m}Fe$ decays by 14.4 keV photon emission. In more than 90% of the decay events internal conversion in the K shell of the iron takes place giving rise to 7.3 keV electrons and 6.45 keV iron X-rays. The latter are however substantially internally converted to produce 5.6 keV electrons and L iron X-ray photons, (See sec.2.2.1). Thus each absorption event leads to the production of 1.53 soft electrons by the absorber.

A simple CEMS counter cum absorber holder is shown in Fig.8.1.

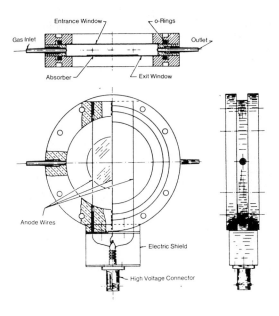

**Fig.8.1**

8.1.1  **Glazes**

Glazes can be investigates non-destructively by attaching a detector of this kind to the object under study. The well known celadon blue and ox-blood glazes on Chinese vases have been explored in this way.

8.1.2  **Depth Resolved CEMS**

A valuable feature of this mode of measuring Mössbauer spectra is that the spectrum obtained relates to the superficial material of the absorber. Measuring the scattered 14.4 keV and the 6.45 keV photons, one samples the first 10.to 20 μm of the scatterer because of the strong absorption of such low energy photons by the absorber.

Similarly if one measures the 7.3 and 5.4 keV conversion electrons the spectrum is determined by the first 300 nm or so below the surface of the absorber, because of the very short range of such soft electrons in the absorber. Using a proportional counter to detect the electrons, and recording only pulses above a selected size, one can obtain spectra from the material in layers at different depths below the surface of the absorber. In particular, by selecting pulses of the size produced by 7 keV electrons, one can record the spectrum of the material in the first 10 or 20 nm from the surface.

With much greater elaboration, feeding the emitted electrons into a soft electron spectrometer and recording spectra due to electrons in a narrow band of energies, one can explore the material at different depths in the absorber.

Fig.8.2 shows a schematic diagram of an apparatus for evaporating a coating onto a target and subsequently measuring the spectrum at different depths in the target.

A similar apparatus has been used for studying corrosion.

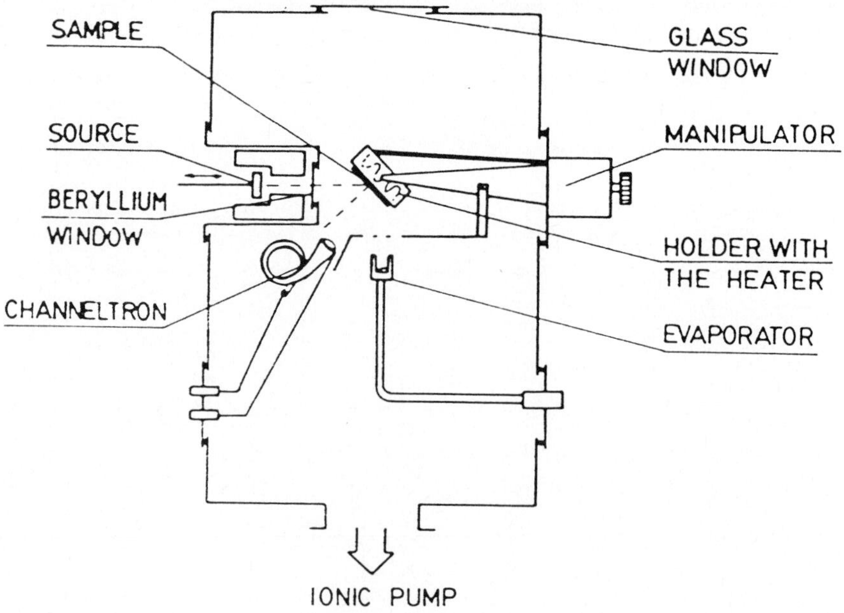

IONIC  PUMP

**Fig.8.2**

Fig.8.3 shows the spectrum of a stainless steel foil coated with different thicknesses of soft iron, recorded for conversion electrons without any pulse size selection. The magnetically split spectrum arises from the soft iron and the single centre line from the stainless steel. The centre line has almost disappeared in the foil with a 300 nm coating of soft iron. It is interesting to note that the $\Delta m_I = 0$ lines in the magnetic spectrum are relatively weak indicating that the magnetic field in the iron is orientated in the plane of the coating.

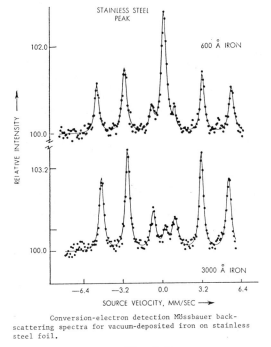

Conversion-electron detection Mössbauer back-scattering spectra for vacuum-deposited iron on stainless steel foil.

**Fig.8.3**

### 8.1.3  Corrosion.

The technique is well suited to the study of the corrosion of iron and steel and of passivation by various treatments. In either case the spectra will identify the compounds formed on the iron surface. In this way the nature of the green rusts and of the compounds formed by phosphoric acid passivation of iron have been explored.

### 8.1.4  Industrial applications

Besides the study of corrosion, the CEMS technique has found industrial application in the study of the case hardening of iron and of the changes taking place due to wear in steel ball bearings.

## 8.2  REACTIONS IN THE SOLID STATE.

Mössbauer spectroscopy is one of a small number of techniques available for the *in situ* study of reactions in solids. Its quantitative sensitivity to the detection of a new phase is not very high; but it has the advantage that rather small domains of the new phase are sufficient. Unfortunately it cannot be used for continuous monitoring of the changes because the recording of a spectrum takes a substantial time. The process must be frozen out during the measurement.

Thermal decompositions of numerous compounds containing elements yielding Mössbauer spectra have been reported. Amongst the most interesting are those relating to iron and steel.

### 8.2.1  Iron and Steel

The body centred cubic $\alpha$ iron gives a magnetically split spectrum with $B_{hf} = 33$ T.

Dissolved carbon, in the martensite phase, occupies tetragonally distorted octahedral sites. This leads to a shorter Fe-C distance for the two axial iron atoms than for the four equatorial atoms in the first coordination sphere of the carbon. The spectrum comprises three sextets, two overlapping considerably. The axial iron atoms give the well resolved sextet with the lower hyperfine magnetic field. The equatorial iron atoms lead to a sextet which broadens the lines in the higher field sextet, which is due to the more remote iron atoms and is essentially the same as the $\alpha$ iron spectrum. The Curie temperature for this phase is about 1500 K.

At about 1200 K the iron changes to the $\gamma$, face centred cubic, phase. This also dissolves carbon to give the austenitic phase which is paramagnetic. On quenching the austenitic phase some part escapes conversion to the martensite phase and a paramagnetic component to the spectrun is found. It comprises a single line due to iron atoms remote from carbon and in cubic sites, together with a quadrupole split pair from iron atoms close to carbon atoms.

Thermal treatment of carbon rich steels leads to the formation of small domains of cementite, $Fe_3C$, $Fe_5C_2$ and other carbides. The contribution of these to the rather complex spectra obtained can be identified. The segregation of cementite in cast iron can readily be seen in its spectrum.

Similar studies have been made on the iron-nitrogen, iron-silicon and other systems. The iron-silicon system gives solid solutions at low silicon content with spectra arising from the disordered iron environments, and ordered iron compounds at higher proportions of silicon.

An interesting application uses the CEMS technique to explore the changes taking place in the surface layers of ball-bearing steel upon plastic deformation and stressing. Spectra showing these effects are shown in Fig.8.4

### 8.2.2   Firing of Pottery

The changes that take place during the firing of clays to produce pottery have been studied extensively by Mössbauer spectroscopy. It was hoped that the spectra might be used to provide evidence of the provenance of ancient pottery, the conditions of firing and, perhaps, the time that has elapsed since firing.

Some of the clay minerals, for example biotite, contain $Fe^{2+}$ with four silicate oxygens and two hydroxyl ligands and the first effect of firing is a dehydroxylation reaction. In air or an oxidising atmosphere

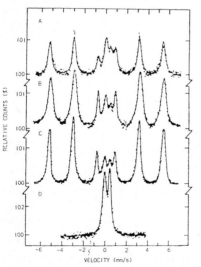

Conversion electron Mössbauer spectra of a ball-bearing steel sample; (A), before plastic deformation; (B), after stressing the sample at 5900MPa for 10 cycles; (C), after 40 cycles at 5900MPa stress; (D), spectrum of the alloyed carbide, $(Fe_{0.85}Cr_{0.15})_3C$, electrochemically extracted from the investigated steel.

**Fig.8.4**

the change is accompnied by oxidation of the iron. Oxidation of the mineral $Fe^{2+}$ is practically complete by 650 K.

Before firing, the haematite in the clay yields a magnetically split spectrum at room temperature. $Fe^{3+}$ in the clay minerals gives a quadrupole split doublet with a small splitting and the $Fe^{2+}$ further doublets generally with larger splittings. During low temperature firing in air the relative intensities of the two $Fe^{3+}$ spectra increase and spectra measured at 4.2 K show a new sextet with broad lines and a smaller hyperfine field. The latter probably arises from superparamagnetic particles of iron oxide or hydroxide.

Above 650 K vitrification begins and the relative intensity of the $Fe^{3+}$ doublet decreases, while that of the haematite component increases. The superparamagnetic iron oxide is aggregating to give larger crystallites. Finally at above about 1450 K crystallisation begins, the haematite spectrum disappears and the iron becomes incorporated in a silicate phase.

The spectral differences consequent to these changes distinguish between oxidising and reducing atmospheres and indicate the firing temperature. However unless the raw material used in producing the pottery is available, it is difficult to give quantitative expression to the interpretation of the spectra..

During firing in the 650 - 1050 K region aggregation of $Fe''O_-$ to give larger crystallites is important. There is evidence that on ageing pottery the reverse reaction takes place, the larger crystallites fragmenting. Such a change should be revealed by a change in the blocking temperature (See section 7.1.4). It was hoped that this effect might provide a means of dating pottery; however measurements of blocking temperatures in these materials are difficult and not very precise.

## 8.3    APPLICATIONS TO MINERALOGY AND PETROLOGY

A large proportion of minerals contain appreciable amounts of iron, and Mössbauer spectroscopy finds extensive application in mineralogy. The oxidation state and coordination of the iron can usually be deduced from the spectra (See Figure 5.1 b).

Although the iron is most commonly present as $Fe^{2+}$ in a six coordinate oxygen environment the quadrupole splittings found cover a wide range of values. The ligand contribution to the EFG is of opposite sign to that of the valence term and generally, the more the environment departs from octahedral symmetry, the smaller the quadrupole splitting. The differences are such that even if there are four or more iron(II) sites they can usually be distinguished by their spectra. Overlapping spectra at one temperature may be resolved at another temperature.

### 8.3.1  **Mineral Identification**

The Mössbauer spectrum is an unique property of the mineral. But care is necessary in the interpretation of the spectra. Minerals seldom possess a fixed composition. The same crystal phase may be found with varying proportions of calcium, magnesium, zinc or other divalent ions occupying the cationic sites. Even with fixed proportions of these cations, if they are randomly distributed the iron(II) will be present in a number of next nearest neighbour environments and, as in the case of the spinels (Sec. 7.2.1), this leads to a series of quadrupole splittings. The differences in EFG are not usually very great but may lead to appreciable line broadening.    In favourable cases the line shape may be analysed in terms of splittings for the different environments, as in Fig.8.5.

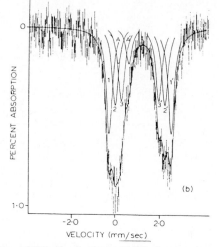

Spectrum of C2/c Omphacite, at 295 K, showing three $Fe^{2+}$ doublets, 1,2,3, arising from the different combinations of $Ca^{2+}$ and $Na^+$ as next nearest neighbours.

## Fig.8.5

In many minerals the iron is present in a number of cationic sites. This complicates the spectra, each kind of site leading to a different spectrum, but it also leads to greater reliability in the identification of the mineral.    In the absence of relaxation effects, as is usually the case for iron(II), and if line broadening due to next nearest neighbour effects is unimportant, the analysis of the spectra is facilitated by assuming equal areas for each line of the quadrupole split pairs.    But this may not be justified if the    $Fe^{2+}$ is present in a highly distorted site, such as in planar coordination.

### 8.3.2 **Analytical applications**.

The Mössbauer spectrum can provide estimates of the proportions of iron(II) and iron(III) in different sites in a mineral.

The area enclosed by the perimeter of an absorption line and the non-resonant base line is determined by the number of iron atoms per unit area of the    absorber occupying the kind of site giving rise to the line. For a line with the ideal Lorentzian shape the area enclosed by the line and the non-resonant absorption base line is given by $\pi h\Gamma/2$. h, the dip, being the separation of the absorption maximum from the base line. The general expression for the transmission integral, which relates this line area to the number of iron atoms per unit area of absorber is complex (see Sec. 2.6.1) but a useful approximations can be derived for restricted conditions of measurement.    In effect the restrictions ensure that the experimental line shape is close to the idealised Lorentzian form.

There must be no substantial overlap of lines and the thickness of the absorber, $t_a = f_a n \sigma_0$, should not exceed 0.2. For an iron compound this implies that the iron content of the absorber should not exceed 0.6 mg.cm$^{-2}$ of iron of normal isotopic composition. With such a thin absorber $\Gamma_e = (2 + 0.027\, t_a)\Gamma_t$, the subscript e indicating the experimental and t the theoretical line width, as determined by the mean life of $^{57m}$Fe.

In practice a moderate amount of line overlap proves to be tolerable.

Under the above conditions $A_i$, the area under line i is given by $\pi f_s f_{ai} \beta_i n \sigma_0 \Gamma_i / 2$, where the f are Mössbauer fractions, s for the source and a for the absorber; n is the number of iron atoms per unit area of absorber; $\sigma_0$ the resonant absorption cross section; $\beta_i$ the fraction of the iron atoms in sites giving rise to the line i and $\Gamma_i$ the line width. The subscript i implies values relevant to the line i. Thus for lines i and j $A_i/A_j = f_{ai}\beta_i\Gamma_i/f_{aj}\beta_j\Gamma_j$. If measurements are made on minerals, which have high Debye temperatures, at low temperatures $f_{ai} \cong f_{aj}$ and in the absence of relaxation effects $\Gamma_i \cong \Gamma_j$, so that $A_i/A_j \cong \beta_i/\beta_j$ and the areal ratio gives the proportion of iron atoms giving rise to the two lines. In practice the absorbers used may have $t_a > 0.2$ and $A_i/A_j = G(t_a)\beta_i/\beta_j$, where $G(t_a) \rightarrow \square 1$ as $t_a \rightarrow 0$.

If the coordination numbers of the iron atoms are different, the approximation regarding the Mössbauer fractions may not be valid. In some cases a mineral may be available in which all the sites occupied by the $Fe^{2+}$ and $Fe^{3+}$ are filled by iron cations. The ratio of the two kinds of site can be obtained from crystallographic data and so a measurement on this mineral will give an areal ratio that can be compared with the crystallographic ratio, and an experimental value for $G(t_a)f_{ai}\Gamma_i/f_{aj}\Gamma_j = F$ obtained.

Although the determination of the ratio of amounts by line areas is of limited accuracy it has various useful applications.

### 8.3.3  **Proportions of $Fe^{2+}$ and $Fe^{3+}$ in Minerals**

The opening up of minerals for conventional chemical analysis tends to lead to oxidation of $Fe^{2+}$. Measurements of line areas can give the proportions of iron(II) and iron(III) more easily and with no worse accuracy than conventional analysis. The procedure is most sensitive for determining small amounts of iron(II) in an excess of iron(III)

The method is useful in establishing oxygen fugacities at the time of formation of the mineral. It was used to set an upper limit to the occurence of iron(III) in the moon samples. With a value from chemical analysis for the total iron content the separate values for the iron(III) and iron(II) concentrations can be obtained.

### 8.3.4  **Site Preferences**

Numerous minerals have more than one kind of octahedral cationic site. The amphiboles have four sites M1, M2, M3, M4 in the ratio 2 : 2 : 1 :2.    The orthopyroxenes have two sites. These sites will be occupied by various divalent cations, $Fe^{2+}$, $Mg^{2+}$, $Ca^{2+}$, $Mn^{2+}$ and others, in different proportions in different mineral samples. The question arises do the different cations occupy the sites at random or does a preference of, for example, $Fe^{2+}$ for one kind of site lead to some degree of cation ordering?

In the amphiboles, in the cummingtonite-grunerite series of minerals, the iron(II) in the more distorted M4 sites gives the lowest quadrupole splitting and the spectrum is well resolved from the iron in the other sites. Ignoring minor cationic components the mineral can be represented:- $(Fe,Mg)_7Si_8O_{22}(OH)_2$. For this series a best value of F is 0.9.

If the $Fe^{2+}$ occupied the cationic sites at random the proportion of the iron in a given kind of site would be the proportion of that kind of cationic site in the lattice.

The spectra in figure 8.6 show clearly that the iron has a preference for the M4 sites. The line areas provide quantitative expression of this preference.

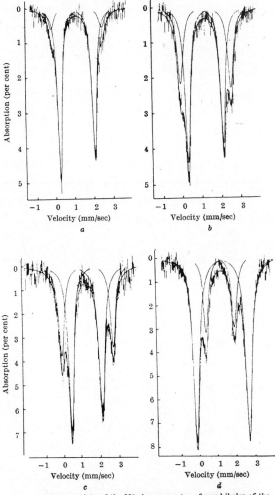

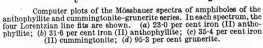

Computer plots of the Mössbauer spectra of amphiboles of the anthophyllite and cummingtonite–grunerite series. In each spectrum, the four Lorentzian line fits are shown. (*a*) 23·0 per cent iron (II) anthophyllite; (*b*) 31·6 per cent iron (II) anthophyllite; (*c*) 35·4 per cent iron (II) cummingtonite; (*d*) 95·3 per cent grunerite.

**Fig.8.6**

One can treat the data in terms of an exchange of magnesium and iron on the different sites. Choosing the simpler case of the orthopyroxenes which have only two kinds of cationic site, M1 and M2, the $Fe^{2+}$ preferring the more distorted M2 site. One can write $Fe_1 + Mg_2 \Leftrightarrow F\underline{e} + Mg_1$. with an equilibrium constant in terms of atomic fractions $K = x_2(1 - x_1)/x_1(1 - x_2)$ where $x_1$ represents the fraction of iron atoms on the M1 sites etc. Expressing this in terms of the experimental quantities $y = Fe_1/Fe_2$ obtained from the areal ratio and $x = (Fe_1 + Fe_2)/(Fe_1 + Fe_2 + Mg_1 + Mg_2)$, which can be obtained by chemical analysis, one has:- $K = [1 + y - 2xy]/[y^2 + y + 2xy]$

One really needs to use activities rather than atomic fractions but the activity coefficients are not available.

Since the entropy change for increasing disorder is positive, as the temperature of equilibration is increased one will approach the disordered state with random occupation of the two sites. It might appear that this would provide a method for determining the prevailing temperature at the time a mineral was formed. This is not generally possible because the speed of equilibration is quite fast down to about 900 K and even down to 750 K is rapid on a geological time scale.

Fig.8.7 shows the ordering in an orthopyroxene equilibrated at different temperatures.

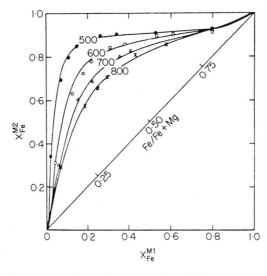

Equilibrium distribution isotherms for orthopyroxenes
**Fig.8.7**

However the kinetics of the approach to equilibrium can be explored for the sample by heating it untill equilibrium at some temperature and then measuring the ordering ratio $Fe_2/Fe_1$ as a function of time at some lower temperature. An energy of activation for the re-ordering process can be calculated and estimates made of the rate at lower temperatures. The possibility of quenching the equilibrium achieved at a high temperature can be assessed.

There are numerous complications in these measurements. Other cationic species, especially $Ca^{2+}$ and $Mn^{2+}$ also show a strong preference for the M2 site in orthopyroxenes and related minerals, so that competitive ordering reactions occur. Iron may be present in other sites in the mineral in which case chemical analysis does not yield a value for x, as defined above.

### 8.3.5  Valence Delocalisation in Minerals

In several minerals containing iron in both main valence states the Mössbauer spectrum shows that valence delocalisation takes place at, or above. room temperature. It is necessary that the two iron sites be closely similar. Such behaviour can range from electron hopping on a time scale that is faster than the mean life time of the excited $^{57m}Fe$ to delocalisation of the d electrons in a conduction band.

Most of these minerals give rather complex spectra but a detailed analysis shows that a component with an isomer shift in the 0.6 to 0.75 region is indicative of delocalisation. This suggests electron hopping between $Fe^{2+}$ and $Fe^{3+}$ on a time scale that is fast compared to the mean life of the $^{57}Fe$ excited state. The electron hopping process may only take place quickly enough above some critical temperature region. To confirm that electron hopping is responsible for such a spectrum it must be shown that it is replaced by spectra from localised valence states at some lower temperature. Conversely it will become more intense at elevated temperatures. An example is shown in Fig.8.8.

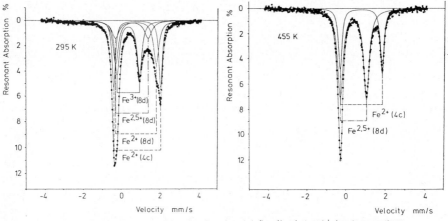

Spectra of Ilvaite showing increased delocalisation at higher temperature.

### Fig.8.8

Minerals showing such effects include ilvaite, deerite, some augites and the orthorhombic form of cubanite. There are several others for which there is less convincing evidence.

### 8.3.6  Sulphidic Minerals

Several iron sulphides occur as minerals. The bonding in these compounds is more covalent than in the oxides and this leads to still greater variety of behaviour.

8.3.6.1 *Troilite,  FeS*. The mineral has an hexagonal close packed structure with the high spin iron(II) in octahedral holes, giving $\delta = 1.02$. It is antiferromagnetic and the spectrum is a single sextet. Troilite was identified in lunar samples using its Mössbauer spectrum.

8.3.6.2  *Pyrrhotite*, $Fe_{1-x}S$ to $Fe_7S_8$. The structure derives from troilite by the introduction of ordered   iron vacancies. A ferrimagnetic hexagonal variety, commonly with small x, gives poorly resolved magnetic spectra. The monoclinic form gives sharp magnetically split spectra comprising three or perhaps four sextets. $T_C \approx$   580 K. A single isomer shift of 0.65 is reported. This indicates valence delocalisation.

8.3.6.3  *Mackinawite,  $Fe_{1+x}S$*. Another nonstoichiometric phase is paramagnetic and gives a single line spectrum down to 4.2 K. The low isomer shift, 0.5, suggests it may have a low spin configuration and perhaps valence delocalisation.

8.3.6.4  *Greigite,  $Fe_3S_4$*. This inverse spinel is analogous to magnetite. It gives a broad doublet at room temperature changing to a broad line sextet at 4.2 K. The spectrum has been analysed in terms of three components, $Fe^{2+}$ and $Fe^{3+}$ in distorted octahedral sites and $Fe^{3+}$ in a regular tetrahedral site.

8.3.6.5 *Pyrite  and  Marcasite,  FeS$_2$*. Both minerals give quadrupole split  doublets and yield the parameters: pyrite  $\delta = 0.31$ with $\Delta = 0.61$ and marcasite $\delta = 0.27$ with $\Delta = 0.51$.   Clearly they both contain low spin iron(II).

Attention has been paid to the use of the spectrum for analytical purposes in coal. In some coals the principal component containing sulphur is pyrite, various treatments being used to oxidise the pyrite to iron oxide to permit the use of the coal in steel production.   Analytical control of the pyrite content is important.

Normal Mössbauer spectrometry takes too long for industrial purposes.  However it is possible to reduce the time to an acceptable level by making a two point measurement. With thin samples of the same thickness a constant narrow line width is obtained so that the area  under  the line is proportional to the dip h, that is, the difference between the photon count at maximum absorption and the count far off resonance. The instrument measures the count with the source travelling at a constant velocity corresponding to one absorption peak of the quadrupole doublet in the absorption spectrum. A second count is made at a velocity known to be well separated from any spectral absorption by the sample. The difference measures the pyrite content of the absorber.

The progress of the de-sulphurisation of coal, for example by washing and aerating or acid treatment, can be monitored using the spectrum. In addition the oxidation products can be identified.

## 8.4  AMORPHOUS  MATERIALS

Mössbauer spectroscopy provides one of the few ways of investigating the environment of atoms in amorphous solids. In silicate glasses and related materials the spectrum will usually identify the oxidation state and the coordination number of the iron. Iron(III) is found in both four and six coordinate sites.

The formation of compounds in alloys can also be detected.  For instance in the aluminium iron system the formation of $Al_5Fe_3$ and $Al_6Fe$ can be detected in a matrix of a solid solution.

Perhaps the most interesting systems to be studied are the amorphous metals. Those containing iron often display magnetic ordering. Their room temperature spectra are magnetically split with broad lines. The line widths are not much less at lower temperatures suggesting a number of kinds of iron site. By subjecting the absorber to a radiofrequency field during measurement the magnetic spectrum collapses and a quadrupole split pair of lines is seen. However the line widths are large, indicating unresolved doublets reflecting a number of iron environments.


## 8.5 APPLICATIONS IN BIOCHEMISTRY AND MEDICINE

Numerous natural products contain iron. They include the iron transport and storage proteins, the haem proteins, the iron-sulphur proteins and various enzymes. A number of them participate in redox reactions in organisms. Their investigation presents many difficuties. A combination of several techniques is needed to elucidate their nature and properties. Mössbauer spectroscopy can make a useful contribution.

Even the preparation of the material for examination is difficult. The compounds concerned commonly contain a few iron atoms in molecules of relative mass $10^3$ to $10^6$, so that it is highly desirable to use material enriched in $^{57}Fe$. Since the compounds cannot, in general, be directly synthesised this involves biosynthesis by some organism supplied with $^{57}Fe$.

In some cases exchange reactions can be used to enhance the $^{57}Fe$ content of a natural product. Such preparations are very expensive. Well crystallised products may not be possible and this complicates interpretation of spectra. Single crystals are rarely obtainable. The products are rather fragile systems and need careful handling.

All the complexities of behaviour of iron compounds are displayed within this group of compounds. Spin cross-over, spin-spin exchange or superexchange, valence delocalisation, and unusual spin states may complicate the spectra of the iron atoms in the low symmetry sites involved. Further there are often relaxation effects leading to broadening of the lines.

Magnetically split spectra generally supply the most information about these compounds. A small external magnetic field yields the most informative spectra. Measurements at or below liquid helium temperature may be needed. Spectra with very low statistical error are necessary and an exhaustive computer analysis of the data is essential. Since measurements are generally made on powders or frozen solutions direction sensitive parameters must be averaged over all directions when simulating spectra.

Using frozen solutions or suspensions the spectra are often pH dependent, because of deprotonation reactions changing the iron environment.


### 8.5.1 Porphyrin complexes.

The environment of the iron in the Haeme proteins is very similar to that of the iron in the porphyrin complexes, as well as having something in common with the phthalocyanines. Since these complexes are rather simpler to handle than the haeme proteins a considerable amount of work has been done using the porphyrin complexes as models of the natural products.

Three kinds of porphyrin complex will be considered. The basic structure is shown in Figs.8.9 a & b. The $\beta$ substituted complexes have R = H. In OEP, octaethyl porphyrin, X = Y = Z = Et. In PP, protoporphyrin IX, X = Me, Y = -CH=CH$_2$ and Z = -CH$_2$CH$_2$COOH. PPM is similar except that the carboxyl group is methylated. In the meso complexes X = Y = Z = H, and R = Ph in TPP, tetraphenyl porphyrin. In the "picket fence" porphyrin, TpivPP, R = o.NH.CO.C$_6$H$_4$.C(CH$_3$)$_3$, one side of the FeN$_4$ ring is stericallly protected from large potential ligands.

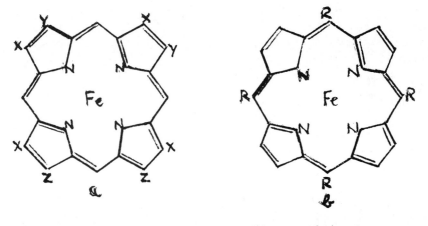

(a) $\beta$ substituents                    (b) meso substituents

**Fig.8.9**

Consider the effect of a small external magnetic field, less than 0.2 T. The spin Hamiltonian operator, $H_s$, which yields the energy levels relevant to the Mössbauer spectra, can be regarded as as being determined by a Hamiltonian comprising two terms $H_e$ and $H_n$. The first of these relates only to orbital electronic factors, the second only to nuclear factors.

$$H_e = D[S_z^2 - S(S+1)/3] + (E/D)(S_x^2 - S_y^2)] + \beta\, S \bullet g \bullet B_{ext}\quad \text{Eq.(i)}$$

The expression in square brackets determines the energy levels of the atom in the absence of an applied field. D is the zero field splitting. It is a measure of the tetragonal distortion from octahedral symmetry at the iron site. E measures the rhombic distortion; E = 0 when the crystal field is axially symmetric and $\eta$ = 0. The final term gives the Zeeman interaction of the applied magnetic field, $B_{ext}$, with the electronic magnetic moment. $\beta$ is the Bohr magneton and g the tensor for the interaction of the field with the electronic magnetic moment. $H_n = S \bullet A \bullet I + g_n\beta_n\, B \bullet I + H_Q.$  Eq.(ii)

Where $\beta_n$ is the nuclear magneton and $g_n$ the nuclear gyromagnetic ratio, the field-nuclear spin interaction being isotropic. I is the nuclear spin. A is the tensor determining the nuclear spin interaction with the electronic spin. The second term is small unless B $\gg$ 0.1 T.

It is generally assumed that the g and A tensors have the same axes.

$$H_Q = eQV_{zz}[3I_z^2 - I(I+1) + \eta(I_x^2 - I_y^2)]/4I(2I-1).$$

For very small fields, Bext. < 0.1 T, both terms in eq.(i) are larger than the hyperfine terms in equation (ii). The nucleus has little effect on the orientation of S, quantisation of S is determined entirely by the electronic system. In these circumstances one can transform eq.(ii) to obtain:     $H_n = -g_n\beta_n (B_{int.} + B_{ext})\bullet I + H_Q$ where the internal hyperfine field at the nucleus $B_{int.} = -<S>_{av} A /g_n\beta_n$  and $<S>_{av}$.is the expectation value of S averaged over the different states.

Now the spectra obtained from ions with half integral spins, that give Kramers doublets, are very different from those with integral or zero spins and it is convenient to consider them separately.

8.5.1.1 *High spin iron (III)porphyrin complexes*. Most complexes of the types FePorph.X and FePorph.LX are high spin.

The ground state of $Fe^{3+}$ is $^6S$, an orbital singlet state, but spin orbit coupling, involving an excited $^4T_1$ state, leads to three Kramers doublets (See Fig.5.6). In large complexes spin-spin relaxation and, at low temperatures, spin-lattice relaxation are both slow (see Section 6.3). So that magnetically split spectra are found at low temperatures even in the absence of an applied field.

But a modest applied field, less than $\approx$ 1 T , yields simpler spectra, with sharper lines, because it decouples spin-spin interactions of the iron with the nuclear spin of the $^{14}N$ nuclei on the ligand.

With many porphyrin ligands there is axial symmetry at the iron so that the **g** and **A** tensors can each be specified by two components, $g_\perp$, $g_{||}$, $A_\perp$ and $A_{||}$. The D values for the porphyrin complexes are rather large, typically D $\approx$ 10 $cm^{-1}$ so that at 4.2 K they are predominantly in the ground state $|\pm 1/2>$. At higher temperatures the $|\pm 3/2>$ and $|\pm 5/2>$ levels are populated and the lines in the magnetic spectrum broaden. Eventually at some higher temperature the magnetic spectrum collapses and an asymmetric broad line quadrupole doublet is obtained. Relaxation effects persist up to 295 K. In stronger magnetic fields mixing of the $|\pm 5/2>$, $|\pm 3/2>$ and $|\pm 1/2>$ states occurs.

The analysis of the spectra is simplified if electron spin resonance data are available, since these will provide D, E/D and the components of the **g** tensor.

The spectrum of FePPCl, in powder form, taken at 2 K and in a magnetic field of 0.05 T perpendicular to the photon beam, can be fitted satisfactorily by the parameters: D = 6 $cm^{-1}$,  E/D $\cong$ 0,  $\delta$ = 0.38, $1/2e^2qQ$ = +0.8 and $B_{int.}$= 48T, and thus the spectrun spans about 9.5 mm.$s^{-1}$. Other high spin $Fe^{3+}$ porphyrin complexes give rather similar parameters, although the D values are usually larger.

For the lowest lying $|\pm 1/2>$ state there is an easy plane of magnetisation, the x-y plane, <S> is highly anisotropic, Bint. aligns with $B_{ext.}$ and if $B_{ext.}$ is perpendicular to the photon beam a 3:4:1::1:4:3 type spectrum is obtained. If the ground state has a simple easy axis of magnetisation, as is the case for the $|\pm 5/2>$ state, a normal 3:2:1::1:2:3 spectrum, characteristic of random orientation of <S>, will ensue.

The temperature dependence of <S> also enables one to calculate D.

**8.5.1.2** *Spin 3/2 complexes*. This unusual spin state has already been mentioned (see Section 5.8.2). Some five coordinate iron porphyrin complexes, such as FeOEP.BF, with the fifth ligand a weak donor, prove to have S = 3/2. They show larger quadrupole splittings than the S = 5/2 or S = 1/2 complexes; for example, $\Delta$ +3.42 for $FeOEP.BF_4$, which changes little with temperature. Isomer shifts are similar to the high spin complexes. The iron is probably in a $^4A_2$ state, $(d_{xy})^2(d_{xz},d_{yz})^2(d_{z^2})^1$. The magnetically perturbed spectrum shows $\Delta$ is positive so that there must be a large ligand contribution to the EFG. $B_{int}$, at about 21 T, is much smaller than for the high spin complexes.

Some iron(III) porphyrin complexes show $3/2 \Leftrightarrow 5/2$ cross over, for example the $SbF_6^-$ and $ClO_4^-$ salts of $FeTPP^+$.

**8.5.1.3** *Low spin S = 1/2 complexes*. These are all of the $FePorph.L_2+.X-$ type, with L a strong donor ligand. They give sharp line symmetric quadrupole split doublets at 77 K and above. Spin relaxation is fast. Isomer shifts are close to 0.15 at 4.2 K and quadrupole splittings around 2.

Spin-orbit coupling and lower than octahedral symmetry reduce the six-fold degeneracy of the $^2T$ configuration to give two, or three doubly degenerate energy levels. These are determined by D and E (See equation (i) above). In a low external magnetic field and at a low temperature only the $|\pm 1/2>$ state is concerned. A number of porphyrin complexes have tetragonal symmetry around the iron, for these $\eta = 0$.

Analysis of the spectra is simplified if electron spin resonance data for the compound are available. Such data yield $D/\lambda$, $E/\lambda$ and k, where $\lambda$ is the spin-orbit coupling parameter and k a covalence factor. The occupation of the energy levels at different temperatures can then be calculated and thence values of $\Delta$. The contribution of each level to <S> can be calculated and hence the magnetically split spectrum.

When E = 0, the $q_{val}$.term should be negative if the orbital doublet lies lowest and positive for the singlet lying lowest; but generally the ligand contribution to $\Delta$ dominates.

Alternatively if $\eta$ and the temperature dependence of $\Delta$ can be determined D and E may be evaluated.

**8.5.1.4** *Spin coupling*. In complexes with two iron atoms of the type $[FeTPP]_2O$ super exchange coupling of the spins on the iron atoms can occur. This introduces another term, $-JS_1 \bullet S_2$ into the spin Hamiltonian. If the coupling is antiferromagnetic, J < 0, permitted spin values range from the difference of $S_1$ and $S_2$ to their sum so that with two $Fe^{3+}$ atoms the ground state may be diamagnetic. With ferromagnetic coupling, J > 0, the ground state will have $S = S_1 + S_2$.

The separation of the different S states must be quite large for the above porphyrin complex since sharp spectra are obtained over a range of temperatures. An external magnetic field shows $\Delta$ is negative and the $B_{hf.}$ is equal to $B_{ext}$. confirming the diamagnetic properties. [FeOEP]$_2$O behaves in a similar way.

Similar complications arise if there is a mixture of spin states. Another term has to be added to the spin Hamiltonian.

Finally, in mixed valence complexes if delocalisation occurs, a spin exchange term allowing for this process must be included. The effective spin then depends on the relative magnitude of this term and the superexchange term. For J negative and its absolute magnitude smaller than the delocalisation term the effective spin of the ground state is 7/2, but for large |J| it is 9/2.

When the ligand is magnetically active a similar spin-spin exchange process term must be included.

8.5.1.5   *High spin $Fe^{2+}$, $S = 2$*.   The porphyrin complexes containing high spin $Fe^{2+}$ are very air sensitive and difficult to handle. In addition, the analysis of the magnetically perturbed spectra is laborious. Spin relaxation is fast down to very low temperatures.

Magnetically split spectra are otained with large external fields, $> 2$ T.

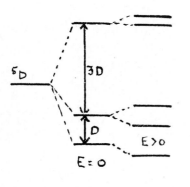

**Fig.8.10**

The degeneracy of the spin quintet from the $^5D$ ground state of the $Fe^{2+}$ is removed by spin-orbit coupling and a low symmetry for the iron, to give singlet energy levels if $E \neq 0$. When $E = 0$ two states are doubly degenerate (See Fig.8.10). With an applied field $<S>$ is no longer zero. Once the energies of the five states are established $<S>_{av.}$, the average value can be calculated.

The Hamiltonian for the hyperfine interaction can be expressed, using equation (ii) above:-

$$H_{h f.} = -\beta_n g_n [B_{ext.} -- <S>_{av}.A /\beta_n g_n] I + H_Q$$

The observed spectrum must be fitted with the levels given by this Hamiltomian varying D, E, g and A, knowing the temperature of the measurement and the direction and magnitude of the applied field.

Clearly this involves considerable computation and demands very high quality spectra to yield accurate values of the unknown parameters.

In the high spin complexes the iron atom lies outside the plane defined by the four nitrogen atoms of the ligand, so that the local symmetry at the iron is $C_{4v}$.

In the absence of an applied field the spectra are sharp quadrupole doublets and can be simply analysed to give $\delta$ and $\Delta$. The temperature dependence of $\Delta$ yields D and E.

The local environment of the iron in the high spin complex FeTPP.(THF)$_2$ is approximately tetragonal and a spectrum taken at 4.2 K could be fitted by the values $\delta = 0.96$; $\Delta = -2.75$; $\eta = 0.4$; $\lambda = 100$ cm$^{-1}$ ; $D = 6.0$ cm$^{-1}$; $E = 0$; $A_\perp /\beta_n g_n = -24.3$ T; $A_{||} /\beta_n g_n = -7.2$ T.

8.5.1.6     **S = 1 complexes.** The spectra of Fe(II)PP IX in frozen aqueous solution are pH dependent. Three different spectra can be distinguished. Above pH 10 two of the spectra have   $\delta = 0.93$ and $\Delta = 0.98$ and must arise from S = 2 complexes. But in the pH range 6 to 9 there appears a spectrum, due to a polymeric species with $\delta = 0.52$ and $\Delta = 1.42$.   These data suggest an S = 1 state for the iron.  Spin coupling is also possible.

8.5.1.7     **S = 0 Complexes.** With such diamagnetic species the spectra readily yield $\delta$ and  $\Delta$.  An applied magnetic field will determine the sign of $\Delta$ and may give a value for $\eta$.

## 8.5.2  **Iron Transport Proteins.**

Amongst the simplest of the iron protein complexes are the compounds involved in the transport of iron. In primitive organisms the siderochromes have relative molecular masses as low as $10^3$, while in higher organisms the transferrin, lactoferrin, and other transport compounds are much larger. All are high spin $Fe^{3+}$ complexes. There may be one or two iron atoms per molecule, but the atoms are well separated and do not interact.

At 4.2 K these compounds show magnetically split spectra, with broad lines and a hyperfine field of from 30 to 40 T. The application of a small external field aligns the iron spins and a sharp spectrum is obtained. The interpretation of the spectra is along the same lines as for the large high spin iron(III) porphyrin complexes, ( See 8.5.1.1) D and E measure the tetragonal and rhombic distortion from octahedral, of the iron environment. A range of small external fields applied perpendicular to the photon beam should be tried and calculated and observed spectra compared for different E, D, **g** and **A** .

Because of magnetic anisotropy effects the spectra will show a 3:4:1 intensity pattern.

Differences in the parameters obtained often reflect differences in the iron environment in transferrins from  different organisms.

## 8.5.3  **Iron storage proteins.**

These behave as a core of iron hydroxide molecules surrounded by a shell of protein. There is usually a small amount of phosphate in the shell. They contain a much higher iron content than the other compounds considered in Section 8.5. Their spectra resemble those found for $\delta$ FeOOH.

Typical iron storage proteins are ferritin and haemosiderin. At 4.2 K these compounds give magnetically split spectra with somewhat broad lines and $B_{hf}$ about 50 T. At higher temperatures a quadrupole split pair appears in the spectrum. The relative intensity of the magnetic spectrum decreases, and that of the quadrupole split  pair increases rapidly as the temperature of measurement rises.

The disappearance of the magnetic spectrum is consistent with a transition to the superparamagnetic state  by  the  iron  hydroxide  core.  (See Sec.  7.1.4).  The rather asymmetric quadrupole  split spectrum seen at the higher temperatures is quite typical of the spectra found for the  polymeric products of the  hydrolysis of iron(III) salts as well as the hydrolytic sediments found on the bottom of lakes and the ocean.

From the blocking temperature it is possible to estimate the size of the FeOOH core.

In this way it has been shown that the ferritins from different organisms vary considerably in the size of their cores. Frozen ferritin suspensions are sufficient for such studies.

However, size estimates made in this way do not agree very well with measurements using electron microscopy.

Some metabolic studies of the transfer of iron from ferritin to hemoglobin have been made using the Mössbauer spectra and highly $^{57}$Fe enriched ferritin.

## 8.6 HAEM PROTEINS AND RELATED COMPOUNDS

In very many natural iron proteins the prosthetic group is an iron atom bound to a porphyrin type ligand. In haemoglobin and myoglobin the porphyrin moiety is the protoporphyrin IX ligand. This prosthetic unit is embedded in a protein shell comprising chains of aminoacids. The iron is attached to the protein chain by bonding to the nitrogen in an imidazole ring of a histidine unit. Thus the iron is present in a five coordinate environment. In the cytochromes and other biologically important compounds the iron in the prosthetic group is also in a porphyrin complex, but the ligand may be differently substituted to protoporphyrin IX, the groups X,Y and Z in Fig.8.9a may be different, and the fifth coordination site may be attached to a different donor atom in the aminoacid chain.

In the natural deoxy form of both myoglobin and haemoglobin, the iron is in the high spin $Fe^{2+}$ state. The iron atom lies out of the plane set by the four nitrogen atoms, about 0.05 nm closer to the nitrogen atom on the histidine. Both these compounds bind oxygen reversibly, the iron preserving its oxidation state of two, in contrast to the $Fe^{2+}$ porphyrin complexes previously described which, as noted above, very readily suffer aerial oxidation in aqueous media to give trivalent iron. The difference can be attributed to the location of the prosthetic groups in hydrophobic pockets in the protein shell.

In the oxygen adducts the iron becomes low spin and is nearer the plane of the nitrogen atoms of the porphyrin ligand. The oxygen is attached to the iron in a non-linear Fe$\rightarrow$O fashion. Myoglobin has one prosthetic group per molecule and the oxygen
      $\diagdown$O   uptake can be treated as a simple equilibrium reaction. Haemoglobin is much larger and the molecule contains four iron atoms. The attachment of oxygen by one iron atom leads to small structural changes in the protein and alters the ability of the other iron atoms to attach oxygen.

Most of these compounds are involved in natural redox processes. Those normally containing divalent iron can be oxidised to give trivalent iron species and conversely. The oxidation products may be high or low spin. In some cases intermediate spin, S = 3/2, compounds are found. Oxidation to iron(IV) may also be possible. A further complication is that an oygen adduct might involve $O_2$ or $O_2^-$ with an increase in the oxidation state of the iron and spin coupling of the two paramagnetic centres.

Other ligands can attach to the vacant sixth coordination site of the deoxymyo- or deoxyhaemo-globin.

## 8.6.1 **Related Compounds**

The peroxidases and catalases are closely related to the above natural products, containing a similar prosthetic group to myo- and haemo-globin, iron(III) protoporphyrin IX. The associated protein is a glycoprotein containing sugar units. The peroxidases usually contain one prosthetic group per molecule, like the myoglobins, while the bigger catalases contain four. In these there are two of each of two kinds of prosthetic centre.

In the cytochromes the porphyrin ligand is different. Referring to Fig.8.9 a, the Y group is - $CH.CH_3S$-, the sulphur bonding to the aminoacid chain. In addition the fifth coordination site on the iron bonds to a sulphur atom of a methionine group in the protein chain.

## 8.6.2 **General features of the spectra**

Many of the measurements reported have been made on frozen solutions or suspensions of the iron proteins. Such spectra are sometimes dependent on the pH of the solution or the presence of other ligands, introducing some uncertainty as to the identity of the molecules responsible for the spectrum. Deprotonation of attached water or, less importantly, of the carboxyl groups of PP IX may occur.

The Mössbauer spectrum will usually decide the oxidation and spin state of the iron. They are likely to be affected by configurational changes in the vicinity of the iron atom and, particularly, to the presence and nature of the axial ligands on the iron. The measurement of $B_{int}$ usually provides a clear indication of the spin state of the iron.

The iron(II) compounds give quadrupole split spectra with reasonable line widths. No magnetic splitting is seen even below 4.2 K. The high spin iron(III) compounds give quadrupole split spectra at room temperature with very broad lines, characteristic of slow relaxation. On cooling a complex magnetically split spectrum develops. The complexity is due to coupling of the nuclear spin of the nitrogen, $I = 1$, and, in the fluorides, with the $I = 1/2$ spin of the fluorine, with the iron. Application of a modest external magnetic field, < 1 T, decouples these interactions and at 4.2 K two six line spectra are resolved. They arise from the two components of the $|\pm 1/2>$ ground state of the complex.

Haemoglobins from different biological sources give very similar spectra since the differences lie in the protein shells, and not in the prosthetic group.

A summary of some data for human haemoglobin and some derivatives is given in Table 8.1. The data is for 298 K.

**Table.8.1**

| Compound | $\delta$ | $\Delta$ | S |
|:---:|:---:|:---:|:---:|
| Hb | 1.05 | 2.3 | 2 |
| $HbO_2$ | 0.21 | 1.9 | 0 |
| HbCO | 0.35 | 0.20 | 0 |
| HbCN | 0.19 | 1.30 | 1/2 |
| $HbN_3$ | 0.20 | 2.20 | 1/2 |
| $Hb^+ H_2O$ | 0.29 | 2.03 | 5/2 |
| HbF | * | * | 5/2 |

* This iron(III) compound gives very broad lines due to relaxation.

### 8.6.3   The Different Iron States

8.6.3.1   *Low spin iron(II) compounds*.  These are the least, rewarding compounds to measure. They are diamagnetic. Their isomer shifts do not change much from compound to compound but the quadrupole splittings cover a wide range.

Some compounds in this class are the carbon monoxide adducts of Myoglobin, Haemoglobin, and the reduced form of various cytochromes, including P450, c and c'.

They give quadrupole-split spectra. The EFG must arise entirely from ligand donation and the sign of $\Delta$ is in most cases positive. These spectra are the least sensitive to the iron environment. Thus the parameters obtained from the spectra of the carbon monoxide adducts of myoglobin and reduced cytochrome P450 are almost the same.

Both haemoglobin and myoglobin can attach carbon monoxide giving low spin complexes. Spectra taken using frozen solutions, with an external magnetic field perpendicular to the photon beam, show that the sign of the quadrupole splitting is positive for the carbon monoxide adduct and negative for the oxygen adduct, Fig.8.11. In both cases the internal hyperfine field was zero confirming their diamagnetic character. Even on raising the temperature no internal field could be detected showing that an excited paramagnetic  S = 1 state must lie considerably above the ground state. The asymmetry parameter was about 0.3 for both adducts.

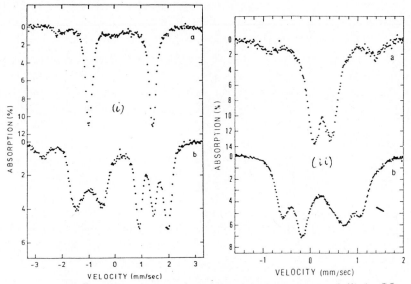

Spectra at 4.2 K of frozen solutions of (i) the oxygen  and (ii) the CO adducts of myoglobin, (a) in the absence of, and (b) in a magnetic field of about 4.5 T applied perpendicular to the photon beam.

### Fig.8.11

The most interesting result has been obtained with single crystal studies using the myoglobin oxygen and carbon monoxide adducts. These showed that the principle axis of the EFG was nearly perpendicular to the $FeN_4$ plane in the CO complex but lay in this plane in the oxygen complex.

The spectrum at 4.2 K of the oxygen adduct of haemoglobin gives $\delta = 0.25$ and $\Delta = -2.23$ with $\eta = 0.2$, although the probable error on the last figure is substantial. Unusually for $S = 0$ compounds the quadrupole splitting is quite temperature dependent. It is hardly likely to reflect electronic excitation, since magnetic data shows the compound is diamagnetic up to room temperature.

It might be due to the onset of rotation about the Fe - O bond or to vibronic coupling effects in the molecule. It is also conceivable that the nature of the iron-oxygen bond changes with temperature.

**8.6.3.2  *Low spin iron(III) compounds*.** The peroxidases and cytochromes contain iron(III) in the natural state. At low temperatures they are generally low spin species.

The combination of electron spin resonance data with the Mössbauer spectrum is particularly effective for this group of compounds, and there are extensive data on such studies. The spectra of frozen solutions of these compounds are often pH dependent because of hydrolysis of the aquo- cations, such as $CPO^+H_2O..$(CPO =Chloroperoxidase).

The spectrum of CPO, measured at 4.2 K in a magnetic field of 0.13 T was fitted satisfactorily with the parameters:

$$g_x = 1.84; \; A_x/g_n\beta_n = -49 \text{ T}; \; \Delta = +2.9.$$
$$g_y = 2.26; \; A_y/g_n\beta_n = +7.9T;$$
$$g_z = 2.63; \; A_z/g_n\beta_n = +28.4T. \; \eta = -2.9.$$

The unusual value for $\eta$ arises because of the difference in the axes for **g** and **A** and for the EFG.

At room temperature CPO gives a quadrupole split spectrum with a much lower splitting. Both CPO and Cytochrome P450 display spin cross-over below room temperature as can be seen in Fig.8.12. From about 190 up to 235 K two quadrupole split pairs appear in the spectrum. $T_{50}$ is about 200 K. (See Sec.7.5)

In a similar way the spectrum of Cytochrome P450 can be fitted with the parameters:-

$$g_x = 2.45; \; A_x = -2.29 \text{ mm.s}^{-1}; \; \Delta = 2.85;$$
$$g_y = 2.26; \; A_y = -1.02 \text{ mm.s}^{-1}; \; \eta = 0;$$
$$g_z = 1.91; \; A_z = -5.24 \text{ mm.s}^{-1}; \; \delta = 0.30.$$

**8.6.3.3  *High Spin Iron(III) compounds*.** At room temperature haem fluoride, and most peroxidases and cytochromes are high spin iron(III) compounds. Their spectra taken in a low magnetic field can be analysed in the same way as the low spin iron(III) spectra.

Electron spin resonance data can also be obtained. Haem fluoride measured at 4.2 K in a magnetic field of 0.2 T yields the information: $D = 7.0 \text{ cm}^{-1}$; $E \cong 0$; $B_{int} = 52.9 \text{ T}$; and $\Delta = +0.7$.

**8.6.3.4  *High Spin Iron(II) compounds*.** The iron(III) compounds previously mentioned can be reduced to give iron(II) species. Some are high spin and unlike the analogous porphyrin complexes are not very air sensitive.

These compounds do not give Kramers doublets, and electron spin resonance data cannot be obtained. Quadrupole split spectra are found down to very low temperatures. The splittings lie in the range 2.0 to 2.8 and are generally temperature dependent. They are readily distinguished by their isomer shifts which range from 0.7 to 1.0.

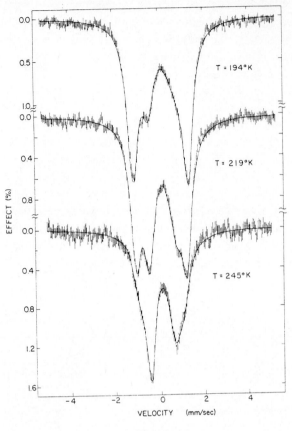

**Fig.8.12**

Magnetically split spectra measured in a magnetic field of a few Tesla at 4.2 K can be analysed in an analogous fashion to the related porphyrin complexes (See Sec 8.5.1.5).

### 8.6.4  High Oxidation State Compounds

Chloroperoxidase and other similar copounds can be oxidised, removing two electrons, to give a green compound. This compound, I, can be reduced in two one electron steps, I -e → II -e → original chloroperoxidase.   At 77 K the oxidized compounds give quadrupole split spectra with the parameters shown in Table 8.2.

**Table 8.2**

| Initial Peroxidase | I | | II | |
|---|---|---|---|---|
| | $\delta$ | $\Delta$ | $\delta$ | $\Delta$ |
| Chloroperoxidase | 0.15 | 1.02 | | |
| Horse Radish Peroxidase | 0.00 | 1.20 | 0.03 | 1.36 |
| Japanese Radish Peroxidase | 0.10 | 1.33 | 0.11 | 1.44 |

The isomer shifts for both I and II are characteristic of iron(IV). This suggests that compound I contains a radical cation ligand, which accounts for its colour.

Similar behaviour has been described in Sec. 5.7.2.

Such species are involved in the biological function of the peroxidases and catalases.


## 8.7   THE IRON-SULPHUR PROTEINS

In a number of iron proteins the prosthetic unit contains between one and four iron atoms, each attached to four sulphur atoms. These compounds have an important role in biological redox processes

In these compounds the iron can display all the complexities of behaviour, spin coupling, valence delocalisation and spin cross over, which have previously been described. In addition the state of the iron seems to be more sensitive to the more remote parts of the molecule than is the case with the iron proteins previously considered. It may be affected by hydrogen bonding and configurational changes in the protein shell.

A thorough understanding of the nature of these compounds and their role in biological processes requires the use of many techniques; the combination of electron spin resonance studies and Mössbauer spectra make a very useful contribution to this difficult topic.

The prosthetic units contain from one to four iron atoms, the donor atoms being sulphur. Some of the iron sulphur proteins molecules contain more than one prosthetic unit.


### 8.7.1  Species  containing  an  [FeS$_4$] unit.

Both natural iron proteins, for example rubredoxin and desulphoredoxin, and synthetic model compounds such as $Fe(SC_6H_5)_4^{2-}$ , containing this unit have been studied. In the natural products the four sulphur atoms are provided by cysteine units in the protein shell

The sulphur atoms are arranged approximately tetrahedrally around the iron.

The iron(II) compounds, with isomer shifts close to 0.7, show large quadrupole splittings usually > 3. In the absence of an applied field simple quadrupole split spectra persist down to below 4,2 K. At 4.2 K application of a magnetic field of 2.6 T gives a magnetically split spectrum with fairly sharp lines. The calculations needed to fit the spectra are more complex than for many iron(II) complexes because the quadrupole interaction is comparable with the magnetic interaction.

For Iron(II) rubredoxin the spectrum yields D = +7.8 cm$^{-1}$, E/D = 0.28 while the desulphoredoxin gives D = -6.0 cm$^{-1}$ and E/D = 0.19. Δ is positive for the latter compound while it is negative for the model compound $Fe(SC_6H_5)_4^{2-}$ . In this model compound **A**, **g** and the EFG have approximately the same axes.


### 8.7.2   Species  containing  an  [Fe$_2$S$_2$] unit.

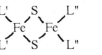

L' = L" = S of cysteine in spinach
and parsley ferredoxins.

L' = S of cysteine and L" = N of histidine in
Rieske protein.

The prosthetic unit is cationic and may carry one or two units of charge. When both atoms are iron(III) and both L the same there is a simple quadrupole split spectrum with the isomer shift $\delta \approx 0.2$ and $\Delta \approx 0.7$. With Rieske protein there is evidence of a second quadrupole-split pair with $\delta = 0.3$ and $\Delta = 0.9$.

There is strong antiferromagnetic spin coupling between the iron atoms by both exchange and superexchange through the sulphur atoms. $J = 380$ cm$^{-1}$. The ground state is $S = 0$, the compound is diamagnetic. This is confirmed by the absence of any $B_{int}$ when the spectrum is measured in a strong external field.

If one of the iron atoms is reduced to iron(II) the valences are localised and spectra arising from each kind of iron are found. There is still antiferromagnetic spin coupling, although the J value is rather smaller, about 200 cm$^{-1}$. The ground state is an $S = 1/2$ state.

### 8.7.3  Species containing an [Fe$_4$S$_4$] unit.

A number of ferredoxins contain a prosthetic unit composed of four iron and four sulphur atoms arranged approximately at alternate corners on a cube. Thus each iron atom is interacting with three sulphur atoms on the cube. This unit is attached to the protein shell by sulphur atoms in the amino acid chains occupying the forth coordination site on the iron atoms

With various proportions of iron(II) and iron(III) these units can have a formal charge of from 0 to +3.

Typical parameters are given in Table 8.3
The isomer shift for the delocalised Fe$^{2..5+}$ lies in the range 0.4 to 0.52 and can easily be distinguished from iron(II) or iron(III). Spin coupling between pairs of iron(II) or iron(III) atoms and iron(II)/iron(III) pairs is strong and finally the antiferromagnetic coupling between the pairs gives rise to the low effective spins of $S = 0$ and $S = 1/2$. Although the spectra are complex, detailed analyses of the spectra at 4.2 K in a weak external magnetic field have been made.

### Table  8.3.

| Cluster | Composition | Species | $\delta$ | $\Delta$ | e.s.r. |
|---------|-------------|---------|----------|----------|--------|
| [Fe$_4$S$_4$] | 3Fe(II) + 1Fe(III) | Fe$^{2.5+}$ | 0.50 | 1.18 | Yes. |
| Reduced Ferredoxin | S = 1/2 | Fe$^{2+}$ | 0.60 | 1.82 | |
| [Fe$_4$S$_4$] | 2Fe(II) + 2Fe(III) | Fe$^{2.5}$ | 0.42 | 1.08 | No |
| Ferredoxin | S = 0. | | | | |
| [Fe$_4$S$_4$] | 1Fe(II) + 3Fe(III) | Fe$^{2.5}$ | 0.49 | 1.03 | Yes |
| Oxidised Ferredoxin | S = 1/2 | Fe$^{3+}$ | 0.29 | 0.88 | |

### 8.7.4  Species containing an [Fe$_3$S$_4$] unit

Mössbauer spectroscopy played a part in discovering this kind of centre. The neutral centre, containing two iron(III) and one iron(II), has a resultant spin of 2. Valence delocalisation is apparent as the spectral parameters show: Fe$^{2.5+}$ $\delta = 0.41$ and $\Delta = 1.47$; Fe$^{3+}$ $\delta = 0.32$ and $\Delta = 0.52$.

## 8.8   PROTEIN DYNAMICS

The iron protein spectra differ from from those of most other compounds in their line shape. Above about $\approx 200$ K the spectral lines are best represented as a superposition of a narrow and a very broad line at the same isomer shift. The mean square displacement $<x^2>$ of the iron rises sharply, as shown in Fig.8.13.

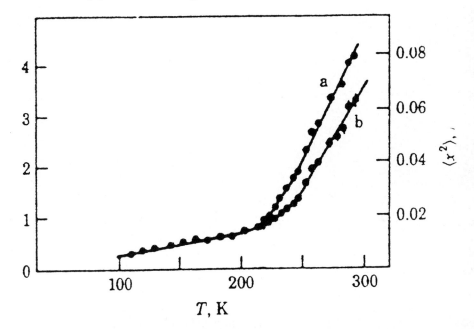

Temperature dependence of -ln f, where f is the recoil-free fraction and $<x^2>$ the mean square displacement, for myoglobin, obtained from (a) the area under the narrow component and (b) the total spectral area.

**Fig.8.13**

Such effects can also be seen, but less clearly, in the Debye Waller factors from X-ray diffraction data. Another method of determining $<x^2>$ uses intensity measurements on the Raleigh scattering of a Mössbauer photon beam by the iron protein complex. This employs a detector, at first protected by and then without, a resonant absorber for the Mössbauer radiation. The results of this method closely parallel the Mössbauer absorption data. The scattering method demands very strong sources.

An interpretation of these results has been developed supposing that conformational changes in the protein shell lead to movement of the iron on a time scale comparable to the mean life of the $^{57m}$Fe, say $10^{-7}$ to $10^{-9}$ s. A model for the process consists of a harmonic oscillator containing the iron atom diffusing in a cage. Diffusion constants calculated on this basis are of a reasonable magnitude.

## 8.9  MAGNETITE AND OTHER IRON COMPOUNDS IN ORGANISMS

A large number of organisms are known to contain inorganic iron compounds. Mössbauer spectroscopy provides a powerful technique for investigating the identity, location and biological mode of deposition of such iron. Growing the organism in a nutrient medium containing a salt made from separated $^{57}Fe$ one can explore the mineralisation process in the organism. In many of these processes the previously described ferritin and hemosiderin produce tiny particles of crystalline goethite, lepidocrocite or, especially, magnetite.

Magnetotactic bacteria orientate themselves and migrate in relation to the direction of a magnetic field. Electron microscopy has shown that they contain iron rich particles 40 to 120 nm in diameter, roughly the size of a magnetic domain in $Fe_3O_4$. The Mössbauer spectra of the organisms show that these particles are crystallites of magnetite. By culturing these organisms in a medium containing $^{57}Fe$ as a soluble iron(III) salt, the

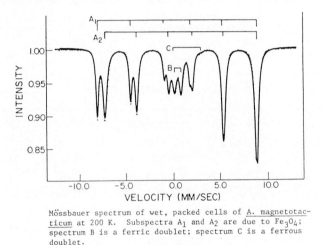

Mössbauer spectrum of wet, packed cells of A. magnetotac-ticum at 200 K.  Subspectra $A_1$ and $A_2$ are due to $Fe_3O_4$; spectrum B is a ferric doublet; spectrum C is a ferrous doublet.

### Fig.8.14

transitions involved in the formation of the magnetite can be explored.   A typical spectrum taken at 4,2 K, see FIg.8.14, shows the presence of two sextets due to the magnetite and two quadrupole split doublets from di- and tri-valent iron compounds.

The iron compound deposited in limpets must fulfil a very different function. In this organism acicular crystallites of goethite, about 1 μm by 50 nm, are deposited on the "teeth" on the radula. The mechanism of formation has been studied by means of the Mössbauer spectrum.

Inorganic iron compounds deposited in a wide variety of other organisms have been investigated in this way.

## 8.10 Comments on Iron Mössbauer Spectroscopy

The extensive application of Mössbauer spectroscopy using $^{57}$Fe is due to both the favourable nuclear characteristics of the isotope and the diversity of behaviour of iron and its compounds. The narrow line width and the large resonance cross section permit the analysis of very complex spectra. The magnetic interactions that can be explored enable one to develop a detailed picture of the interaction of the iron atom with its surroundings. The ubiquitous nature of iron compounds has lead to an extraordinary variety of applications.

# Acknowledgements

Fig.8.1  Reproduced with permission from Fenger,J. (1969) *Nucl.Instrum.Methods,* **69**, 268.

Fig.8.3  Reproduced with permission from Spijkerman,J.J. (1971) *Mössbauer Effect Methodology* , Ed.Gruverman,I.J. Plenum Press, N.Y., **7**, 91.

Fig.8.4  Reproduced with permission from Skrzypek,S.,Kalawa,E.,Sawicki,J.A.,and Tyliszczak,T. (1984) *Material Sci,& Eng.,* **66**, 145.

Fig.8.5  Reproduced with permission from Aldridge,L.P., Bancroft,G.M., Fleet,M.E. and Herzberg,C.T. (1978) *American Mineralogist,* **63**, 1107.

Fig.8.6  Reproduced with permission from Bancroft,G.M., Maddock,A.G.,Burns,R.G. and Strens,R.G.J. (1966) *Nature,* **212**, 913.

Fig.8.7  Reproduced with permission from Virgo,D. and Hafner,S.S. (1970) *American Mineralogist,* **55**, 201.

Fig.8.8  Reproduced with permission from Amthauer,G. and Rossman,G.R. (1984) *Phys.Chem.Minerals,* **10**, 250.

Fig.8.11 Reproduced with permission from Maeda,Y., Harami,T., Morita,Y.Trautwein,A. and Gonser,U. (1981) *J.Chem.Phys.,* **75**, 1.

Fig.8.12 Reproduced with permission from Champion,P.M., Munck,E.,Debrunner,P.G. Hollenberg,P.F. and Hager,L.P. (1973) *Biochemistry,* **12**, 426.

Fig.8.13 Reproduced with permission from Bauminger,E.R. Cohen,S.G., Nowick,I. Ofer,S. aqnd Yariv,J. (1988) *Proc.Nat.Acad.Sci.U.S.A.,* **80**, 736.

Fig.8.14 Reproduced with permission from Frankel,R.B., Papaefthymiou,G.C., Blakemore,R.P. and O'Brien,W. (1983) *Biochim.Biophys.Acta,* **763**, 147.

# 9

# Mössbauer Spectroscopy of Elements Other Than Iron and Tin

Although the majority of the elements with atomic number greater than eighteen have isotopes with excited states displaying the Mössbauer effect, rather less than half are suitable for studies based on the hyperfine structure in their spectra. In favourable cases information can be obtained that parallels that obtained from iron and tin spectra.

A new feature however is that the spin state combinations involved often differ from the $3/2 \Leftrightarrow 1/2$ combination found with iron and tin. This leads to more complex spectra with more lines when there is an EFG or a magnetic field at the nuclei.

Provided the lines can be resolved the spectrum will carry more information than do the iron and tin spectra. With a quadrupole split spectrum one can extract $\eta$, and the sign and magnitude of $V_{ZZ}$.

A short account of the Mössbauer spectroscopy of some other favourable elements follows, drawing attention to their special features.

As in previous chapters all numerical data on Mössbauer parameters is given in $mm.s^{-1}$ unless otherwise indicated.

## 9.1 MAIN GROUP ELEMENTS

### 9.1.1 Iodine

An unusual feature of the Mössbauer spectroscopy of iodine is that useful information can be obtained with two different isotopes, $^{127m}I$ and $^{129m}I$. The ground state of the latter species is a very long-lived purely beta active species formed as a fission product and available in macroscopic amounts. The absorbers must all be made using iodine containing a reasonable proportion of this isotope. The radioactivity of $^{129}I$ and the biological characteristics of iodine demand that it is handled in such a way that there is no possibility of the user ingesting or inhaling any of the $^{129}I$ or its compounds. Well designed glove boxes are suitable. Because the beta particles emitted by $^{129}I$ are of low energy, a centimeter thickness of perspex provides adequate screening from the beta radiation.

Excepting this troublesome aspect, in all other respects $^{129}I$ is much more satisfactory to use than $^{127}I$. The theoretical line width for $^{129}I$ is only about a quarter that for $^{127}I$, the cross section for resonant absorption is rather larger and the Mössbauer photon energy smaller for the $^{129}I$, so that the Mössbauer fraction is more favourable: however with either species spectra must be measured with the absorber at liquid nitrogen temperature or below. The value of $\Delta R/R$ for $^{129}I$ is of rather larger magnitude and opposite in sign to the negative value for $^{127}$I.  Both species give M1 transitions. (See Table 2.1)

Both the iodine excited states are conveniently fed by the decay of moderately long lived tellurium isomeric states, $^{127m}$Te and $^{129m}$Te, with half lives of 109 and 34 days, respectively. Only a small proportion of the $^{127m}$Te decays to feed the $^{127m}$I state and larger activities are needed than with $^{129m}$Te. Zinc telluride is a satisfactory source material,

Because there is a substantial body of data on quadrupole couplings for iodine compounds from nuclear quadrupole spectroscopy, $\Delta$ values are often expressed as frequencies.
1 mm.s$^{-1}$ = 22.4 MHz for $^{129}$I and 46.4 MHz for $^{127}$I, and quadrupole couplings measured in mm.s$^{-1}$ with $^{129m}$I can be converted to the equivalent for $^{127m}$I in MHz by multiplying by 31.8.

### 9.1.1.1 *Extraction of Mössbauer parameters from the spectral data*

The transitions occurring with $^{129}$I in an EFG with axial symmetry round the iodine bond are shown in Fig.9.1a and a stick diagram giving relative intensities in Fig.9.1b

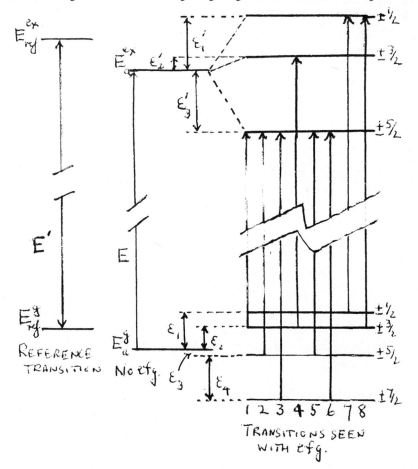

**Fig.9.1  a**

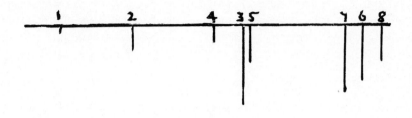

**Fig.9.1 b**

A spectrum of methyl iodide is shown in Fig.9.2. For $^{127}I$ the sequence of transitions shown in Fig.9.2 would be reversed and the resolution of the lines much inferior.

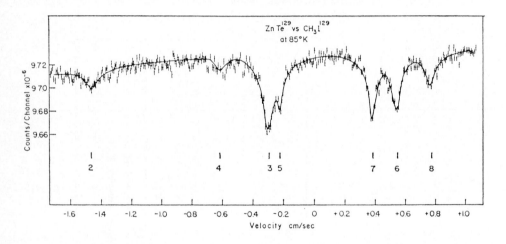

**Fig.9.2**

These rather complex spectra yield more informatiom than do the quadrupole split spectra from $3/2 \Leftrightarrow 1/2$ transitions, although more computation is needed. The simple relation between $\Delta$, $\eta$ and $V_{ZZ}$ derived in Section 1.4.4 is no longer valid. But the spectrum enables one to calculate $\delta$, $V_{ZZ}$, $\eta$ and the ratio of the quadrupole moments of the excited and ground states of the Mössbauer nucleus.

The Hamiltonian for the quadrupole interaction may be written:

$$H = \mu_I[3I_Z^2 - I(I + 1)] + \mu_I\eta/2(I_+^2 + I_-^2) \quad \text{where } \mu_I = eQV_{ZZ}/4I(2I-1).$$

The determinant this Hamiltonian yields reflects across the diagonal so that the energy levels obtained are doubly degenerate, Kramers doublets. The EFG does not remove the $\pm$ degeneracy. When $\eta = 0$ the eigenvalues of the energies are the diagonal elements of the determinant, only the first term of the Hamiltonian is involved. When $\eta > 0$, $m_I$ is no longer a good quantum number and the spectrum for an M1 transition may contain more than eight lines.

The secular equations the above Hamiltonian yields are:

For I = 5/2,   $[E^3 - 28\mu_{5/2}^2(3+\eta^2)E - 160\mu_{5/2}^3(1-\eta^2)]^2 = 0$

and for I=7/2,   $E^4 - 126\mu_{7/2}^2(3+\eta^2)E^2 - 1728\mu_{7/2}^3(1-\eta^2)E + 945\mu_{7/2}^4(3+\eta^2)]^2 = 0$

A numerical approach to the extraction of the Mössbauer parameters from the spectrum can be made as follows:

Assuming that $\eta$ is not very large and the lines are well resolved, the eight line spectrum can be fitted with eight Lorenzians and the positions of the lines established. The differences in the energies of all the pairs of lines are calculated. Some differences will be the same. Referring to Fig. 9.1a it will be seen that differences of the lines labelled 6 and 2 and of lines 5 and 1 are the same and equal $\varepsilon_3$. Similarly from 5 and 4 and 8 and 7 one obtains $\varepsilon_1 - \varepsilon_2$. With some guidance from the intensities, Fig.9.1b and Table 3.7, and also bearing in mind that the EFG does not change the centre of gravity of the split levels, so that $\sum_{n=1}^{3}\varepsilon_n = \sum_{n=1}^{3}\varepsilon_n' = 0$, one can obtain $\varepsilon_1, \varepsilon_2 \varepsilon_3$ and $\varepsilon_4$ as well as $\varepsilon_1', \varepsilon_2'$ and $\varepsilon_3'$. These $\varepsilon$ values must be the solutions of the secular equations, given above, giving the E values for the different states the EFG produces. Thus one can write:

$(E - \varepsilon_1')(E - \varepsilon_2')(E - \varepsilon_3') = 0$ for the I = 5/2 levels and

$(E - \varepsilon_1)(E - \varepsilon_2)(E - \varepsilon_3)(E - \varepsilon_4) = 0$ for the I = 7/2 levels.

Comparing coefficients with the secular equations above one has for I = 5/2 :

$$-28\mu_{5/2}^2(3 + \eta^2) = (\varepsilon_1'\varepsilon_2' + \varepsilon_1'\varepsilon_3' + \varepsilon_2'\varepsilon_3')$$

$$\text{and } 160\mu_{5/2}^3(1 - \eta^2) = \varepsilon_1'\varepsilon_2'\varepsilon_3'.$$

Since the $\varepsilon'$ values are known, these two equations enable one to calculate $\eta$ and $\mu_{5/2}$. In a similar way $\mu_{7/2}$ can be obtained. Now $\mu_{5/2} = eQ_{ex}V_{zz}/4I(2I-1) = eQ_{ex}V_{zz}/40$ and $\mu_{7/2} = eQ_{gr}V_{zz}/84$, and $Q_{gr}$ is known from other data, for example, nuclear quadrupole spectroscopy, so that $V_{zz}$ and $Q_{ex}$ can be calculated.

Using the symbols on Fig.9.1 a, the position of the line corresponding to the 3/2 $\Leftrightarrow$ 1/2 transition must be $(E_{ex} + \varepsilon_2') - (E_{gr} - \varepsilon_1) - (E_{ref}^{ex}. - E_{ref}^{gr}) = \varepsilon_1 - \varepsilon_2' + \delta$

Hence the isomer shift may be calculated.

Alternatively, but more elaborately and demanding of computer time, one may calculate spectra using the Hamiltonian given above, varying the input parameters $V_{zz}$ and $\eta$ until one gets the best fit to the observed spectrum. This approach is essential if $\eta$ is substantial and if the resolution is imperfect, as will be the case for [127]I spectra. It also leads to better estimates of the probable errors in the calculated parameters.

9.1.1.2 *Interesting aspects of iodine Mössbauer spectroscopy*

Many of the features of iron spectroscopy discussed in Chapters 5 and 7 reappear in iodine spectroscopy, but magnetic splitting leads to such complex spectra that overlapping lines are inevitable, and the spectra are not very useful as a source of information about the iodine absorbers. The magnetic fields are either external or arise from transferred fields from magnetically active atoms, for example in $CrI_3$.

Solid iodine gives a spectrum with $\Delta = 0.12$ suggesting interaction between the iodine molecules in the lattice. This is confirmed by measurements on frozen solutions of iodine in inert solvents which give $\Delta = 0$. Values for hs and hp provide evidence of charge transfer interactions in frozen solutions in aromatic solvents.

The results are sometimes surprising. The phosphine adducts of copper(I) iodide, $CuI(PPh_3)_3$, $(CuI)_2(PPh_3)_3$, $[CuI(PH_2Ph)_2]_2$ and $[CuI(PPh_3)]_4$ all show very small quadrupole couplings, less than 500 MHz. The isomer shifts are also small.

The last of the above compounds is known from crystallographic studies to have two iodine sites, the iodine atoms being two coordinate in one site and three coordinate in the other. When the spectra are analysed in terms of two iodine environments isomer shifts of -0.65 and +0.42 with quadrupole couplings of 620 and 690 MHz are found. These results suggest that the iodine in these compounds is only very slightly perturbed by the neighbouring atoms.

### 9.1.1.3 *Interpretation of the spectral parameters.*

It is useful to have some simple, roughly quantitative, way to interpret the Mossbauer parameters in terms of the bonding in the iodine compound.

For main group elements it seems reasonable to suppose that the spectral parameters are determined predominantly by the occupation of the valence shell s and p orbitals on the atom concerned. As stated in 4.1 one might expect that:

$\delta = \kappa[ -h_s + (h_p + h_s)(2 - h_s)]$ in terms of the number of "holes" in the orbitals specified, thus $h_s + n_s = 2$, $h_p = h_x + h_y + h_z$, $h_x + n_x = 2$ etc. The x,y and z referring to the $5p_x$, $5p_y$ and $5p_z$ orbitals, respectively. The $\kappa$ includes $\Delta R/R$ and will be different for $^{127}I$ and $^{129}I$. While $h_s$ is small expression can be approximated by:-

$\delta = ah_s + bh_p + c.$

The quadrupole splitting is determined entirely by the imbalance of the occupation of the 5p orbitals, $U_p$. $U_p = -n_z + 1/2 (n_x+n_y) = h_z - 1/2 (h_x+h_y)$. If we can suppose that for the singly bound iodine atom $h_x = h_y = 0$, then $h_z = h_p = -U_p$. For such iodine containing molecules $e^2q_{mol}.Q$, which can be derived from the spectrum, is equal $e^2q_{at}QU_p$, since atomic iodine has one 5p vacancy. The value of $e^2q_{at}Q$ is available from nuclear quadrupole spectroscopic measurements, being equal to 2293 MHz. Thus $U_p$ for other iodine compounds containing singly bonded iodine can be obtained from their quadrupole coupling in MHz by dividing by 2293. The quadrupole coupling for the compound can be calculated from its spectrum.

This approach can be extended to $^{129}I$, $^{125}Te$, $^{121}Sb$ and $^{119}Sn$ in the following way. The quadrupole couplings are measured for frozen solutions of iodine in an inert solvent using first $^{127m}I$ and then $^{129m}I$. Then if $e^{129}QV_{zz}/e^{127}QV_{zz} = k$, the frequency corresponding to the loss of one 5p electron for $^{129}I$ spectral data will be 2293 k.

A similar procedure can be used to calibrate the quadrupole couplings with the other elements.

Results are shown in Table 9.1.

### Table 9.1

| Species | $e^2qQ$ for $h_z = 1$ mm.s$^{-1}$ | $e^2qQ$ for $h_z = 1$ MHz | $E_-$ keV | conversion factor mm.s$^{-1}$. |
|---|---|---|---|---|
| $^{127}I$ | -49.4 | -2293 | 57.60 | 46.47 |
| $^{129}I$ | -71.7 | -1616 | 27.80 | 22.43 |
| $^{125}Te$ | -26.9 | -771 | 35.48 | 28.62 |
| $^{121}Sb$ | -26.4 | -791 | 37.15 | 29.96 |
| $^{119}Sn$ | -8 | -154 | 23.83 | 19.22 |

The couplings for tin and antimony are dependent on the estimated value of -8 for $^{119}Sn$. Clearly since the spectra provide only $\delta$, $V_{zz}$ and $\eta$, one cannot hope to evaluate $h_s$, $h_x$. $h_y$ and $h_z$, even supposing a, b and c are known.

A first approximation to a, b and c can be made in the following way. The constant c is introduced to change the reference for the isomeric shifts from $^{129}I$ in ZnTe equals zero to $^{129}I^-$ equals zero. The different alkali iodides give different isomer shifts and these are assumed to arise entirely from small differences in $h_p$. Values for $h_p$ in these compounds are available from other sources, e.g. dynamic quadrupole coupling and nuclear magnetic resonance measurements. Assuming $h_s$ is zero for these compounds this enables the constant b to be determined.

Evaluation of a is more difficult. An early method made the probably doubtful assumptions that the bonding in $KIO_3$ involved only p orbitals so that $h_s = 0$ and $\delta = bh_p + c$. Now $h_p/3$ will be the number of electrons removed from the iodine per I - O bond. It was then assumed that the same transfer of electrons per bond took place in $KIO_4$. One quarter of this was attributed to $h_s$. Then using the $h_p$ value calculated from the observed quadrupole coupling in $KIO_3$, together with its isomer shift, a value for a could be obtained.

Estimates made in this way have been refined and alternative methods and theoretical treatments of the data developed. Best estimates of a = -9.20, b = 1.50 and c = -0.54 for $^{129}I$ data in mm.s$^{-1}$, have been obtained.

It is still necessary to make further assumptions to reach useful results. For the compounds $CH_nI_{4-n}$ it is reasonable to suppose $h_x = h_y = 0$, when one obtains the results shown in Table 9.2.

### Table 9.2.

| n | $e^2qQ$ * | $h_p$ | $h_s$ | Charge on iodine | $\delta$ |
|---|---|---|---|---|---|
| 0 | -2132 | 0.93 | - | -0.07 | - |
| 1 | -2051 | 0.89 | 0.029 | -0.11 | 0.53 |
| 2 | -1920 | 0.84 | 0.043 | -0.16 | 0.32 |
| 3 | -1757 | 0.77 | 0.06 | -0.23 | 0.06 |

* All quadrupole coupling data converted to $e^2q\,^{127}Q$ values.

### 9.1.1.4  *Iodide as a ligand*

Iodine forms many compounds in which it is bound to only one other atom, including numerous metal complexes in which it is present as one of the ligands.  It is quite informative to treat the spectral data for these compounds on the assumption that $h_x$ and $h_y$ are zero, so that $U_p = h_p$ and $h_p$ and $h_s$ can be obtained from $\delta$ and $\varepsilon$ the quadrupole coupling. For all the data treated in this way $\eta$ must be zero.  The sum $h_p + h_s$ will be a measure of the donation by the iodine to the metal in the complex.  This approach is analogous to that used in the analysis of the data on $SnCl_3^-$ adducts in 4.6.3.3.

Data of this kind, shown in Table 9.3, give evidence of cis and trans effects in four coordinate platinium complexes.

It will be seen that $h_s$ is nearly constant at 0.05.  The absolute errors in $h_s$ and $h_p$ may be large because of the approximations in this treatment, but the relative values are probably significant.

**Table  9.3**

Complexes of the type $PtI_2L_2$

| Compound | | hs | hp | Total donation by iodine |
|---|---|---|---|---|
| $PtI_2(PEt_3)_2$ | cis | 0.04 | 0.44 | 0.48 |
| "    " | trans | 0.07 | 0.51 | 0.58 |
| $PtI_2(PMe_2Ph)_2$ | cis | 0.05 | 0.45 | 0.50 |
| "    " | trans | 0.06 | 0.51 | 0.57 |
| $PtI_2(\gamma\ Pic)_2$ | cis | 0.06 | 0.47 | 0.53 |
| "    " | trans | 0.05 | 0.46 | 0.51 |
| $PtI_2(Py)_2$ | cis | 0.05 | 0.41 | 0.46 |
| " | trans | 0.05 | 0.45 | 0.50 |
| $PtI_2(NH_3)_2$ | cis | 0.04 | 0.42 | 0.46 |
| " | trans | 0.05 | 0.43 | 0.48 |

The interaction of iodine or the iodine halides with the substrate in intercalates, for example in graphite, has been explored in this way.

### 9.1.2  **Antimony**

In some respects the Mössbauer spectroscopy with $^{121m}Sb$ resembles that described for iodine and tin in previous sections.  The photon emission involved is of M1 type and the transitions, like $^{127}I$, are between an I = 7/2 excited state and an I = 5/2 ground state.

There are a number of favourable aspects of $^{121m}Sb$. The best source, $Ca^{121m}SnO_3$, lasts for several decades, but the initial expense is high (see Section 2.2). The Mössbauer photon energy, 37.15 keV, is rather high, and measurements must be made at 80 K or lower temperature. The source emits antimony X-rays at about 26 keV and the detection system must distinguish the Mössbauer photons from this emission. The value of $\Delta R/R$ is negative and its magnitude is very large, so that isomer shifts extend from about +4 to

- 21 with respect to the $CaSnO_3$ source. Quadrupole splittings also span a large range from 0 to about 25 mm.s$^{-1}$

The dominant unfavourable factor is the large theoretical line width, 2.1 mm.s$^{-1}$. Although $Ca^{121m}SnO_3$ sources give line widths close to the theoretical value, the eight lines arising in a quadrupole split spectrum are never resolved. The spectra obtained are always envelopes of overlapping lines, generally appearing as two broad, rather asymmetric, lines. This complicates the extraction of the usual parameters from the spectra. Even the extraction of the isomer shift is not a simple matter.

To obtain reasonably accurate spectral parameters one needs to make measurements at liquid helium temperature and to count long enough so that the statistical error attached to each point on the spectrum is very small. Further, unless the absorbers are "thin" in the sense discussed in Section 2.6.1 it is necessary to introduce transmission integral corrections to obtain reliable isomer shifts and quadrupole couplings.

A computer comparison of calculated spectra with the observed spectrum is then made varying the parameters until the best fit is obtained.
The quadrupole moments of both $^{121m}Sb$ and $^{121}Sb$ are negative so that excess electron density on the z axis of the EFG leads to a positive $eQV_{ZZ}$.

### 9.1.2.1 *Interpretation of results*.

The large value of $|\Delta R/R|$ leads to an easy differentiation of Sb(V) and Sb(III). For the former the isomer shifts run from about +4 to -7 and for the latter from -8 to -20. Thus the two states of antimony in $Rb_4Sb(III)Sb(V)Cl_{12}$ are readily distinguished, $\delta$ for the Sb(III) = -19 and $\delta$ for Sb(V) = -2.4.

There is a close relationship with the data for tin compounds, although the 5s orbitals play a lesser role in the bonding of the antimony. One can find several pairs of tin(IV) and antimony(V) compounds which are both isoelectronic and isostructural, their quadrupole couplings are linearly related and $e^{121}QV_{ZZ} / e^{119}QV_{ZZ}$ is about 3.3: indeed this is the way the $e^{121}QV_{ZZ}$ for $h_p = 1$ has been derived.

The minor role of the 5s electrons is shown by the observation that the isomer shift for a numerous group of compounds can be represented by an expression of the form $\delta = a + bh_p$. More accurately, using molecular orbital calculations of the 5s populations in different compounds, it was found that for $\delta = ah_s + bh_p$, the best fit is with a = -14.4 and b = 0.61. Like all these estimates of a and b there may well be large absolute errors in the values but, they are significant for comparing one compound with another.

The $SbR_3$ have a trigonal pyramidal configurations and yield positive $e^2qQ$. Since both $Q_{gr}$ and $Q_{ex}$ are negative for $^{121}Sb$, the greater electron density must be on the trigonal axis suggesting substantial 5p character in the lone pair.

There are extensive data on compounds with $SbR_3$ ligands and they can be treated in the same way as the derivatives of $SnCl_3^-$.

One has $h_p^* = [eQV_{ZZ} (complex) - eQV_{ZZ} (ligand)]/26.0$,

($eQV_{ZZ}$ for $h_p = 1$ for $^{121}Sb$, is 26 ). and $h_s^* = [\delta (complex) - \delta(ligand) -b h_p]/a$ where a = 15.0 and b = -0.80. For these compounds one finds $h_s^*$ lies between 0.15 and 0.21 and $h_p^*$ lies between 0.13 and 0.3. These starred quantities represent the electron density lost by the antimony on forming the complex. The results can be contrasted with those for the $SnCl_3$ adducts; the $h_s^*$ for $SbR_3$ span a much narrower range of values.

With some compounds the Mössbauer spectrum provides useful information about their structure. The spectral parameters, δ and $e^2qQ$, for $Ph_4SbF$ are -4.56 and -7.2, for $Ph_4SbCl$ -4.58 and -6.0, while for $Ph_4SbBr$ they are -5.52 and -6.8. Their sensitivity to the identity of the halogen and their substantial quadrupole couplings exclude the presence of $Ph_4Sb+$. However the perchlorate, $Ph_4SbClO_4$, shows no quadrupole coupling and presumably contains a tetrahedral $Ph_4Sb+$.

The substantial quadrupole couplings for the pyridinium and 4.methyl pyridinium salts of the $SbBr_4^-$ anion preclude a tetrahedral structure.

The quadrupole couplings in the series of trigonal bipyramidal compounds $Ph_3SbX_2$ increase in the sequence I < Br < Cl < F, while δ becomes less negative implying the s electron density decreases. This reflects 5s participation in the bonding and the greater tranfer of 5s density to the axial halogens along the series.

### 9.1.3  Tellurium

The nuclear characteristics of $^{125m}Te$ have been treated in Chapter 2 and in Table 2.1. The Mössbauer transition is of M1 type, from an excited I = 3/2 to an I = 1/2 ground state. Suitable sources have also been described. The Mössbauer fraction is big enough to allow satisfactory spectra to be recorded at 80 K. The line width is very large, 5.02. ΔR/R is positive but the range of isomer shifts for compounds of tellurium in the same oxidation state is less than the line width, so that the spectra are not sensitive to differences in the environment of the tellurium atoms in the same oxidation state. Isomer shifts are usually referred to an $^{125}I/Cu$ source, although this constitutes a slightly variable reference. Sources can also be made from the $^{125m}Te$. A very satisfactory source is $\beta^{125m}TeO_3$.

With the $^{125}I/Cu$ reference base Te(VI) compounds give shifts in the range -1.5 to -1.0, and can be distinguished from the lower oxidation states. Isomer shifts for Te(IV) compounds lie between +0.2 and +1.8, and Te(II) compounds give values between +0.3 and +1.0. Quadrupole splittings for the compounds with tellurium in its lower oxidation states are generally large. Notwithstanding the great line width, quite extensive studies of tellurium spectra have been reported.

#### 9.1.3.1  *Interpretation of the spectral data.*

As for the previous elements the quadrupole coupling supplies a value for $U_p$. The value of $eQV_{zz}$ for $h_z = 1$ given in Table 8.1 was established by measuring the quadrupole couplings for $^{125}Te$ in $TeO_2$ and, using emission spectroscopy, for $^{129m}I$ produced in the same site by $^{129m}Te$ decay.

Isoelectronic and isostructural compounds of tellurium and iodine have linearly related

isomer shifts, and quadrupole resonance spectra for Cl, Br and I in salts of the $TeX_6^{2-}$
anions give estimates of the bond ionicities. The $Te^{2-}$ in CaTe can be assumed to have
$h_s = 0$. Together these data enable estimates of a and b in $\delta = ah_s + bh_p$ to be made.
One obtains a = -2.7 and b = 0.45. As for other elements the errors in these quantities
may be considerable.

There are data for numerous complexes in which $TeR_2$ acts as a ligand. These data can
be treated in the same way as were the $SbR_3$ and I$^-$ data. The average value of $h_s^*$ for
these compounds, about 0.02, is less than for the $SbR_3$ or I$^-$ complexes, but the spread of
values is rather greater.

The participation of the 5s orbital in the bonding, or in a stereochemically active lone
pair is less important for tellurium. Thus the halotellurate(IV) salts show no measurable
quadrupole splitting, nor does trans $Te(tmtu)_2Cl_4$. The last compound exists in two
crystal modifications, orthorhombic with $\delta = 1.69$ and monoclinic with $\delta = 1.52$.
These data suggest three centre p orbital bonding. (tmtu = tetramethylthiourea)

A selection of data for tellurium compounds is given in Table 9.4, listing $\delta/\Delta$ values.

Overall the data show the importance of the lone pairs in the lower oxidation state
compounds.

The small effect of changing the anionic component in the compounds containing
$Tetu_4$, Table 9.4.D, suggest this planar species is effectively a doubly charged cation.

The very large quadrupole splittings show that the lone pairs are not in pure 5s or 5p
orbitals, but probably in s-p hybrids lying around the fourfold axis above and below the
plane of the molecule.

**Table 9.4**

| | | | |
|---|---|---|---|
| A. $Na_2TeO_4$ | -1.41/0 | $Na_2TeO_3$ | 0.17/5.94 |
| $\beta\,TeO_3$ | -1.19/0 | $TeO_2$ | 0.73/6.63 |
| $Na_2H_4TeO_6$ | -1.12/0 | | |
| B. $TeCl_4$ | 1.9/4.7 | $Me_2TeCl_2$ | 0.44/9.97 |
| $TeBr_4$ | 1.4/4.4 | $Me_2TeBr_2$ | 0.65/8.50 |
| $TeI_4$ | 1.4/3.5 | $Me_2TeI_2$ | 0.52/7.6 |
| C. $Ph_2TeBr_2$ | 0.60/7.9 | $PhTeBr_3$ | 0.91/7.8 |
| $(p.EtOC_6H_4)_2TeCl_2$ | 0.79/9.2 | $p,EtOC_6H_4TeCl_3$ | 0.91/9.1 |
| D. $Me_2Te$ | -0.14/10.5 | $Tetu_4(SCN)_2$ | 0.81/16.1 |
| $Ph_2Te$ | -0.02/10.5 | $Tetu_4Cl_2$ | 0.93/15.6 |
| $(p.MeC_6H_4)_2Te$ | 0.5/10.1 | $Tetu_4I_2$ | 0.80/15.4 |
| $Ph_2Te_2$ | 0.17/10.7 | $Te(Et_2NCS_2)_2$ | 0.59/15.2 |
| E. $(C_6H_5)_4As^+TeCN^-$ | 0.15/12.2 | $PhTeCN$ | 0.43/14.1 |
| $Ph_3PTeCN$ | 0.08/12.5 | $t.Bu_3PTeCN$ | 0.02/10.3 |

tu = thiourea.

The rather small isomer shifts in the compounds in 9.4.E suggests some 5s
participation in the bonding.

## 9.1.4 Xenon

Xenon might not seem a promising element to study by Mössbauer spectroscopy, but the rather few results present some interesting features. Two isotopes of xenon display the Mössbauer effect, $^{129m}Xe$ and $^{131m}Xe$, but the former species, emitting a lower energy γ photon, gives the better spectra. In either case spectra must be measured at or below the temperature of liquid helium. Like tellurium the line width is very large, 6.85. Other nuclear details can be found in Table 2.1.

The $^{129m}Xe$ state is fed by $^{129}I$ beta decay. A satisfactory source consists of $Na_3H_2$ $^{129}I/^{127}IO_4$. The recoil following the emission of the beta particle by the iodine is small. The nascent $^{129m}Xe$ remains in its lattice site and any change due to radiolysis by the beta particle is negligible, so the source emission is a line of not much more than the theoretical width. The xenon clathrate in hydroquinone can be used as a single line reference absorber, although it probably has some unresolved quadrupole splitting. $XeF_6$ is a more satisfactory alternative.

Assuming that hs = 0 and η = 0 for these compounds, then changes in the isomer shift and quadrupole splitting must arise entirely from changes in the population of the 5p orbitals. Thus Up = hp, except for the square planar complexes where Up = ´hp. A few results are given in Table 9.5.

### Table 9.5

| Compound | $\delta$ | $\Delta$ | $U_p$ | $h_p$ |
|---|---|---|---|---|
| $XeF_4$ | 41.0 | 0.40 | 1.50 | 3.00 |
| $XeF_2$ | 39.0 | 0.10 | 1.43 | 1.43 |
| $XeCl_4$ * | 25.6 | 0.25 | 0.94 | 1.88 |
| $XeCl_2$ * | 28.2 | 0.17 | 1.03 | 1.03 |
| $XeO_3$ | 10.9 | | 0.40 | |

These results confirm our chemical intuition that the ligand fluorine removes more 5p electron density than does chlorine.

Comparing with the data for analogous iodine compounds, it is seen that much more 5p electron density is removed by chlorine ligands fron iodine than from xenon.

The most interesting results are those obtained for the compounds marked with an asterisk,*. These cannot be isolated and studied as absorbers but can be made as sources by the decay of suitable $^{129}I$ compounds and studied by emission spectroscopy.

This is only possible because the $^{129}I$ decay disturbs the lattice so little.

Thus $XeCl_4$, which cannot be prepared in macroscopic ammounts, is formed by beta decay in $K^{129}ICl_4$. In a similar way $XeCl_2$ is found in $K^{129}ICl''$. It will be noted that charge defects must form in the parent lattices to balance the negative charge lost by the $ICl_4^-$ or $ICl_2^-$ upon decay, but it cannot be close enough to modify the Mössbauer parameters of the emission.

Fortunately the validity of these conclusions about the formation of $XeCl_4$ and $XeCl_2$ can be confirmed by the results for $XeO_3$. This species can also be formed by $^{129}I$ decay in $KIO_3$. The quadrupole splitting found for such a source, 11.0, agrees with the value for an $XeO_3$ absorber in Table 9.5 above.

Mössbauer spectroscopy provides a means of characterising species that can only survive if produced in a solid matrix.

### 9.1.5 Other main group elements

A number of other main group elements have isotopes with excited states displaying the Mössbauer effect.    For various reasons they are not much used.    The appropriate excited states are not fed by radioactive decay in the case of $^{40m}$K or $^{73m}$Ge.    $^{83m}$Kr has a low photon energy, $\Gamma_t$ about 0.2 and can be fed by decay of the 83 d $^{83}$Rb, but this rare gas offers limited opportunities for study.

## 9.2 OTHER TRANSITION ELEMENTS

A substantial proportion of the second and third families of transition elements have isotopes with suitable excited states to permit Mössbauer spectroscopy and indeed many of them have been exploited.

### 9.2.1 Iridium

Both $^{191m}$Ir and $^{193m}$Ir display recoilless emission, and indeed Mössbauer discovered the effect using the former  species.    There are two excited states giving recoilless emission for each of the above species, but virtually all the Mössbauer spectroscopy on iridium has been conducted using the 73 keV state of $^{193}$Ir.    The transition between $I_{ex} = 1/2$ and $I_{gr} = 3/2$ is of M1/E2 type (see Table 2.1).    A suitable source has been described in Section 2.2.    Source line widths close to the theoretical 0.6 can be obtained.    $\Delta R/R$ is positive and Iridium metal is a convenient reference for the isomer shifts.

Quadrupole splittings cover a wide range, from zero to about 9, but the nuclear magnetic moment is too small for the magnetic perturbation method to be used to determine the sign of $\Delta$.    However the orientated single crystal absorber method has provided signs for a few compounds.

Disadvantages are that spectra have to be measured at liquid helium temperature or below and that the Mössbauer emission cannot be separated from the K X-rays from the source.

Data for a collection of iridium compounds, excluding those with ligands encouraging $\pi$ bonding and compounds with the iridium in an oxidation state less than three, give a simple picture: the isomer shift increasing by about 1 for the loss of each d electron, see Fig.9.3.

However, as is shown on the figure the $\delta$ for $K_3Ir(CN)_6$ corresponds to that found for iridium(IV) compounds. For the numerous iridium compounds, in formal oxidation states lower than three, no such correlation is found.    Increased covalence in these compounds leads to the formal oxidation state no longer relating closely to the electronic environment of the iridium nuclei. The data included in Fig.9.3 for $IrO_4$ and $IrO_4^+$ are based on emission spectra from $^{193}OsO_4^-$ and $^{193}OsO4$ decay. (see Chap.10)

### 9.2.1.1  *Iridium(III) compounds*.

In a formal way Ir(III), a $d^6$ species, is analogous to the low spin iron(II) and it is interesting to enquire how far the p.i.s. and p.q.s. treatments prove to be applicable to Ir(III) compounds. Although sets of p.i.s. and p.q.s. values can be calculated, they are much less reliable at predicting isomer shifts and quadrupole splittings than in the case of iron(II).

The fact that fac and mer isomers give notably different isomer shifts shows that such a treatment will be very approximate. The ratio of the magnitudes of the quadrupole splittings of the cis and trans forms of $MA_2B_4$ complexes is also not as close to two as it is with the iron(II) complexes.

Some data are given in Table 9.6. Considering the conditions the molecular orbital approach sets for a successful p.q.s. analysis, this could indicate that the localisation and cylindrical symmetry of the iridium bonds is no longer valid, or that the bonds are no longer practically independent of each other. The latter possibility is hardly surprising in view of the trans and cis effects, the former being well known in iridium compounds. Both factors are likely to be involved.

Although the sequence of ligand p.i.s. or p.q.s. values according to magnitude, is practically the same as for iron(II), the values for the two elements are not linearly related. For both elements the p.i.s. parallels the spectrochemical series for the ligands.

The bonding of one ligand appears to be influenced by that of another in such a way as to minimise the change in the electronic environment of the iridium nucleus.

### 9.2.1.2  *Lower oxidation state Iridium compounds*

These effects become still more important when one examines the spectra of the numerous Ir(I) complexes. These, formally, $5d^8$ often give isomer shifts more positive than those found for Ir(III) complexes, which are formally $5d^6$. Clearly the population of the 6s orbital must have increased.

A large number of square planar compounds of the general formula $(PPh_3)_2IrCOX$ are known.

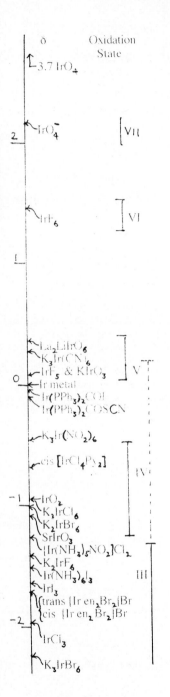

**Fig.9.3**

**Table 9.6**

| Compound | isomer | δ | Δ | isomer | δ | Δ | Ratio of Δ's |
|---|---|---|---|---|---|---|---|
| $[IrCl_2en_2]Cl$ | cis | -1.76 | 2.01 | trans | -1.72 | 3.62 | 1.80 |
| $[IrBr_2en_2]Br$ | cis | -1.70 | 1.95 | trans | -1.69 | 3.47 | 1.78 |
| $[IrPy_4Cl_2]Cl$ | cis | -1.30 | 1.50 | trans | -1.30 | 2.60 | 1.73 |

A large number of square planar compounds of the general formula trans $(PPh_3)_2IrCOX$ are known. Spectral data for several of these compounds are given in Table 9.7

**Table 9.7.**

| X | δ | Δ | X | δ | Δ | X | δ | Δ |
|---|---|---|---|---|---|---|---|---|
| 1. Cl | -0.06 | 6.66 | 5. OH | +0.28 | 7.17 | 9. SCN | +0.08 | 7.42 |
| 2. Br | +0.01 | 6.92 | 6. SH | +0.15 | 7.31 | 10. $PPh_3$ | 0.00 | 8.31 |
| 3. I | -0.02 | 6.65 | 7. $O_2CCF_3$ | 0.06 | 7.26 | 11. NCMe | -0.22 | 7.90 |
| 4. F | +0.28 | 7.22 | 8. N3 | 0.10 | 7.51 | | | |

Considering the variety of donor power shown by these ligands, both isomer shifts and quadrupole splittings change very little. Now the isomer shifts of iridium(I) compounds must be determined partly by the extent of donation by the ligands into the 6s orbital and to a smaller extent by any reduction in occupation of the 5d orbitals by π donation from the metal. The quadrupole splitting must be determined by the electronic imbalance in the 5d orbitals; the data in the table show that the imbalance only changes with ligand to a modest extent. Changes in the carbonyl to iridium bond greatly reduce changes in the Mössbauer parameters upon introducing different ligands.

The spectra show that according to the nature of X, the extent of donation into the iridium 6s orbital and of π donation to the ligands by the iridium will change. These conclusions are substantiated by infra red data on the trans carbonyl group stretching frequencies in the different complexes.

The Mössbauer data on these and related compounds provide some of the most convincing evidence of the role of π bonding in iridium chemistry. The isomer shift for $[Ir(PPh_3)_2CO(OH)]$ is 0.28 while that for $[Ir(PPh_3)_2NO(OH)]PF_6$ is 0.73. The increase must reflect a reduced population of the 5d orbitals because of π donation by the iridium to the NO+.

There is some possibility that imbalance in the 6p orbitals may be involved in determining the magnitude of the EFG. The calculated values of $<r^{-3}>_{5d}$ and $<r^{-3}>_{6p}$ are probably not very accurate. In gold(I) compounds the EFG certainly arises from 6p imbalance. Towards the completion of the 5d shell, the 5d orbitals are becoming more diffuse and at the same time the 6p are contracting, so that the difference between the two expectation values must be decreasing.

The oxidative addition reactions in which a molecule XY adds to a planar four coordinate Ir(I) species to produce a quasi-octahedral Ir(III) compound has been investigated. One might expect that electron density would be removed from the $d_{z^2}$ orbital and added to the 6s orbital so that δ of the Ir(III) species produced should be larger and Δ should be smaller than for the Ir(I) parent. This has been verified for a number of $Ir(diphenylphosphinoethane)_2^+$ complexes. The reduction in $\Delta$ is much more marked than the change in $\delta$.

### 9.2.2 Gold

Very satisfactory Mössbauer spectra can be obtained with $^{197m}Au$. The excited state is conveniently fed by the decay of $^{197}Pt$. This species has a half life of only 18 h, but it is produced by neutron irradiation of platinum and a piece of platinium foil can be repeatedly irradiated producing a source with a line width not much more than the theoretical value of 1.88.

The Mössbauer transition is from an $I_{ex.} = 1/2$ to the $I_{gr.} = 3/2$ ground state, and it has mixed M1/E2 character. The photon energy $E_\gamma = 77.3$ keV gives a rather large recoil and spectra have to be recorded at liquid helium temperature.

Like iridium, the nuclear magnetic moments are too small to permit the magnetic perturbation method to be used to determine the sign of $\Delta$. The orientated crystal method has been used for a few compounds.

The line width is rather large but both isomer shifts and quadrupole splittings cover a wide range of values. $\Delta R/R$ is positive. Isomer shifts are referred to gold metal.

The isomer shift alone will not distinguish Au(I) from Au(III), or even Au(V). This last gives an isomer shift of 3.5 in $AuF_6^-$ salts. Cs (or Rb)Au, containing the unusual $Au^-$, yield a positive value of about 8.3. The combination of $\delta$ and $\Delta$ however will distinguish Au(I) from Au(III). (See Fig.9.4).

### 9.2.2.1 *Gold(I) complexes.*

Supposing the linear gold(I) complexes have 6s/6p hybrid bonds the stronger donor the ligand is, the greater $\delta$. This is because although increased electron density in the 6p orbital will reduce $\delta$, the increase in density in the 6s orbital will be more important. The increase in $6p_z$ will increase the imbalance in the 6p orbitals and $\Delta$ will rise. On this basis the EFG should be negative and this has been confirmed for $KAu(CN)_2$. One can expect that $\delta$ and $\Delta$ will be very roughly linearly related.

The data in Fig.9.4 show this to be the case. They also show that the data for ClAuL and for $Ph_3PAuL$ complexes lie on lines of different slope. This effect is not due to lattice contributions to the EFG, since $[Ph_3PAuL]^+$ and neutral $Ph_3PAuL$ complexes show no systematic differences. Similar difficulties arise if one tries to derive tables of p.i.s. and p.q.s. for these complexes.

This behaviour can be explained by changes in the proportions of 6s and 6p in the bonding hybrids. The data show that $\Delta$ rises more quickly than $\delta$ for the $Ph_3PAuL$ complexes, which suggests more donation to the $6p_z$ orbital. This is in agreement with Bent's rule that the bond to the least electronegative element has the greater s character, because this implies that the Au-L bond has the greater 6p character. This supports a changing hybridisation interpretation.

On changing from two to three coordinate gold with $sp^3$ bonding, the isomer shift

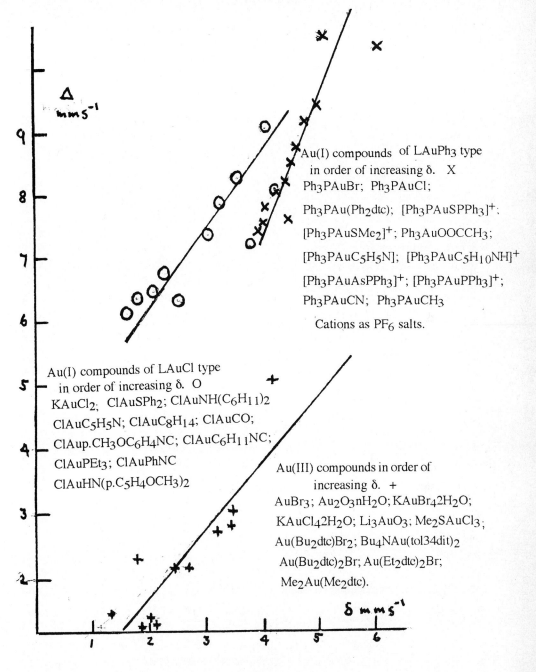

Au(I) compounds of LAuPh$_3$ type
in order of increasing δ.　X
Ph$_3$PAuBr;　Ph$_3$PAuCl;

Ph$_3$PAu(Ph$_2$dtc);　[Ph$_3$PAuSPPh$_3$]$^+$;

[Ph$_3$PAuSMe$_2$]$^+$;　Ph$_3$AuOOCCH$_3$;

[Ph$_3$PAuC$_5$H$_5$N];　[Ph$_3$PAuC$_5$H$_{10}$NH]$^+$

[Ph$_3$PAuAsPPh$_3$]$^+$;　[Ph$_3$PAuPPh$_3$]$^+$;

Ph$_3$PAuCN;　Ph$_3$PAuCH$_3$

Cations as PF$_6$ salts.

Au(I) compounds of LAuCl type
in order of increasing δ.　O
KAuCl$_2$;　ClAuSPh$_2$;　ClAuNH(C$_6$H$_{11}$)$_2$
ClAuC$_5$H$_5$N;　ClAuC$_8$H$_{14}$;　ClAuCO;
ClAup.CH$_3$OC$_6$H$_4$NC;　ClAuC$_6$H$_{11}$NC;
ClAuPEt$_3$;　ClAuPhNC
ClAuHN(p.C$_5$H$_4$OCH$_3$)$_2$

Au(III) compounds in order of
increasing δ.　+
AuBr$_3$;　Au$_2$O$_3$nH$_2$O; KAuBr$_4$2H$_2$O;
KAuCl$_4$2H$_2$O;　Li$_3$AuO$_3$;　Me$_2$SAuCl$_3$;
Au(Bu$_2$dtc)Br$_2$;　Bu$_4$NAu(tol34dit)$_2$
Au(Bu$_2$dtc)$_2$Br;　Au(Et$_2$dtc)$_2$Br;
Me$_2$Au(Me$_2$dtc).

**Fig.9.4**

decreases sharply, but the quadrupole splitting hardly changes. A ratio of Δ/δ > 3
identifies a three coordinate gold(I) complex, but the converse is not always true.

If the effect is to be mainly associated with increased p shielding of s electron density
at the nucleus, it is larger than would be expected. It is interesting to find on going to a
four coordinate complex, (Ph$_3$P)$_3$AuSnCl$_3$, that the isomer shift falls again to the low

value of 1.64, suggesting the p screening is indeed substantial. The four coordinate $Au(I)L_4$ complexes are easily identified by the absence of quadrupole splitting.

It may be noted that, if one can apply a p.q.s. approach and the donation by a given ligand is independent of the coordination number, the quadrupole splitting should be the same in $Au(I)L_2$ and $Au(I)L_3$ complexes.

### 9.2.2.2 *Gold(III) compounds.*

As can be seen in Fig.9.4 the combination of the δ and Δ values will generally distinguish Au(III) from the Au(I) compounds. The EFG for the planar $Au(III)X_4$ compounds should however be positive and this has been verified for $KAu(CN)_4$.

A curious feature is the sensitivity of the Mössbauer parameters to the identity of the cation in $MAuX_4$ salts.

### 9.2.3 **Ruthenium**

Although two isotopes of ruthenium possess excited states showing recoil free resonance, $^{99}Ru$ and $^{101}Ru$, the former is more satisfactory for Mössbauer spectroscopy. A source containing the parent $^{99}Rh$ can be made by cyclotron bombardment of a ruthenium foil, $^{99}Ru(d.2n)^{99}Rh$ or $^{99}Ru(p.n)^{99}Rh$. The parent half life is 15 d. Although ruthenium metal is hexagonal, such a source gives a single line of width close to the theoretical value of 0.15. (Table 2.1).

$E_\gamma$ is rather large, 89.4 keV, and spectra must be recorded at 80 or better 4.2 K. The transition is between an excited state with $I_{ex.} = 3/2$ and a ground state with $I_{gr.} = 5/2$. The emission is of mixed E2/M1 type.

### 9.2.3.1 *Isomer shifts in ruthenium compounds.*

Isomer shifts, expressed with respect to ruthenium metal, span about 2 $mm.s^{-1}$. $\Delta R/R$ is positive and the narrow line width permits differentiation of the oxidation states of ruthenium from Ru(VIII), $d^0$, with the most positive δ, to Ru(II), $d^6$, with the most negative δ. (See Fig. 9.5)

Thus in "ruthenium red", $[(NH_3)_5Ru(III)-O-Ru(IV)(NH_3)_4-O-Ru(III)(NH_3)_5]$, the Ru(III) and Ru(IV) can be distinguished. Isomer shift data for the Ru(II) compounds show:

δ of $[Ru(CN)_5NO_2]^{2-}$ < δ of $[Ru(CN)_6]^{4+}$ < δ of $[Ru(CN)_5NO]^{2-}$ and δ of $[RuCl_5NO]^{2-}$ < δ of $[Ru(NH_3)_5NO]^{3+}$ < δ of $[Ru(CN)_5NO]^{2-}$. The sequences reflect the varying π acceptance from the ruthenium by the different ligands.

### 9.2.3.2 *Quadrupole splitting in ruthenium compounds.*

The E2/M1 type transition permits $\Delta m_I = 0, \pm 1, \pm 2$ so that a six line spectrum should ensue. The excited state has much the larger quadrupole moment and as a result the spectrum seen with an EFG usually appears as two broad lines.

The salts of $[RuNCl_5]^-$ and $[RuNBr_5]^-$ give very large quadrupole couplings and incipient resolution of the six lines is seen. A computer analysis of the quadrupole split spectra, similar to that used in the case of $^{129}I$, yields δ, ε the quadrupole coupling and η.

### 9.2.3.3 *Magnetically split spectra.*

The narrow line width permits analysis of the multi-line spectra magnetically split spectra

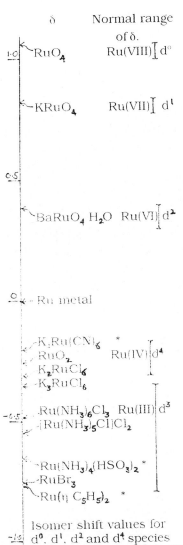

$\delta$     Normal range
of $\delta$.

RuO$_4$      Ru(VIII) $d^0$

KRuO$_4$     Ru(VII) $d^1$

BaRuO$_4$ H$_2$O   Ru(VI) $d^2$

Ru metal

K$_4$Ru(CN)$_6$   *
RuO$_2$       Ru(IV) $d^4$
K$_2$RuCl$_6$
K$_3$RuCl$_6$

Ru(NH$_3$)$_6$Cl$_3$   Ru(III) $d^5$
[Ru(NH$_3$)$_5$Cl]Cl$_2$

Ru(NH$_3$)$_4$(HSO$_3$)$_2$ *
RuBr$_3$
Ru($\eta$ C$_5$H$_5$)$_2$ *

Isomer shift values for
$d^0$, $d^1$, $d^2$ and $d^4$ species
do not overlap. * $d^6$ Ru(II)
compounds overlap both
$d^4$ and $d^5$ values.

**Fig.9.5**

Several ruthenium mixed oxides order ferromagnetically at low temperatures with $B^o_{hf}$ in the range 10 to 40 T. The E2 character of the transition leads to an 18 line spectrum and most of these lines can be resolved. Because of the M1 admixture the intensities of these lines do not correspond to those expected on the basis of the Clebsch- Gordan coefficients for an E2 3/2 ⇔ 5/2 transition. The differences due to the M1 admixture can be used to obtain the E2/M1 mixing ratio.

### 9.2.4 Tungsten

The Mössbauer effect has been observed in several isotopes of tungsten. In all cases the life times of the Mössbauer excited states are around 1 ns so that the line widths are rather large. In addition the $E_\gamma$ are about 100 keV so that spectra have to be recorded at liquid helium temperature.

Most work has been done with $^{182m}$W. Neutron irradiation of a tantalum foil produces the 115 d $^{182}$Ta, which feeds the $^{182m}$W excited state. Such a source gives a line width not much greater than the theoretical value, about 2.1. The main limitation is that $\Delta R/R$ is very small, so that isomer shifts vary very little for different tungsten compounds.

Information is restricted to quadrupole splittings and the recoil free fractions. Fortunately the quadrupole splittings are often very large, up to 19, and different tungsten environments can sometimes be distinguished by differing $\Delta$, for instance two tungsten sites in Na$_2$W$_2$O$_7$.

Another useful feature is that the transition is from an $I_{ex.} = 2$ to an $I_{gr.} = 0$ ground state. This means that in the presence of an EFG, the magnitude, sign of $\Delta$, and the asymmetry parameter can all be determined. (See Fig.9.6). This provides an easy way of deciding the structure of W(CN)$_6^{2-}$ salts. In a dodecahedral configuration, $D_{2d}$, the b$_1$ ( $d_{x^2-y^2}$ ) will lie lowest, while in the square antiprismatic $D_{4d}$ form the a$_1$ ( $d_{z^2}$ ) will be

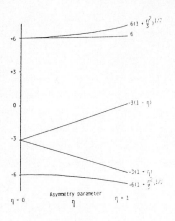

**Fig.9.6**

lowest. Taking into account that Q is negative for $^{182}W$, the sign of $\Delta$ will be negative for the $D_{2d}$ form and positive for the $D_{4d}$ form. The $D_{2d}$ form will give a 2.2.1 intensity pattern and the $D_{4d}$ the reverse pattern. See Fig.9.7.

### 9.2.5 Tantalum

Tantalum Mössbauer spectroscopy has considerable potential importance but presents formidable practical difficulties. The theoretical line width is only $6.5 \times 10^{-3}$ mm.s$^{-1}$ so that fortuitous vibrations of the spectrometer can ruin the spectra. In addition impurities and, or, defects in the absorbers will modify spectra. So far the best experimental line widths have been almost an order of magnitude greater than the theoretical value. It seems likely that these difficulties can be overcome.

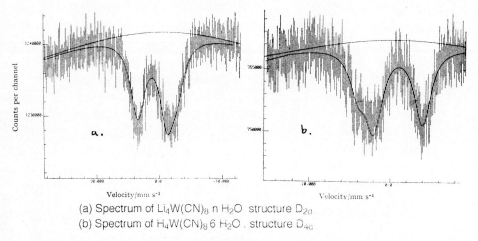

(a) Spectrum of $Li_4W(CN)_8$ n $H_2O$  structure $D_{2d}$
(b) Spectrum of $H_4W(CN)_8$ 6 $H_2O$ . structure $D_{4d}$

**Fig.9.7**

There are also some very favourable aspects, $E_\gamma = 6.23$ keV, and the recoil is small enough to permit measurements at quite high temperatures. $\Delta R/R$ is extremely large in relation even to even the present experimental line widths. Since tantalum is near the beginning of the 5d elements both $\delta$ and $\Delta$ are determined by the occupation of the 5d orbitals. Both changes in oxidation state and cf ligand produce very large changes in Mössbauer parameters as the following data show. $NaTaO_3$  $\delta = -15.45$,  $\Delta = +18.2$, $\eta = 0.5$; $TlTaSe_4$  $\delta = +3.85$, $\Delta = 0$; $TaS_2$  $\delta = +70.2$, and $\Delta = -240$.

### 9.2.6 Some others

Several other transition elements have been investigated. In most cases there is some deterrent to their more extensive use, although $^{190}Os$ and $^{195}Pt$ have attracted attention.

The unfavourable features include one or more of the following: (i) Cyclotron or accelerator necessary to produce rather short lived sources, $^{61}$Ni, $^{67}$Zn, $^{178}$Hf and $^{189}$Os. (ii) Very small absorptions even at < 4.2 K. $^{178}$Hf, $^{189}$Os and $^{199}$Hg. (iii) Line width inconveniently large, $^{178}$Hf, $^{189}$Os and $^{199}$Hg. (iv) Extremely narrow line width, $^{67}$Zn.

## 9.3  RARE EARTHS AND ACTINIDES

A high proportion of the rare earth elements have isotopes which display the Mössbauer effect. The interesting magnetic properties of the rare earths has stimulated numerous studies using Mössbauer spectroscopy.

### 9.3.1  Europium

The greater amount of work has concerned europium. Both $^{151}$Eu and $^{153}$Eu have suitable excited states but $^{151m}$Eu has the more favourable characteristics. The $E_\gamma$ for $^{151m}$Eu is 21.6 keV, so that the Mössbauer fraction tends to be high and spectra can be measured at room temperature. The excited state can be fed by either the long lived $^{151}$Sm (87 y.) or $^{151}$Gd (120 d). The latter has the advantage that a higher proportion of decays populate the $^{151m}$Eu state; but it is a cyclotron product.

The theoretical line width is 1.44, but a lot of work has been done using oxide or fluoride sources. Neither of these provide a cubic environment for the rare earth and such sources cannot give a monochromatic emission: there is always some unresolved quadrupole splitting and line broadening.

However both the isomer shifts and quadrupole splittings are large and for some purposes the large line width is unimportant. $\Delta R/R$ is positive and $Eu^{2+}$ is readily distinguished from $Eu^{3+}$. This has enabled some very interesting work on electron delocalisation in the mixed valence $Eu_3S_4$.

### 9.3.2  Neptunium

Several actinide elements have isotopes with excited states showing the Mössbauer effect. By far the most useful is $^{237}$Np, indeed its properties make it one of the best elements for Mössbauer spectroscopy.

$^{237}$Np is an alpha emitter with a half life of about $2.2 \times 10^6$ years. It is formed during the operation of uranium fuelled reactors. Ample amounts for all kinds of Mössbauer studies are commercially available. Suitable precautions during its use must be taken regarding its radiotoxicity, but it does not emit much penetrating radiation. The source materials are more dangerous and must be kept sealed.

The excited state of $^{237}$Np at 59.5 keV has a half life of $\approx$ 63 ns. giving a theoretical line width of 0.073. The excited state can be fed by $\beta$ decay of $^{237}$U, which has a half life of 6.75 d.; by orbital electron capture in $^{237}$Pu, $t_{1/2}$ = 45.6 d., or by alpha decay in $^{241}$Am, $t_{1/2}$ = 432.5 y. The first two of these, as expected, provide sources with narrow line widths, but the uranium isotope is rather short lived and the plutonium isotope hard to obtain.

One would not expect an amerecium source to be very satisfactory; for a monochromatic emission. The $^{237m}$Np, which suffers a substantial recoil upon formation, must thermalise and reach a cubic site before Mössbauer emission. Indeed the sources so far produced, generally Am - Th alloys, give line widths about an order of magnitude greater than the theoretical value.

However the isomer shifts, quadrupole and magnetic splittings are all so large that this is not too disturbing. Most studies have used such sources.

The transition is $I_{ex.}= 5/2+$, $I_{gr.} = 5/2-$, an E1 type. $\Delta R/R$ is negative and rather large. $Q_{ex.}$ and $Q_{gr.}$ are to a good approximation equal and $\mu_{ex.}$ is about 0.54 times $\mu_{gr.}$, all these terms are large.

The isomer shifts reflect the screening of the 6s electron density by the occupation of 5f orbitals as can be seen in Table 9.8.

Other compounds with neptunium in these oxidation states have isomer shifts lying within bands up to about 15 wide centred on the values in Table 9.8. There is no appreciable overlap. Quadrupole splittings are also considerable. With the slight approximation $Q_{ex.} = Q_{gr.}$ an axial EFG gives rise to a five line spectrum as shown in Fig.9.8

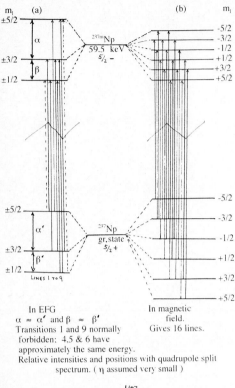

In EFG
$\alpha \approx \alpha'$ and $\beta \approx \beta'$
Transitions 1 and 9 normally forbidden: 4.5 & 6 have approximately the same energy.
Relative intensities and positions with quadrupole split spectrum. ( $\eta$ assumed very small )

In magnetic field.
Gives 16 lines.

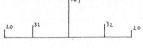

Relative intensities and positions for magnetically split spectrum.

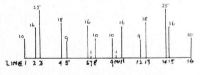

**Fig.9.8**

**Table 9.8**

| Species (No.of 5f) | Np$^{3+}$ (4) | Np$^{4+}$ (3) | NpO$_2^+$ (2) | NpO$_2^{2+}$ (1) | NpO$_6^{5-}$ (0) |
|---|---|---|---|---|---|
| $\delta$ | +34 | -9 | -37 | -63 | -77 |

Isomer shifts are given with respect to NpAl$_2$.

If $\eta > 0$ weak $\Delta m_I = \pm 2$ lines may appear at higher and lower velocities than the above five line spectrum. The isomer shift is given by the position of the centre line, line 3 numbering the five lines from 1 to 5. The asymmetry parameter $\eta$ and the quadrupole

coupling, eqQ/4, are related to the line positions by the functions F(η and f(η) such that
F(η)  = $(v_5 - v_1)/(v_4 - v_2)$, and eQV$_{zz}$/4 = $(v_4 - v_1)$/ f(η)  ≅ $(v_5 - v_2)$/ f(η). The functions
F(η ) and f(η ) are shown in Fig.9.9.  When η = 0 the intensity pattern of the spectrum is
20 : 32 : 106 : 32 : 20.  A quadrupole split spectrum is shown in Fig. 9.10.

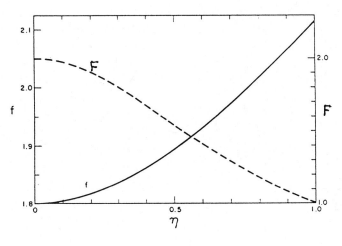

**Fig.9.9**

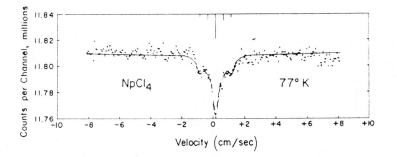

Fig.9.10.  Quadrupole split spectrum of NpCl$_4$ at 77 K.

**Fig.9.10**

As would be expected the linear NpO$_2^+$ and NpO$_2^{2+}$ give very substantial quadrupole
splittings.

In the absence of an EFG the magnetically split spectrum with 16 lines can appear. ( See
Fig.9.8 ) An example is shown in Fig. 9.11.

### 9.3.3  **Other actinide elements**.

The Mössbauer effect has been observed with nearly all the elements in the first half of
the family (Z < 96). But with the possible exception of $^{243}$Am none  seem very suitable
for extended study.

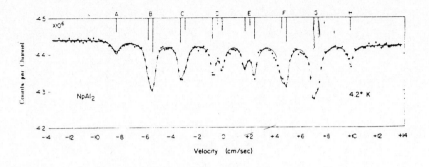

Spectrum of NpAl₂ at 4.2 K.  Some lines overlap.
**Fig.9.11**

## Acknowledgements

Fig.9.2   Reproduced with permission from Bukshpan,S. and Sonnino,T.  (1968)
*J.Chem.Phys.*, **48**, 4441.

Fig.9.7   Reproduced with permission from Clark,M.G.,Garrod,J.R.,Maddock, A.G. and
Williams, A.F.  (1975)  *J.Chem.Soc., 1975*, 120.

Fig.9.9   Reproduced with permission from  Pillinger,W.C.and Stone,J.A. (1968)
*Mössbauer effect Methodology*, **4**, 222.

Fig.9.10 Reproduced with permission from  Pillinger,W.C. and Stone,J.A.  (1968)
*Mössbauer effect Methodology*, **4**, 217.

Fig.9.11 Reproduced with permission from  Pillinger,W.C. and Stone, J.A. (1968)
*Mössbauer effect Methodology*, **4**, 331.

# 10

# Mössbauer Emission Spectroscopy

## 10.1  Mossbauer Emission Spectroscopy

The environment of $^{57m}$Fe, or other Mössbauer emitting atoms, in a source can be investigated by measuring the spectrum from the source using a single line absorber. Either the source or the absorber can be the vibrating part of the spectrometer. The change in sign of the shifts when the source is the vibrator should be noted.

Line widths in emission spectra are generally larger than for absorption spectra, probably because of the disturbance of the lattice by the formation of the Mössbauer emitting atom. Reproducibility of spectra is also less consistent.

Two kinds of measurement are possible. The source may consist of a compound of the parent species, for example a $^{57}$Co labelled compound, or it may be a substance doped with the parent species. The chemical amount of the parent species needed is extremely small and it is often possible to introduce it into a solid without much disturbing the lattice, and thus a source composed of $MnCl_2$ containing $^{57}$Co can easily be made.

Even if the site and charge state of the dopant Mössbauer parent are known, the question arises how far are these changed by the decay processes preceding Mössbauer emission. In the example of $^{57m}$Fe emission spectra, how far is the site and charge state of the atom altered by the orbital electron capture by the $^{57}$Co and the de-excitation of the higher excited state of $^{57m}$Fe ?   (See decay scheme Fig.2.1)

If the decay process leads to sustantial recoil, sufficient to eject the atom from its lattice site, effectively more than about 20 eV, it is difficult to predict the site and chemical state of the atom at the time of Mössbauer emission.. Fortunately for $^{57m}$Fe, and most other Mössbauer nuclei, the parent decay process is orbital electron capture, beta decay or isomeric transition, all processes where the direct recoil is unlikely to eject the daughter species from the site occupied by its parent in the lattice.

The question of the charge state of the nascent Mössbauer emitter is more difficult to predict. All the above decay processes can change the charge state of the decaying atom and may produce chemical changes in the affected molecule and its surroundings.

## 10.1  EMISSION SPECTRA FROM LABELLED COMPOUNDS

Pure beta decay produces the smallest disturbance of the lattice, although oxidation will normally ensue. The formation, of an atom of one unit greater atomic number than the parent species, implies an increase of one unit in the oxidation state of the daughter; parent and daughter are isoelectronic. But charge compensation demands that the increased charge on the daughter be balanced by an appropriate lattice defect. A trapped electron, an increase in the charge on an anion, or a reduction in the charge on a cation, in the matrix lattice can all serve for this purpose. Since the mole fraction of radioactive atoms is extremely small such compensation can usually occur.

## 10.2 **SYNTHESIS IN A MATRIX**

For the reasons given above one finds the emission from $^{129}$Te labelled $K_2TeCl_6$ arises from $^{129}ICl_6$ , a compound not available for measurement by absorption. Similarly a source of an $^{193}$Os labelled osmium compound gives a spectrum corresponding to that of the iridium compound of the next higher oxidation state; Os labelled $^{193}Os(C_5H_5)_2$ gives a spectrum attributed to $Ir(C_5H_5)^+$.

However with some compounds either the initial product is unstable or neutralisation rather than charge compensation takes place; thus $^{193}$Os labelled $K_2OsCl_6$ yields spectra due to Ir(IV), Ir(III) and even Ir(II) species. The interpretation of such emission spectra is difficult.

Noteworthy examples of such synthesis are found with xenon as described in Section 9.1.4.

## 10.3 **DECAY INVOLVING AN AUGER CASCADE**

Both isomeric transition and orbital electron capture decay usually lead to the creation of a low lying K or L shell vacancy. In the case of isomeric transition this is due to the internal conversion of a low energy photon emission. Such a vacancy initiates an **Auger cascade**, low energy electrons are ejected and the atom acquires several units of positive charge.

The subsequent fate of the affected atom, before emission of the Mössbauer photon, depends on how quickly and completely electrons lost in the Auger cascade are replaced. For an isolated molecule the positive charges migrate to the periphery of the molecule and fragmentation of the molecule due to Coulombic repulsion takes place. In a solid, greater dispersion of charge and more rapid neutralisation may prevent such catastrophic events. However some bond rupture and chemical change may still occur. Besides bond rupture, from a chemical point of view, the Auger cascade is essentially an oxidising event.

The consequences of an Auger cascade in a solid are determined by how easily the ejected electrons return to the affected atom, and how the energy released on neutralisation is dissipated. In metals return is very rapid. In some compounds the ejected electrons may become delocalised in low lying conduction bands and rapid return may also be possible. The low energy Auger electrons travel at most a few lattice units. In insulators, if the source compound is susceptible to radiolytic decomposition, they may be trapped and unable to return to the positively charged parent atom. In such a source the nascent Mössbauer emitter will be formed in a nest of radical species.

This has been shown in an elegant Mössbauer experiment using $^{57}$Co labelled β -diketone complexes. The acetylacetonate is rather susceptible to radiolytic decomposition while the dibenzoylmethane complex is not. It has already been observed that in such large iron(III) complexes both spin-spin and spin-lattice relaxation processes are rather slow. The emission spectra of the two cobalt complexes at 78 and 4.2 K are shown in Fig.10.1.

In the acetylacetonate complex the $^{57m}$Fe is formed in a nest of radicals which facilitate the spin-spin relaxation, so that the relaxation is still fast enough to exclude a well formed magnetically split spectrum. But with the dibenzoylmethane complex the radical density is much lower, relaxation is much slower and a clear magnetic spectrum is obtained.

When the matrix compound is easily radiolysed the emission spectra are usually complex with a number of components. If it is resistant to radiolysis and especially if it has a rather low lying conduction band the principal component of the spectrum corresponds to the absorption spectrum of the compound of the daughter species analogous to the parent compound.

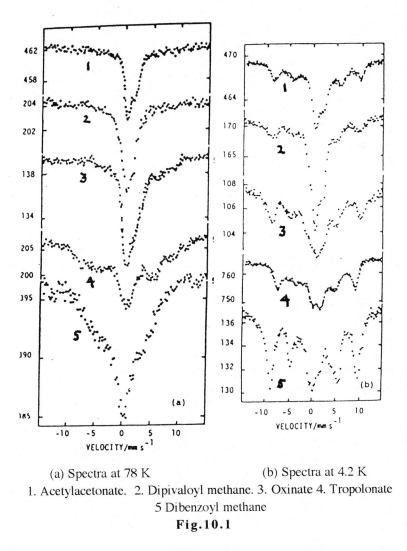

(a) Spectra at 78 K                    (b) Spectra at 4.2 K
1. Acetylacetonate.  2. Dipivaloyl methane. 3. Oxinate 4. Tropolonate
5 Dibenzoyl methane

**Fig.10.1**

$^{57}$Co complexes of highly conjugated ligands generally survive the Auger cascade, and yield a spectrum corresponding to the absorption spectrum of the iron complex.

### 10.3.1 Time Resolved Mossbauer Spectroscopy

By using time resolved Mössbauer spectroscopy one can learn something about the events immediately following the decay generating the Mössbauer excited state. In the case of $^{57}$Fe spectroscopy, in 90% of the decay events the formation of $^{57m}$Fe is preceeded by the emission of a 122 keV photon. The detection of this photon denotes a zero time and the recording of the Mössbauer photons is then controlled, by gating access to data collecting system, so that only photons received after a chosen delay following zero time are recorded.

In this way one can explore how far the environment and charge state changes during the delay. One can investigate such changes, or "after effects" as they are sometimes called, in the time interval from about $10^{-9}$s to $10^{-6}$s.

### 10.4  $^{57}$Co DECAY IN METAL HALIDES

From the above description of the consequences of the Auger cascade, one might expect $^{57}$Co decay in a transition metal halide of the $MX_2$ type would generally lead to emission from iron in a higher oxidation state than two, perhaps from iron(IV) or even iron(VI). In fact emission spectra arising from the higher oxidation states are rare, although $K_2CoF_6$ gives an iron(IV) emission.

Generally the emission spectrum shows iron(II) and iron(III) components. The proportion of the iron(III) component increases as the ionic radius of the metal M decreases. It is very small if $M = Fe^{2+}$, $Mn^{2+}$ or $Zn^{2+}$. For a given fluoride $MF_2$ it decreases as the temperature is raised, presumably because neutralisation proceeds more rapidly at the higher temperature. No iron(III) emission is found with $MCl_2$ or $MBr_2$.

An iron(III) emission is only found when the lattice energy of the $MF_2$ matrix exceeds about 2.7 MJmol$^{-1}$. These observations provide a clear demonstration of the importance of the matrix in determining the state of the $^{57m}$Fe at the time of Mössbauer emission.

Hydrated compounds, including the halides, give a strong iron(III) component in their emission. This is due to the local radiolysis of the water by the Auger electrons. The OH produced by radiolysis traps electrons and inhibits their return to the nascent iron atom. Typical emission from $CoF_2$ and its hydrate are shown in Fig.10,2.

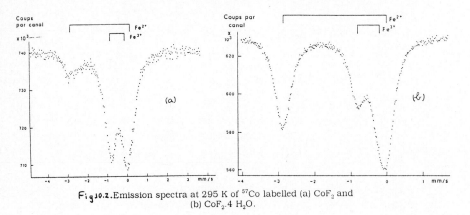

Fig.10.2. Emission spectra at 295 K of $^{57}$Co labelled (a) $CoF_2$ and (b) $CoF_2.4 H_2O$.

**Fig. 10.2**

In some cases new spectra, not seen in absorption, are found, but the oxidation state of the iron emitter can be identified by the isomer shift.

### 10.4.1 Aliovalent compounds

$^{57}$Co can be introduced homogeneously into solids such as sodium or silver chloride and there is an extensive literature on the emission spectra from such sources. Its interpretation is difficult. Firstly the situation of the $^{57}$Co in the crystals is usually uncertain. Very pure materials must be used, sodium chloride commonly contains hydroxide and superoxide ions and these will tend to associate with the aliovalent cobalt ions.

The spectra observed are often affected by annealing the source, suggesting that the defect character and density affect the state of the $^{57m}$Fe before photon emission.

The spectra, whose reproducibility is poor, supply evidence of several environments for the iron including, surprisingly, iron(I) with an isomer shift of 1.98.

## 10.5  OXIDE SYSTEMS

There are also extensive data on the emission spectra from $^{57}$Co labelled oxides, including the cobalt oxides and both homovalent and aliovalent oxides. These compounds are generally markedly nonstoichiometric and display a wide variety of defects. For this reason reproduceability of the spectra is poor, depending on the precise previous treatment of the source and conditions of measurement.

## 10.6  IMPLANTATION SYSTEMS

$^{57}$Co can be introduced into most refractory solids by ion implantation, but similar difficulties in interpretation of the emission spectra to those outlined above usually arise. Other parent species have also been used. If the decay process is isomeric transition or orbital electron capture, the Mössbauer species will be produced in the site at which the implanted atom came to rest. However, except in semiconductors and metals, its charge state is still questionable. $^{125m}$Te has been used in this way.

**Acknowledgements**

Fig. 10.1  Reproduced with permission from Sakai,Y.,Endo,K.and Sano,H. (1980)
*Bull.Chem.Soc.Japan*, **5 3**, 1317.

Fig. 10.2  Reproduced with permission from Friedt,J.M. and Adloff,J.P. (1969)
*Comp.rend.*, **268C**, 1342.

# Appendix 1

## Sources of data on Mössbauer spectroscopy

Mössbauer spectroscopy is fortunate in that, from its beginning in the early 1960's, it has been well abstracted and indexed.

In the U.S.A. the Mössbauer Effect Data Center publishes the Mössbauer Effect Reference and Data journal. the current volume is number 20. In addition it supplies a variety of software data bases ( 3.5" or 5.25" discs ) covering individual elements and areas of application. Some 60000 entries are recorded.

[ Mössbauer Effect Data çenter, University of North Carolina, Asheville, North Carolina, 28804-3299. U.S.A. ]

In the U.K., the Royal Society of Chemistry and the Manchester Computing Centre have implemented Mössbauer Effect Center's data base, which can be accessed through JANET. It contains data on individual compounds, and can be searched by formula, author, keyword etc.. Further details can be obtained from Prof. R.V.Parish, UMIST, Manchester M60.1QD. Email: R.V.Parish@UMIST.ac.uk) or Mrs.S.J.Davnall, MCC, Email: Sarah Davnall@mcc.ac.uk).

# Appendix 2

# Table of Symbols

A  Mass number of nucleus. Tensor determining magnetic coupling.

B  Magnetic field flux. $B_{hf}$ Hyperfine magnetic field at the nucleus. $B_F$ Fermi contact contribution to $B_{hf}$. $B_o$ Contribution to $B_{hf}$ from orbital motion of 3d electrons. $B_D$ Contribution to $B_{hf}$ from dipolar interaction with electron spin. $B_{int}$ internal magnetic field at nucleus. $\overset{\circ}{B}_{hf}$ the saturation hyperfine magnetic field.

c  Velocity of light in vacuum. C Clebsch-Gordan coefficient.

D  Change in energy due to Döppler effect. D determines tetragonal distortion. Misfit parameter used in testing suitability of model to data.

E  Energy, subscripts indicate origin e.g. $E_\gamma$ denotes photon energy. Parameter related to rhombic distortion.

e  Charge on ptoton. Doubly degenerate energy level.

f  Mössbauer fraction.

g  Gyromagnetic ratio. $g_n$ nuclear g, $g_e$ electronis g.

h  Planck's constant. $\hbar = h/2\pi$

H  Hamiltonian operator

I  Nuclear spin. $I_x$ etc. components of I along axes. $I_+$ and $I_-$ nuclear raising and lowering operators.

J  Coulomb exchange integral, components $J_x$ etc. along axes.

k  Boltzmann's constant

K  Absolute temperature.

l and m Angular momentum quantum numbers.

M  Misfit parameter for testing suitability of model. See p.43.

m  Component of nuclear spin. $\Delta m$ Change in m in process. M Relative atomic or molecular mass n    Notional number of electrons in an orbital, e.g. $n_p$ number in p orbital.

P and p  Momenta. p.i.s. Partial isomer shift. p.q.s. Partial quadrupole splitting.

Q    Quadrupole moment of nucleus.    $Q_{ii}$ defined in text on p.12.

$q_i$    Charge q located at $r_i$, $\theta_i$, $\varphi_i$ . $q_{lat}$, $q_{val}$, $q_{fi}$, $q_{mo}$, Equivalent charges producing lattice, valence, free ion and molecular orbital contributions to the EFG.

R    The gas constant. Recoil energy. R and r radii. r  polar coordinate.
    R   Sternheimer anti-shielding factor for charges within atom.

S    Electronic spin. Misfit parameter see p.43.

T    Kinetic energy. $T_C$ Curie temperature. $T_N$ Néel temperature. $T_{50}$ Temperature for 50% spin cross over. t triply degenerate energy level. $T_B$ Blocking temperature. $T_V$ Verwey transition T. $T_M$ Morin transition T.

$T_a$    Dimensionless thickness of absorber. $t_a$  thickness of absorber. See p.37.

V    Potential. v velocity of source relative to absorber. $V_{xx}$ etc. components of EFG tensor.

$Y_{LM}$  A spherical harmonic function.

Z    Atomic number.

Subscripts ∥ and ⊥ components parallel and perpendicular to a defined axis.

<x>  Expectation value of quantity within bra and ket.

α    **Exponential in Debye model for vibrational frequencies in a solid.** Internal conversion coefficient.

β    **Defined on p.14. Mode of radioactive decay.**

χ    Electronegativity.  Wave function.  $\chi^2$ Measure of closeness of fit of data to model used.

δ    **Isomer or chemical shift.**

Δ    Quadrupole splitting.    $\Delta^0$ and $\Delta^T$ values at 0 and T K.

ε    Quadrupole coupling defined on p.18. $\varepsilon_0$ Vacuum permittivity.

φ    Wave function

γ    Photon   Sternheimer anti-shielding factor for lattice charges.

η    Asymmetry parameter of EFG.

φ    Polar coordinate

λ    Spin-orbit coupling constant

Γ    Line width. $\Gamma_t$ Theoretical line width calculated from $\tau_n$.

μ    Mass absorption coefficient for electronic absorption. Subscripts s and a refer to source and absorber.

$\mu_o$   Vacuum permeability. μ Nuclear magnetic moment. Nuclear magneton $\mu_n$
    = $\mu$ / I $g_n$. $\mu_B$ Bohr magneton.

υ    Photon frequency.

θ    Polar coordinate.  Angle between field and principle axis of EFG, or other defined axis. Debye temperature of solid.

Ψ   Wave function.

ρ    Density  Subscripts s and a refer to source and absorber.

$\rho(r)$ Charge density distribution.

σ    Cross section for resonant absorption.

τ    Mean life-time.  With subscripts for various relaxation times e.g. $\tau_L$ the Larmor precession time, $\tau_Q$ quadrupole precession time

ω    Angular velocity $= 2\pi\upsilon$.

# Index

Bold type signifies -- and following pages

**ANTIOXIDANTS in Science, Technology, Medicine and Nutrition**
**GERALD SCOTT, Professor Emeritus in Chemistry, Aston University, Birmingham**

ISBN 1-898563-31-4          350 pages          1997

The use of antioxidants is widespread throughout the rubber, plastics, food, oil and pharmaceutical industries. This book brings together information generated from research in quite separate fields of biochemical science and technology, and integrates it on the basis of the common mechanisms of peroxidation and antioxidant action. It applies present knowledge of antioxidants to our understanding of their role in preventing and treating common diseases, including cardiovascular disease, cancer, rheumatoid arthritis, ischaemia, pancreatitis, haemochromatosis, kwashiorlor, disorders of prematurity, and diseases of old age. Antioxidants deactivate certain harmful effects of free radicals in the human body due to biological peroxidation, and thus provide protection against cell tissue damage.

The book is of considerable interest to scientists working in the materials and foodstuff industries, and to researchers seeking information on the connection between diet and health, and to those developing new drugs to combat diseases associated with oxidative stress. It is important not only to scientists and practitioners in the world of medicine and nutrition, but also to all scientists and technologists seeking to improve the manifold spectrum of human and industrial life.

*Contents:*
Peroxidation in Chemistry and Chemical Technology; Biological Effects of Peroxidation; Chain-breaking Antioxidants; Preventive Antioxidants and Synergism; Antioxidants in Biology; Antioxidants in Disease and Oxidative Stress.

**"Drives the reader through mechanisms of peroxidation and antioxidant effect to discuss possible interventions in polymer technology...contains more on rubber and polymers....life scientists would hardly get in a single volume such an impressive amount of information. Particularly worth appreciation is the constant effort to define and to describe mechanism and kinetics of reactions discussed, of relevance for the antioxidant effect....chemical aspects are described in depth....the body of chemical information is of relevance for the biologist interested to rationalise the effect of enzymes, drugs and food components interacting with hydroperoxides."**

*Society For Free Radical Research Newsletter*
(Professor Ursini, Padova University, Italy)

# FUNDAMENTALS OF INORGANIC CHEMISTRY
**An introductory text for degree course studies**
JACK BARRETT, Imperial College of Science, Technology and Medicine, London University *and* MOUNIR . MALATI, Mid-Kent College of Higher and Further Education, Chatham

ISBN: 1-898563-38-1 *ca.* 320 pages 1997

This text, from two well-known and experienced teachers, offers a foundation course for 1st and 2nd year undergraduate inorganic chemists. It covers the main underlying theoretical ideas for an understanding of inorganic chemistry, taking account of the lower level of mathematical ability among present-day students commencing university study. The necessary mathematics, clearly explained where appropriate, is deliberately non-rigorous. Undergraduates reading chemistry will find much benefit from these teachers' proper and kindly approach which will launch them into their more advanced part of the inorganic chemistry degree course. The book will be helpful also to those reading any of the sciences where chemistry forms a significant part.

The authors' broad treatment of inorganic chemistry and the elements and their compounds establishes a sound basis for further study. Each chapter provides "worked example" problems, supported by additional problem-exercises which test comprehension and serve for revision or self-study. Solutions and hints are given at the end of the book.

# CHEMISTRY IN YOUR ENVIRONMENT
JACK BARRETT, Imperial College of Science, Technology and Medicine, London University

Hardback: ISBN: 1-898563-01-2 250 pages 1994
Paperback: ISBN: 1-898563-03-9

**"A super book which I thoroughly recommend. I suggest you take a large dose of *Chemistry in Your Environment*."**
John Emsley in *IC Reporter*

**"A sound understanding .... clear description of those branches of modern chemistry that are directly involved in our lives and in the world around us"**
Robin Turner in *Chemistry & Industry*

# SYMMETRY AND GROUP THEORY IN CHEMISTRY
MARK LADD, DSc (Lond) FRSC FInstP
Department of Chemistry, University of Surrey

ISBN 1-898563-39-10                    300 pages                    1998

This introductory undergraduate text, for a module course of about 20 lectures, presents a readable account of symmetry and group theory which has become so fundamental in the teaching (and learning) of chemistry and chemical physics.  It clarifies this area which undergraduates often find difficulty in understanding when meeting it for the first time.  Students embarked on a chemistry degree course will find much benefit from this experienced teacher's proper and kindly approach.  The book will be helpful also to those reading any of the sciences where chemistry forms a significant part.

Copiously illustrated, including many stereoviews of molecules, it stems from successful courses given by the author, and tested, honed and proven over many years.  Reference is made to programs developed by the author and currently on the Internet, which aid the derivation, study and recognition of point groups, and provide for other germane procedures. Rigorous mathematics is avoided, the pre-requisite being an A-level standard: other mathematical topics are developed within the text or the appendices  Each chapter concludes with a set of problems designed to enhance the reader's appreciation of the subject matter: brief answers are provided at the end of the book, with  detailed solutions accessible on the Internet.

*Contents:*   Symmetry Everywhere; Symmetry Operations and Symmetry Elements; Group Theory and Point Groups; Representations and Character Tables; Group Theory and Quantum Mechanics; Group Theory and Chemical Bonding; Group Theory and Molecular Vibrations; Crystal Symmetry; Space-group Theory

## PRACTICAL INORGANIC/PHYSICAL CHEMISTRY:

MOUNIR A. MALATI, Mid-Kent College of Higher & Further Education, Chatham, Kent

ISBN 898563-40-3                *ca.*300 pages                1998

This approach to practical inorganic and physical chemistry combines instrumental and radiochemical techniques and qualitative and quantitative (volumetric and gravimetric) analysis, with preparation of compounds. This strengthens both analytic and preparative skills. A feature is the emphasis on project investigative work, an important current trend in modern undergraduate chemistry courses.

Each chapter commences with a general introduction on the underlying principles of the chemistry of each group with also a section providing all equations and calculations. In this way the practical aspect in linked with underlying theory. The book explains physico-chemical measurements and reaction kinetics, and the handling of chemical data appears in an appendix. All the main elements and groups of the periodic table are covered, and. most experiments in the text have been published and proven., All can be carried out using readily available instruments and equipment and inexpensive chemicals, and safety is emphasised throughout.. Adequate references are given and techniques are described wherever appropriate.

Dr Malati is a well-known and experienced teaching chemist with a high involvement in project work. He has published extensively at undergraduate level, and much original research in journals; and has supervised many higher degree programmes. This is an ideal practical manual for undergraduates reading chemistry, at least in their first two years. HND and HNC (Higher National Diploma and Certificate) students and their tutors will find useful project ideas and tested experiments, and assignment work.

*Contents*: Preliminary Experiments; Alkali Metals; The Alkaline Earth Metals; Boron and Aluminium; The Carbon Group; The Nitrogen Group; The Oxygen Group; The Halogens; Titanium; Vanadium; Chromium; Manganese; Iron; Cobalt; Nickel; Coinage Metals; The Zinc Group; Other Experiments; Properties of Radiation; Neutron Activation Analysis; Szilard Chalmers Processes; Exchange Reactions; Radiometric Titration..

## ORGANIC CHEMISTRY:

**a comprehensive degree text and source book**

Professor Dr. Beyer, formerly at Institute of Organic Chemistry Ernst-Moritz-Arndt University, Greifswald, Germany.

Professor Dr. Walter, Institute of Organic Chemistry, University of Hamburg.

*Translator and Editor:* Dr. Douglas Lloyd, University of St. Andrews, St.Andrews, Fife, Scotland

ISBN 1-898563-37-3        1038 pages        1997

This book fills a need felt for several decades by providing a standard text from European sources. It provides a background book in support of lecture courses serving for reference and consultation. It will also be kept for postgraduate use and as a reference source for practising chemists in industry and academia.

**"Contains a considerable amount of up-to-date chemistry, a delightful blend old and new. The significant feature is the vast amount of material it contains. There are sections: general, aliphatics, alicyclics, carbohydrates, aromatics, isoprenoids, heterocyclics, amino acids** *etc*,** nucleic acids, enzymes, and metabolic processes. Each Chapter divided logically into sub-sections, complete with literature references. Compounds confined to specialist areas are covered, eg carbonic and cyanic acid derivatives, and compounds such as thioethers, thiocyanates** *etc.* **Will become an invaluable reference. The chemical equivalent of a sedate gentlemen's club."**

*Chemistry in Britain*

**"Primarily aimed at degree students, it is also a valuable reference for professional chemists in industry or academia."**

*Chemistry in Industry*

**"A prime example of the European Lehrbuch at it's best: thoroughly up-to-date, comprehensive, lucidly written. To my knowledge there is no comparable book for students with basic knowledge of organic chemistry; it is also a reference book for chemists at any stage of their career."**

*J.Am.Chem.Soc.*